Green Catalytic Hydrogenation of Phthalate Plasticizers

Zhongyi Liu

Green Catalytic Hydrogenation of Phthalate Plasticizers

科学出版社
Science Press

Springer

Zhongyi Liu
High-Tech Zone
Zhengzhou University
Zhengzhou, Henan, China

ISBN 978-981-16-9788-3 ISBN 978-981-16-9789-0 (eBook)
https://doi.org/10.1007/978-981-16-9789-0

Jointly published with Science Press, Beijing, China
The print edition is not for sale in China (Mainland). Customers from China (Mainland) please order the print book from: Science Press.

This Springer imprint is published by the registered company Springer Nature Singapore Pte Ltd.
The registered company address is: 152 Beach Road, #21-01/04 Gateway East, Singapore 189721, Singapore

Preface

The benzene ring structure in phthalate compounds is harmful to humans, animals, plants, and the environment. These compounds are a family of substances that can accumulate in the human body, are potentially mutagenic, carcinogenic, and teratogenic, and can cause damages to livers and kidneys after long-term exposure. They are also considered as a type of environmental estrogens or endocrine-disrupting chemicals that have reproductive toxicity and developmental toxicity, a serious threat to ecological security and human reproductive health.

In 2002, BASF SE in Germany successfully achieved the benzene ring hydrogenation for phthalate compounds. After 2005, the European Union (EU) and the United States of America (USA) published a series of laws and regulations, including REACH, RoHS 2.0, CPSIA, etc., to restrict or prohibit the use of phthalate plasticizers in human contact products such as food and drug packaging, medical devices, children's toys, and electronics products. Because phthalates are the major plasticizers used in China, the products are facing serious technical and trade barriers when penetrating the international market.

Since 2001, to tackle the technological monopoly by foreign countries, the author's research team have carried out the investigation of catalytic selective ring hydrogenation of benzene. This resource-saving and environmentally benign technology has been promoted in China since 2010 and has achieved high social and economic benefits. In recent years, to solve the plasticizer challenges China is facing, our team have started the research of catalytic ring hydrogenation of phthalate plasticizers. These two topics are basically in succession to solve the key technological problems during our economic and social developments and facilitate the realization of sustainability.

The research work discussed in this book was supported by China Petroleum and Chemical Industry Federation (CPCIF), the Development and Reform Commission of Henan Province, Department of Science and Technology of Henan Province, the Wisdom Zhengzhou—1125 Talent Plan, Qitaihe Longao Environmental Technology Co., Ltd., and Henan KeMingJunDe Chemical Industry Co., Ltd. Professors Shouchang Liu and Zhongjun Li, and Drs. Weidong Li, Shuaihui Li, Yuebin Lian, Zhihong Wei, Haiyang Wang, and Chao Lv participated in the research work. Senior

Engineer Qiangyong Zheng carried out the pilot testing and commercialization. We would like to thank all the above-mentioned government agencies, companies, and persons.

Your suggestions on how to further improve this book will be greatly appreciated.

Zhengzhou, China
October 2018

Zhongyi Liu

Contents

Chapter 1
Introduction

1.1 Background

Phthalate plasticizers, such as dioctyl phthalate (DOP), di(iso-octyl) phthalate (DIOP), di(2-ethylhexyl) phthalate (DEHP), dibutyl phthalate (DBP), di(iso-butyl) phthalate (DIBP), and di(iso-nonyl) phthalate (DINP), have the earliest use, the best performance, the broadest application, and the most production globally. Phthalate compounds have achieved an annual production capacity of 5 million tons in China and a consumption capacity of 3.6 million tons, about 1/5 of the global numbers. They are also one of the fundamental raw materials for a number of strategic industries.

A plasticizer is a manufacturing additive for polyvinyl chloride (PVC) resins, 40–50% in the final soft PVC articles. It can modify the PVC resin structure, weaken the interactions between polymer chains, decrease the glass transition temperature, reduce the tensile strength, increase the elongation at break, improve the plasticity, soften the articles, and harvest better low-temperature performance, thus enhancing the processability of PVC resins. As a type of commodities, plasticizers are widely applied in hundreds of areas in our economics and society, including tubes, plastics, adhesives, celluloses, lubricants, inks, paints, toys, food wraps, personal care articles (such as nail oils, hair sprays, soaps, and shampoos), etc.

Despite the increasingly wider uses, the toxicity of plasticizers has attracted a great deal of global attention with the focus on their side effects on the embryos, kidneys, hearts, livers, and reproductive systems of mammals [1–6]. Many years of experimental data and patient cases have indicated the harmful effects of the benzene ring structure on humans, animals, plants, and the environment. Phthalate compounds are a family of substances that can accumulate in the human body, are potentially mutagenic, carcinogenic, and teratogenic, and can cause damage to livers and kidneys after long-term exposure to them [7–14]. They are also considered as a type of environmental estrogens or endocrine-disrupting chemicals that have reproductive toxicity and developmental toxicity, a serious threat to ecological security and human reproductive health [15–17].

Z. Liu, *Green Catalytic Hydrogenation of Phthalate Plasticizers*,
https://doi.org/10.1007/978-981-16-9789-0_1

In 2005, the EU issued an official law to prohibit the use of phthalate plasticizers in human contact articles including food, drugs, and toys. The "Registration, Evaluation, Authorisation and Restriction of Chemicals (REACH)" regulation, effective since 2007 in the EU, lists phthalate compounds as substances of very high concern (SVHC) that may cause serious and irreversible damages to human health and the environment. The Consumer Product Safety Improvement Act (CPSIA) of the USA issued in 2008 prohibits the sale, manufacturing, and import of any articles containing >0.1% phthalates. The Restriction of Hazardous Substances (RoHS) implemented by EU in 2006 officially lists the following 6 phthalates as restricted substances in electronics and cables: DOP, DIOP, DEHP, DBP, DIBP, and DINP.

In China, the toxicity of phthalates has attracted broad attention among the communities of the press, drug, food science, etc. In 2011, a "plasticizer incident" happened in Taiwan. In 2012, a "Chinese spirits plasticizer incident" was reported. In 2013, a "toxic capsule plasticizer incident" was noticed; the China Central Television (CCTV) reported the use of a prohibited plasticizer in 15 types of PVC food wraps. In 2014, the CCTV reported an incident of the plasticizer content 368 times over the standard number in "toxic toys". In 2015, the plasticizer incident was labeled as one of the annual top-ten food contamination incidents. Dr. James-Todd from Brigham and Women's Hospital of Harvard Medical School discovered that women with higher phthalic acid ester (PAE) levels are more susceptible to diabetics. Later in October 2016, Dr. Fan Qu's research team from the Department of Traditional Chinese Medicine at the Women's Hospital of Zhejiang University Medical School found that long exposure to DEHP causes significant changes to the transcription of the key granulosa cell-related signaling pathway factors. Such changes will lead to dysfunctions in the anabolism, cell cycle, proliferation, and apoptosis of granulosa cells, and thus failure of ovulation will occur for women, i.e., polycystic ovary syndrome (PCOS). Simply saying, extensive exposure to plastics will cause ovulation dysfunction and infertility. Meanwhile, DEHP also greatly affects the fertile functions of men's sperms and causes gene mutations.

In November 2016, according to the interview of Mr. Xin Liu who was a member of National Committee of the Chinese People's Political Consultative Conference (CPPCC) and also the Chair of the Department of Health Food in the Chinese Institute of Food Science and Technology (CIFST), the China Food Daily published a report titled "High Hopes for Environmentally Benign Plasticizer Industry in China". According to the report, environmentally benign plasticizers are only 15% of the overall production of industrial plasticizers and the environmental estrogen contaminations caused by traditional plasticizers have attracted intensive attention. Data have indicated a rapid increase of the infertility percentage from 3% 20 years ago to the current 12% in China. The relationship between the increasingly serious soil contaminations by environmental estrogens and the decrease of human fertility is being highly concerned among the domain experts. In 2016, the annual production of plasticizers in China has reached 450 million tons, being 56% of the global production, and phthalate plasticizers are the dominant type. However, the greatly harmful phthalate plasticizers have yet to be listed among the national or regional environmental monitoring chemicals and the environmental estrogen pollutants have yet to be seriously assessed, causing new challenges to food security.

For many years, Mr. Xin Liu has kept bringing the attention of the general public to the development of environmentally benign plasticizer industry as well as the importance of environmental estrogen pollutant research. Data from the sample survey indicate that the PAE contaminations of soil, water, and produces are quite serious in economically developed areas of China such as the Zhu River Delta Area in Guangdong Province. Mr. Liu has submitted a proposal entitled "Attention Should Be Paid to The Environmental Estrogens That Damage Human's Fertility" (below abbreviated as The Proposal) to the Fourth Meeting of the 12th National Committee of the Chinese People's Political Consultative Conference. The foresighted, constructive recommendations in The Proposal, including the prohibition of environmental estrogen production from the original sources, the development of environmentally benign plasticizers, the establishment of standards for the use of phthalate plasticizers, and the inclusion of environmental estrogen-type organic pollutants into environmental monitoring chemicals, have received high attention from the Ministry of Industry and Information Technology (MIIT). According to Mr. Liu, the release of relevant inspection standards will promote the research and development of the environmentally benign plasticizer industry, form a world-class platform for scientific and technological innovations, and improve the overall economic competitiveness. It is foreseen that the spring for the development of environmentally benign plasticizers will come in the near future.

Currently, the major problem is that the PAE pollutants have not received sufficient attention. PAEs are important plastics additives for industrial PVC articles that can improve their plasticity, flexibility, and dilatability. Common PAEs include DEHP, dimethyl phthalate (DMP), diethyl phthalate (DEP), DBP, etc. The PAE content in industrial PVC articles can reach as high as 25–65% and even 75%. During the manufacturing process, the low-molecular-weight PAEs are not polymerized with PVC leading to high migration of PAEs in the use of PVC articles. In addition to mutagenicity, carcinogenicity, and teratogenicity, some phthalates such as DBP and DEHP are also environmental estrogens or endocrine-disrupting chemicals that have reproductive toxicity and developmental toxicity. In the areas of food processing, agricultural plastics, and medical devices, the plasticizer residues can migrate to enter the human body by swallowing, inhaling, and skin contact, causing damages to the endocrine system, liver, kidney, and lung. This will greatly increase the risks of cardiovascular diseases and threaten productive health.

The investigations by Mr. Liu show that, in the Zhu River Delta that hosts many printing, dyeing, tannery factories, the untreated discharge of a part of city sewage and industrial wastewater, and the rain flushing of ground surfaces and building tops are the major sources of PAE pollutants in urban rivers and lakes. The wide use and disposal of agricultural thin films and plastic bags cause the migration, conversion, and deposition of plasticizers in the soil, water, and air. Inspection data show that the soil contaminations of PAEs are particularly serious. Six types of PAEs including DEHP, DEP, DBP, DMP, etc. are found in farm-raised fishes. The

15 lake surface deposition samples taken from the Baiyun, Tianhe, Yuexiu, Liwan, and Haizhu Districts of Guangzhou City generally contain 16 PAEs especially DMP, DEP, DBP, di(2-ethoxyethyl)phthalate (DEEP), and DEHP. PAEs are found in the groundwater in many areas of Zhu River Delta; 40% of the agricultural irrigation uses the groundwater bringing the PAEs into the food chain with the adsorption and transportation of crops. PAEs are also found in a great deal of surface water and drink water. Furthermore, PAEs are widely used in the fields closely associated with drug and human life, including plastic gloves, food wraps, food storage bags, packing bags, trash bags, plastic tubing, plastic seal rings, plastic gaskets, plastic containers, and plastic drink and food bottle caps, in which surprisingly high contents of PAEs are detected. So far, the PVC articles containing no PAE additives are rarely seen in China, while the use of environmentally benign plasticizers is very common in foreign countries with the established policy and regulations that define strict inspections and restrictions regarding the use of PAEs.

Environmentally benign plasticizers have a large space of industrial demands. In recent years, the global demand and market of PVC articles is rapidly increasing and so are the additives especially the environmentally benign plasticizers. The fast development of environmentally benign plasticizer industry and PAE-free PVC manufacturing industry is underway. Currently, the commonly used environmentally benign plasticizers include biodegradable citric acid esters, epoxy fatty acids, diisononyl cyclohexane-1,2-dicarboxylate (DINCH), and mixed esters of isooctyl acid and benzoic acid polyol. In China, the production of environmentally benign plasticizers is still in the infancy stage. With the increasing social awareness of food safety and environmental protection, foreign and domestic laws are becoming stricter with the content of PAEs or other additives in PVC articles. Numerous policies and regulations have been enforced to inspect and restrict the use of PAEs in medical devices, maternal and infant articles, food wraps, and other human-contacting PVC articles. Environmentally benign plasticizers with non-toxicity, biodegradability, high plasticization performance, and low risk of food security are demonstrating their high values and thus wide market prosperity.

Regarding the current status of plasticizer development, Mr. Liu points out that the industrial foundation of non-toxic, environmentally benign plasticizers is rather weak in China, as well as their limited types compared with foreign counterparts. Because of the low concern of PAE pollution, the laws that restrict the use of PAEs have not been enforced for the PVC manufacturing industry and plasticizer industry. Meanwhile, due to the lack of government guidance and consumer need, most types of environmentally benign plasticizers have limited market penetration. Currently, the production of environmentally benign plasticizers is still in the infancy stage in China; biodegradable, biomass-sourced plasticizers are even rarely seen due to the high cost and limited competitiveness.

Analysis has indicated that promoting the innovation of environmentally benign plasticizers can eliminate the origin of PAE pollutions. According to Mr. Liu's opinions, with the forthcoming Fourth Industrial Revolution, China should execute strategic planning for the development of environmentally benign plasticizers based on the vast market demand and resource availability, prioritizing the industrial arrangement and project execution to promote the industrial chain growth and economic advancement through government guidance and policy support. It is estimated that the global demand for PVC will increase by 50% to 49,500,000 tons in 2020, and so is the plasticizer. Currently, the global development of PVC and plasticizer exhibits a general trend of larger scale production, higher product specification, multifunction, and better environmental benignness. China should think forward and plan coordinately the industrial development of environmentally benign plasticizers to meet the immense market demand, trigger a new cycle of sustainable advancement of economics, and place China at a globally leading position of plasticizer innovation. Meanwhile, the reality is that the development of environmentally benign plasticizers in China obviously lags behind that of functional plastics articles and fails to fully meet the need, and therefore long-term strategic plans are required to optimize the high-level arrangements and designing. More research support should be granted to achieve core technologies for industrializing new materials and products of environmentally benign plasticizers. Novel catalysts and synthetic protocols will be employed to develop non-toxic, biodegradable, high-performance, and specificated environmentally benign plasticizers and to establish a green industrial chain to satisfy global legal requirements. Promoting the R&D of environmentally benign plasticizers will help form a world-class sci-tech center and advance the competitiveness of the industrial and economic structures. At the same time, the origins of PAE pollution will be eliminated improving food safety. Furthermore, the Chinese government should list the PAEs among the environmental monitoring target substances and achieve real-time analysis, tracking, detection, and assessment. In reference to the USA and EU and also based on our own actual situations, China should establish the regional and national standards of PAE residues in food, nutritional supplements, drug and food additives, as well as their manufacturing parts.

Encouragingly, the Chinese government has approved all three suggestions in The Proposal, including "Suggestion Regarding the Development of Industrial Chain Groups of Environmentally Benign Plasticizers", "Suggestion Regarding the Establishment of Regulations and Standards for Phthalate Plasticizers", and "Suggestion Regarding the Enforcement of Phthalate Monitoring". They have endorsed the development of safe environmentally benign plasticizers as the key to sustainability in the Thirteenth Five-Year Planning and have completely agreed on the high attention to environmental hormone pollutants that may damage human reproducibility. China will increase the policy support to environmentally benign plasticizers. A number of Chinese government agencies including MIIT will investigate the suggestions regarding the overall planning, demonstration, establishment of standards, and construction of environmental monitoring and detection, and make corresponding acts.

In August 2017, a report released by Tencent titled "Shanghai Textbook Cover Sampling Mostly Unqualified" raised a great deal of social concerns especially the parents of young students. Shanghai Quality and Technical Supervision Bureau announced the quality inspection results of the plastic textbook covers and cases produced and sold in Shanghai: 25 of the collected 30 samples contain phthalate plasticizers higher than the standards. This may increase the secretion of estrogen and cause sexual precocity in children.

China promulgated the Plastic Limit Order in 2007 but the effectiveness has been rather unobvious for more than 10 years. To expedite the construction of ecological civilization, the National Development and Reform Commission (NDRC) opened a column of "My Suggestions for Reduction of Plastics Trashes and Pollutants" on its official website on January 5, 2018, focusing on our long, wide use of disposable plastic articles in various forms. In recent years with the growing new businesses such as e-commerce, express delivery, and food take-out, the consumption of plastic food boxes and merchandise wraps has increased rapidly, which has caused new pressures on the resources and environment because of careless disposals of the "white" plastic trashes. To make the policies more reasonable, more specific, and more executable, the NDRC invite a wide audience to offer comments and suggestions based on the requirements for plastic articles of different applications.

In "Industrial Classification for National Economic Activities", the plasticizer is categorized into the high-tech fine chemicals, one of the emerging strategically basic materials. The MIIT proposed the establishment of an independent, innovative, sustainable new materials industry system by 2020 that closely integrates production, education, research, and application. The purposes are to develop materials technology as the pioneering industry in our economics, satisfy the need for economic development and national defense construction, reach the globally leading position, upgrade our materials industry, and realize the transition from "quantity" to "quality".

Based on the urgently critical scientific and technological problems, this book will carry out the research of catalytic hydrogenation of phthalates to facilitate the upgrading transition and green development of the plasticizer industry and increase our voice in international policy making.

1.2 Fundamentals of Plasticizers

1.2.1 Symbols and Abbreviations

See Table 1.1.

1.2.2 Major Functions of Plasticizers

The plasticizer is a type of substance that increases plasticity while not affecting the properties of plastics. Here are the major functions of plasticizers: weakening van der Waals interactions between the polymer chains; increasing the mobility of polymer chains (i.e., plasticity); enhancing the extensibility, flexibility, and flexurity;

Table 1.1 Symbols and abbreviations of plasticizers

Abbreviations	IUPAC name	CAS#
AES	alkanesulfonates or alkyl alkanesulfonates	Unknown
BAR	butyl(*R*)-12-acetoxyoleate	140-04-5
BBP	Same as left	85-68-7
BCHP	Same as left	84-64-0
BNP	Same as left	Unknown
BOA	benzyl 2-ethylhexyl adipate	3089-55-2
BOP	butyl 2-ethylhexyl phthalate	85-69-8
BST	Same as left	123-95-5
DBA	Same as left	105-99-7
DBEP	bis(2-butoxyethyl) phthalate	117-83-9
DBF	Same as left	105-75-9
DBM	Same as left	105-76-0
DBP	Same as left	84-74-2
DBS	Same as left	109-43-3
DBZ	Same as left	2917-73-9
DCHP	Same as left	84-61-7
DCP	bis(1-methylheptyl) phthalate	131-15-7
DDP	Same as left	84-77-5
DEGDB	oxydiethylene dibenzoate	120-55-8
DEP	Same as left	84-66-2
DHP	Same as left	3648-21-3
DHXP	Same as left	84-75-3
DIBA	Same as left	14 1-04-8
DIBM	Same as left	14,234-82-3
DIBP	Same as left	84-86-9
DIDA	Same as left	27,178-16-1
DIDP	Same as left	26,761-40-0
DIHP	Same as left	41,451-28-9
DIHXP	Same as left	71,850-09-4
DINA	Same as left	33,703-08-1
DINP	Same as left	28,553-12-0
DIOA	Same as left	1330-86-5
DIOM	Same as left	1330-76-3
DIOP	Same as left	27,554-26-3
DIOS	Same as left	27,214-90-0
DIOZ	Same as left	26,544-17-2

(continued)

Table 1.1 (continued)

Abbreviations	IUPAC name	CAS#
DIPP	Same as left	605-50-5
DMEP	bis(2-methyloxyethyl) phthalate	117-82-8
DMP	Same as left	131-11-3
DMS	Same as left	106-79-6
DNF	Same as left	2787-63-5
DNM	Same as left	2787-64-6
DNOP	dioctyl phthalate	117-84-0
DNP	Same as left	14,103-61-8
DNS	Same as left	4121-16-8
DOA	bis(2-ethylhexyl) adipate	103-23-1
DOIP	bis(2-ethylhexyl) isophthalate	137-89-3
DOP	bis(2-ethylhexyl) phthalate	117-81-7
DOS	bis(2-ethylhexyl) sebacate	122-62-3
DOTP	bis(2-ethylhexyl) terephthalate	6422-86-2
DOZ	bis(2-ethylhexyl) azelate	2064-80-6
DPCF	diphenyl *x*-tolyl orthophosphate	DPCF
DPGDB	No existing	Unknown
DPOF	2-ethylhexyl diphenyl orthophosphate or octyl diphenyl orthophosphate	1241-94-7
DPP	Same as left	84-62-8
DTDP	Same as left	27,253-26-5
DUP	Same as left	3648-20-2
ELO	No existing	8016-11-3
ESO	No existing	8013-07-8
GTA	Same as left	102-76-1
HNUA	No existing	Unknown
HNUP	No existing	68,515-42-4
HXODA	No existing	Unknown
HXODP	No existing	68,515-51-5
NUP	No existing	Unknown
ODA	decyl octyl adipate	110-29-2
ODP	decyl octyl phthalate	68,515-52-6
ODTM	decyl octyl hydrogen benzene-1,2,4-tricarboxylate	Unknown
PO	No existing	8012-95-1
PPA	Same as left	Unknown
PPS	No existing	Unknown

(continued)

Table 1.1 (continued)

Abbreviations	IUPAC name	CAS#
SOA	surcros octaacetate	126-14-7
TBAC	Same as left	77-90-7
TBEP	tris(2-butoxyethyl)orthophosphate	78-51-3
TBP	tributyl orthophosphate	126-73-8
TCEF	tris(2-chloroethyl)orthophosphate	6145-73-9
TCF	tri-*x*-tolyl orthophosphate	1330-78-5
TDBBP	tris(2,3-dibromopropyl)orthophosphate	126-72-6
TDCPP	tris(2,3-dichloropropyl)orthophosphate	78-43-3
TEAC	Same as left	77-89-4
THFO	Same as left	5420-17-7
THTM	triheptyl benzene-1,2,4-tricarboxylate	1528-48-9
TIOTM	tris(6-methylheptyl)benzene-1,2,4-tricaboxylate	27,251-75-8
TOF	tris(2-ethylhexyl)orthophosphate	78-42-2
TOPM	trtrakis(2-ethylhexyl)benzene-1,2,4,5 tetracarboxylate	3216-80-5
TOTM	tris(2-ethylhexyl)benzene-1,2,4-tricarboxylate	89-04-3
TPP	triphenyl orthophosphate	115-86-6
TXF	tri-*x,y*-xylyl orthophosphate	25,155-23-1

decreasing the hardness, modulus, softening temperature, and brittle temperature; and improving the mechanical properties.

The plasticizers can be categorized as either internal or external.

Internal plasticizers: for the first type, in the polymerization process a second monomer is introduced and copolymerized to form the final polymer chains, leading to a lower degree of crystallinity. For the second type, side chains are introduced onto the polymer backbone via substitution or grafting to reduce the inter-chain interactions and increase the plasticity. An example is the copolymerization of vinyl chloride and vinyl acetate. Because of the covalent bonding with polymer backbones, the plasticizer can resist external extraction. But the monomers must be added during the polymerization process increasing the synthetic complexity and cost. And internal plasticizers have rather narrow application temperature ranges and can be used only in low-flexibility plastic articles. The mechanism for the increase of plasticity lies in the block or graft copolymerization of a different monomer leading to reduction of inter-chain attractions.

External plasticizers: a high-boiling-point, non-volatile liquid or a low-melting-point solid, typically organic esters, is added to the polymer to form a solid-state solution via swelling, instead of chemical reactions, at elevated temperatures. The mechanism for the increase of plasticity relies on the penetration of small solvating molecules into the polymer to increase inter-chain spacings and decrease the attractions between them, leading to softer resins and lower processing temperatures. The external plasticizers have higher overall performance, more convenient production and use, and wider applications. Therefore, the so-called "plasticizers" often mean external plasticizers.

Plasticizers are widely used not only in PVC resins but also in celluloses, polyvinyl acetates (PVAc), ABS resins, polyamides (PA), polyacrylates, polyurethanes (PU), polycarbonates (PC), unsaturated polyesters, epoxy resins, phenolic resins, alkyd resins, melamine resins, certain rubbers, etc.

PVC can be categorized as either hard or soft, with the hard PVC resins primarily used in the construction materials industry. The plasticizers are mainly used in soft PVC resins with an average loading of 45–50% with respect to PVC resin. Globally, 80–85% of the plasticizers are used in soft PVC plastics and the rest is used in rubbers, cellulose resins, paints, etc. Therefore, the development of plasticizers is closely related to that of PVC. Currently, PVC is still one of the most used plastics. 40% of soft PVC is consumed in developed countries and 60% in developing countries.

The plasticizer industry in China was originated in the 1950s, basically at the same pace as the PVC industry. Since the 1980s, the production volume of PVC resins has increased rapidly leading to an increased demand for plasticizers, which has greatly advanced the plasticizer industry.

1.2.3 Major Performance-Determining Factors for Plasticizers

1. **Van der Waals forces**

When the plasticizer is added to a polymer, the intermolecular interactions between them greatly affect the plasticity. There are two types of intermolecular interactions: van der Waals forces and hydrogen bonding. Van der Waals forces include London dispersion forces, dipole-induced dipole forces, and dipole–dipole forces.

London dispersion forces exist between all types of polar or non-polar molecules and result from attractive forces between weak temporary induced dipoles. London dispersion forces are dominant only in non-polar molecules such as benzene, polyethylene (PE), and polystyrene (PS).

Dipole-induced dipole force is a weak attraction that results when a molecule with a permanent dipole induces a temporary dipole in a non-polar molecule. This force is particularly strong for aromatic compounds because of the high polarizability of their π electrons.

Dipole–dipole forces form between the permanent dipoles of polar molecules when they approach each other. A typical example is an interaction between ester plasticizers and PVC resins.

Van der Waals forces are weak, short-range attractive interactions between molecules or polymer chains, and can become so strong, because of the additivity, that the penetration of plasticizer molecules into the polymers can become difficult.

2. **Hydrogen bonding**

Hydrogen bonding is formed between molecules bearing –OH or –NH– groups such as polyamide and poly(vinyl alcohol) (PVA). Hydrogen bonding is a relatively strong intermolecular force and can affect the penetration of plasticizer molecules into the polymer chains, especially for those bearing high degrees of hydrogen bonds. At elevated temperatures, hydrogen bonding becomes weaker as a result of the thermal energy preventing the organization of polymer chains.

The magnitude of hydrogen bonding depends on the functional groups in polymer structures. Polar groups lead to stronger hydrogen bonding and non-polar groups lead to weaker bonding. The polarity of common polymers follows this trend: PVA > PVAc > PVC > polypropylene (PP) > PE.

3. **Crystallinity of polymers**

Full crystallization of polymer chains is not possible. Polymers are generally composed of crystalline grains distributed in amorphous regions. Plasticizer molecules are more different to penetrate into crystalline regions than into amorphous regions. Those that can insert into both crystalline and amorphous regions are called solvent-type or primary plasticizers; and those that can insert only into amorphous regions are called non-solvent-type or secondary plasticizers.

1.2.4 Mechanism for Plasticity Increase by Plasticizers

A number of mechanisms for the plasticity increase via using plasticizers have been proposed, all of which have faults but can well explain certain observations. The major ones include the lubrication effect, gelation effect, and free volume effect [18–21].

1. **Lubrication effect**

Plasticizers function as interfacial lubricants. The presence of plasticizers can break the attractions between polymer chains, decrease the inter-chain "frictions", and thus increase their slip mobility, even when the polymer chains are partially entangled into gel networks. Without any plasticizer, the surface irregularity of PVC chains prevents their relative slips, leading to the formation of a rigid network structure. As molecular lubricants, the plasticizers can soak between the polymer chains and separate them out, resulting in more free relative slips and thus increasing materials flexibility and extensibility. This mechanism can explain the reduction of viscosity, increase of fluidity, facile processing, and minimal property changes, but fails to elucidate the complex plasticization process [21].

2. **Gelation effect**

Plasticization opens up the polymer chains, while the inter-chain attractions re-aggregate these chains. These two processes are in a dynamic "open–close" equilibrium that yields a number of physical "nodes" between polymer chains at certain temperatures and concentrations. The plasticizers function to separate polymer chains by solvating and then breaking these "nodes". This mechanism can explain the plasticization of polar polymers. On the other hand, due to the weaker inter-chain attractions in nonpolar polymers, the addition of plasticizers decreases the number of such "nodes". The plasticizer molecules diffuse into the spacings between polymer chains. Thus, part of the above-stated inter-chain "nodes" are replaced by the newly formed polymer–plasticizer "nodes". Different loadings of the plasticizer increase the irregular mobility of polymer chains to different extents.

This mechanism provides a reasonable explanation for the phenomenon that external plasticizers can soften the polymer more than the internal counterparts as the temperature increases. It can also explain the PVC softening by secondary plasticizers in that they increase inter-chain spacings and reduce the entanglement between polymer chains [19].

3. **Free volume effect**

The free volume is the usable free space in polymers at the absolute temperature (0 K), and generally refers to the space between polymer chains that is proportional to the free-moving space of polymer chains. The origins of free volume include the free moves of the end-tails, of side functionalities and of the middle portions of the polymer [22]. According to this mechanism, plasticizers can separate the polymer chains and increase the free volume inside the polymer, thus leading to improvement of spatial chain mobility and polymer flexibility.

The addition of plasticizers can increase the free volume in the polymer and inter-chain spacings, considering the fixed free volume below the glass transition temperature for all polymer materials. The results of plasticization include the increase of melt mobility, decrease of hardness after cooling, and enhancement of extensibility and flexibility. Obviously, the degree of plasticization is proportional to the plasticizer loading. However, this mechanism fails to explain the antiplasticization phenomenon at low plasticizer loadings for many polymers.

The above three mechanisms have elucidated plasticization to some extent, but so far a comprehensive mechanism that can explain the complexity of plasticization is still lacking. Currently, the widely accepted theory is that plasticization is a result of the weakening of inter-chain interactions in polymer materials. The plasticizer molecules enter the space between polymer chains, decrease the inter-chain attractions, improve the chain mobility, reduce the polymer crystallinity, and thus increase the plasticity of the polymer. When a plasticizer is added to a polymer, there exist the following interactions: (1) polymer–polymer interactions; (2) plasticizer–plasticizer interactions; and (3) polymer–plasticizer interactions [23–25]. In general, plasticizers are small molecules and the plasticizer–plasticizer interactions are negligible. The polymer–polymer interactions are the key. For nonpolar polymers, the polymer chains have weak interactions with other chains; thus, the plasticizers are easy to penetrate into the polymer and increase the inter-chain spacings and weaken the inter-chain interactions, leading to effective plasticization. On the other hand, for polar polymers, the polymer–polymer interactions are too strong for plasticizer molecules to insert into the polymer. The use of plasticizers bearing polar functional groups can form stronger plasticizer–polymer interactions, thus weakening the polymer–polymer interactions for plasticization.

1.2.5 Property Requirements for Plasticizers

1. **Non-volatility**

The volatility of a plasticizer is closely related to its molecular weight, compatibility with PVC, and the functional groups it has. In general, plasticizers with low molecular weights or poor PVC compatibility are more volatile; those with high molecular weights, good PVC compatibility, or bulky functional groups are less volatile. Macroplasticizers such as polyesters have high molecular weights and are generally non-volatile.

Non-volatile heat-resistant plasticizers, such as polyesters, epoxy oleate, diisotridecyl phthalate (DTDP), trimellitates, and dipentaerythritol esters, are mostly found in articles used at high temperatures such as cables and car decors.

Among phthalate plasticizers, low-molecular-weight DBP is the most volatile, while high-molecular-weight DIDP and DTDP are significantly non-volatile. Straight-chain alkyl phthalates are less volatile than branched-chain alkyl phthalates.

Among epoxy plasticizers, the volatility follows the trend: epoxidized triglycerides < epoxy trahydrophthalate < epoxy fatty acid monoesters.

Among aliphatic dicarboxylates, the volatility follows the trend: dioctyl sebacate (DOS) < diisodecyl adipate (DIDA) < dioctyl azelate (DOZ) < dioctyl adipate (DOA).

2. **Atomization resistance**

Fogging of a plasticizer is related to its volatility and is one of the required tests for car decors. BASF carry out the anti-fogging tests following the SAE J1756-5006 standard. The main examination is to detect the change of glossiness and the amount of target materials attached to transparent decors after the co-existence of decors and plasticizers under certain conditions.

3. **Extraction resistance**

Extraction means that, when a PVC article is soaked in a liquid medium such as water, soap water, oil, and chemical solvents, the plasticizer tends to migrate from inside the plastics to the medium. Such a tendency depends not only on the nature of the plasticizer (e.g., the structure, polarity, and molecular weight of the polymer and the plasticizer) but also on the physico-chemical properties of the liquid medium. Anti-extraction properties include oil resistivity, solvent resistivity, water resistivity, and soap water resistivity.

Generally, plasticizers tend to be extracted by gasoline or other oily solvents, but those with benzene, ester, or branched alkyl groups are opposite due to the difficulty of plasticizer molecule diffusion. Plasticizers with bulky alkyl groups in their structures are facile to extraction with gasoline or other oily solvents.

The water and soap water resistivities are opposite to oil resistivity. Plasticizers with bulky alkyl groups have better water and soap water resistivities. Due to the resistivity to water extraction, PVC articles that often contact water or are washed with water use normal plasticizers. But those with frequent oil contact must use oil-resisting polyester plasticizers. High-molecular-weight polyesters have good non-volatility, anti-extraction, and anti-mobility properties, yet relatively poor cold tolerance and plasticity. Polyester plasticizers have good durability and are commonly used in articles with high oil and heat tolerances.

Two factors contribute to the anti-extraction property of plasticizers:

(1) The diffusion rate of plasticizer molecules in plastic articles. Extraction swells the plastic article due to the absorption of the liquid medium by the plastics, which enhances the internal diffusion of the plasticizer molecules.
(2) The solubility of the plasticizer in the liquid medium. In the case of low solubility, the medium dictates the extraction rate by affecting the rate of plasticizer diffusion from the article surface to the medium. Plasticizers are easy to extract by polar media but not by non-polar or weakly polar media. High-molecular-weight plasticizers are easier to be extracted than low-molecular-weight counterparts. Polymeric plasticizers have better than small molecule plasticizers. Epoxy plasticizers have good water and soap water resistivities but poor weather tolerance.

4. **Migration resistance**

Migration refers to the mobility of plasticizer molecules from inside the plastics to the surface and then to the inside of the substances in contact. Because of the migration, PVC articles often become softened, sticky, or cracked. Meanwhile, the migration causes contamination of the plastic articles.

Migration is related to the structure of plasticizers. The migration of phthalate plasticizers rapidly decreases with the alkyl chain length. The introduction of ether or aromatic groups will increase the migration resistance of ester plasticizers. Plasticizers with straight alkyl chains have lower anti-migration than those with branched alkyl chains of the same number of carbons. Aliphatic acid-based plasticizers have higher migration than epoxy fatty acid monoesters. DOS have good cold tolerance but high migration and may cause a decrease of the electrical insulation.

5. **Cold tolerance**

The cold tolerance of plasticizers is closely related to their structures. Generally, plasticizers with good PVC compatibility have inferior cold tolerance. Especially, those plasticizers bearing cyclic rings have significantly low cold tolerance. Phthalate plasticizers with straight alkyl chains have good cold tolerance; branched chains decrease the cold tolerance. The longer the alkyl chains, the better the cold tolerance.

Currently, cold-tolerant plasticizers are mainly based on aliphatic dicarboxylates. Straight-chain alcohol-based phthalates, diol-based aliphatic acid esters, and epoxy aliphatic acid monoesters, all have remarkable low-temperature performance. Representative cold-tolerant plasticizers include dioctyl adipate (DOA), diisodecyl adipate (DIDA), dioctyl azelate (DOZ), and dioctyl sebacate (DOS).

Generally, cold-tolerant plasticizers have inferior compatibility with PVC, and thus can only be used as secondary plasticizers with a content of 5–20% of the primary plasticizers to improve the cold tolerance.

Increasing the content of plasticizers leads to lower glass transition temperature for the plastics and better low-temperature flexibility. However, an excess plasticizer damages the room temperature physical properties of the plastics, especially the strength and weather tolerance. Therefore, these factors should be considered when finalizing the formulations.

6. **Aging resistance**

Improvement of the anti-aging performance of plastics mainly relies on the effects of thermal stabilizer, anti-oxidant, and light stabilizer. For soft PVC, the anti-aging property is closely related to the plasticizer due to its high content. Different plasticizers will result in different anti-aging performances. The addition of anti-oxidant can greatly improve the anti-aging and anti-volatility properties. Epoxy plasticizers can also serve as a stabilizer to improve the anti-aging performance of PVC articles.

7. **Weather tolerance (stability to light, heat, and oxidant)**

The weather tolerance of a plasticizer material is the degree of its deterioration, typically expressed as the weight loss, as a function of time under mimicking or natural conditions. This property reflects the stability to a combination of light, heat, oxidants, humidity, and even pollution.

Because the tertiary hydrogen is more attracted by $C = O$ and decomposed via oxidation, DNOP has better thermal stability than DIOP. That is, alkyl groups with more side chains yield lower thermal stability for plasticizers.

Besides the structural effect, the purity of plasticizers also greatly affects their thermal stability. The high the purity, the higher the thermal stability.

By comparing the pH value changes after introducing O_2 to phthalate plasticizers, straight-chain alkyl groups result in good thermal stability.

Moreover, anti-oxidant additives greatly improve the thermal stability of DIDP and DIOP. Anti-oxidants can prevent the generation of water and thus the hydrolysis of esters by blocking the oxidation reactions that produce peroxide groups.

Epoxy plasticizers, such as epoxidized soybean oil, butyl epoxy oleate, octyl epoxy oleate, and decyl epoxy oleate, can afford good weather tolerance to plastic articles.

8. **Durability (resistivity to volatility, extraction, and migration)**

Plastics, especially soft PVC articles, have high contents of plasticizers, which requires the long-term stay of plasticizers in the plastics, i.e., good durability. The volatility-, extraction-, and migration-based losses of plasticizers include three elementary steps: (1) diffusion of plasticizers to plastics surface; (2) transition to a "recumbent" state at the inner surface; and (3) diffusion away from plastics surface.

The durability of plasticizers is closely associated with their molecular weight and structure. Only with molecular weights above 350 can the plasticizers have good durability. Polyester and benzene multi-carboxylate plasticizers with molecular weights > 1000 have excellent durability and are commonly used in the so-called permanent articles such as electrical cables, refrigerators, and car interiors. Moreover, 1,2-propylene glycol adipate polyester and phosphate plasticizers also have good durability.

9. **Electrical insulation**

Soft PVC articles have a high requirement for electrical insulation, especially those used in cable sheaths. Polar, compatible primary plasticizers, such as chlorinated paraffin, phthalates, phenyl alkylsulfonates, and phosphates, must have high electrical insulation.

Plasticizers containing ionic impurities such as the catalyst residues from esterification reactions have lower electrical insulation. But the electrical resistance of the plasticizer cannot be used as a decisive indicator of its electrical insulation and other factors need to be considered as well.

Plasticizers used for cable sheaths require not only good electrical insulation but also high thermal stability and anti-aging property. Therefore, heat-tolerant plasticizers are often used in high-temperature cables.

Polyester plasticizers are also widely used in wires and cables because of their low volatility and good durability. In the case of underground cables, the plasticizer extraction by soil often causes the hardening of the sheath plastics. Special attention should be paid to their anti-extraction and anti-mildew properties. Common plasticizers in cable sheath formulas include DIOP, DIDP, DTDP, DNP, TOTM (trioctyl trimellitate), chlorinated paraffin, polyesters, etc.

Besides the electrical insulation, the dielectric properties of PVC plasticizers are also very important. The dielectric loss depends on the types and contents of the polymer, plasticizer, filler, or other manufacturing additives.

10. **Flame retarding**

The use of plastic articles in the fields of construction, transportation, electricity, and particularly electrical cables, mining transport belts, and home appliances requires flame retardancy and no toxic gas release when burning. PVC resins have Cl contents as high as 56% and thus are flame retardants by nature, which can be further improved if combined with flame retarding plasticizers. Otherwise, the PVC plastics become more flammable if improper plasticizers are added.

Three factors can affect the flame retardancy of plasticizers: (1) the relative volatility compared to PVC, with higher volatility yielding less flame retardancy; (2) the decomposition products of combustion, which should not be flammable; (3) the chemical structure of the plasticizer, with P, Cl, or aromatic groups leading to better flame retardancy.

Currently, the widely used flame-retarding plasticizers include phosphates, chlorinated paraffin, and chlorinated aliphatic acid esters. Phosphate plasticizers have a unique advantage of strong flame retardancy and are commonly used in PVC and cellulose. Chlorinated paraffin is cost-effective and is mainly used as secondary plasticizers. The performance of chlorinated paraffin is closely related to its Cl content, the increase of which will improve the flame retardancy and compatibility but rapidly worsen the cold tolerance. Thus, the common content of chlorinated paraffin plasticizer is 40–50%. Due to the ester group, chlorinated aliphatic acid esters such as methyl 5-chloro-stearate have better PVC compatibility than chlorinated paraffin. A combination with Sb_2O_3 makes Cl-containing plasticizers even more fireproof. Generally, plasticizers with an oxygen index higher than 28 are considered flame retarding.

11. **Non-toxicity**

Plastics such as films, containers, and straws have been widely used for the storage and packing of food and drug, and thus are required to be non- or low-toxic. Most plastics themselves have no toxicity. For PVC, due to the continuous improvement of synthetic protocols, the content of vinyl chloride monomers has dropped below 5 ppm (1 ppm $= 10^{-6}$). World Green Organisation (WGO) has no longer rejected the use of PVC because of its toxicity. However, many of the additives in plastics can enter the human body via water or oil extraction. Therefore, the toxicity of these additives especially of plasticizers needs high attention. Both their own toxic nature and the possibility of entering the human body should be taken into consideration.

12. **Mildew resistance**

Certain plastics (such as wire cables, agricultural films, and construction materials) are in contact with microbes in nature during their use. The plasticizers inside often become the nutrients for these microbes and are attacked by mildew and bacteria, leading to a decrease in performance.

Long-chain aliphatic acid esters are the most susceptible to mildew attacks, and so are dialkyl dicarboxylic acid esters. In contrast, phthalates and phosphates have strong resistance to mildew, especially those based on phenol-type raw materials such as tricresyl phosphate (TCP) and triphenyl phosphate (TPP). Epoxy soybean oil is also easy to be attacked by mildew.

An ideal plasticizer needs to satisfy the following basic requirements: (1) good compatibility with the resin; (2) high plasticization effect; (3) stability to heat and light; (4) low volatility; (5) good cold tolerance; (6) low migratability; (7) resistance to extraction by water, oil, and solvents; (8) good electrical insulation; (9) flame retardancy; (10) no toxicity, color, or odor; (11) resistance to mildew; (12) resistance to contamination; (13) being non-sticky; (14) low cost.

The plasticizers in PVC are in 80% of the total use of plasticizers. The plasticization effect depends on the following factors: (1) the chemical structure and composition of plasticizers; (2) the average molecular weight of plasticizers; (3) the functional groups in plasticizers. In general, plasticizers with lower molecular weights or less polar groups provide higher flexibility and better plasticization.

The selection of a proper plasticizer needs to consider the following details: (1) whether it has good compatibility with the polymer; (2) whether it is easily manufacturable; (3) whether it has sufficient thermal, electrical, and mechanical stability; (4)

whether it has good resistance to water, corrosion, ultraviolet, aging, dust pollution, and microbial attacks; (5) its effect on the rheology of the polymer; (6) its toxicity; (7) its cost. Therefore, most plasticizers are high-boiling-point organic resins or liquids, or low-melting-point solids.

1.2.6 Inspection Standards for Plasticizers

Here is a list of the major inspection standards of plasticizers:

(1) Plasticizers—Determination of color; GB/T 1664–1995.
(2) Plasticizers—Determination of acid value and acidity; GB/T 1668–2008.
(3) Plasticizers—Determination of flashing point; GB/T 1671–2008.
(4) Plasticizers—Determination of density and relative density; GB/T 4472–2011.
(5) Plasticizers—Determination of refractive index; GB/T 6488–2008.
(6) Plasticizers—Determination of water content; GB/T 11,133–2015.
(7) Plasticizers—Determination of purity; GB/T 9722–2006.
(8) Plasticizers—Determination of composition; DZ/T 0223–2001.
(9) Plasticizers—Determination of Shore hardness A; GB/T 2441–2004.
(10) Plasticizers—Determination of tensile properties; GB/T 1040.3–2006.
(11) Plasticizers—Determination of Brittleness Temperature; ASTM D746-2007.
(12) Plasticizers—Determination of fogging characteristics (gravimetric); DIN 75,201 B.
(13) Plasticizers—Determination of Compatibility in PVC Plastics; ASTM D3291-1997(2003).
(14) Plasticizers—Determination of aging under UV Exposure; ASTM G154-2006.
(15) Plasticizers—Determination of metal content; DZ/T 0223–2001.

1.2.7 Types of Plasticizers

1. Benzoate plasticizers

Benzoate plasticizers include phthalates, terephthalates, trimellitates, etc. The common feature is a benzene ring in its molecular structure. 90% of the total plasticizers are phthalates such as DOP, DBP, DINP, and so on. Table 1.2 lists their commonly seen types.

The currently most used phthalate plasticizers have the following advantages: (1) no chemical reactions with polymer resins at room temperature; (2) ease of melting; (3) good electrical insulation when added to polymers; (4) enabling high flexibility and strength at low temperatures; (5) low volatility in air; (6) low cost.

Table 1.2 The commonly used phthalic plasticizer

Abbreviations	Molecular formula	English name	CAS#
DINP	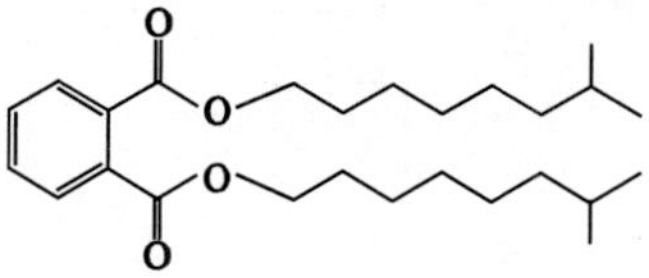	diisononyl phthalate(GB)	28,553-12-0
DIDP	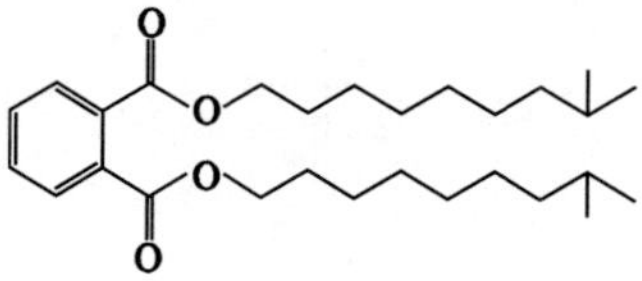	diisodecyl phthalate(GB)	26,761-40-0
DOP	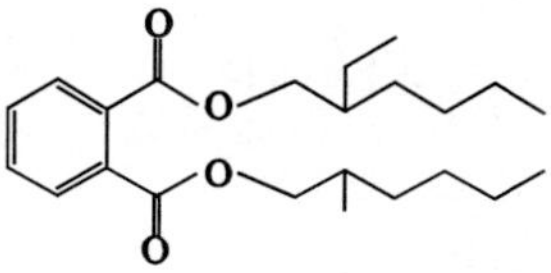	dioctyl phthalate(GB), bis(2-ethylhexyl) phthalate(IUPAC)	117-81-7
DNOP	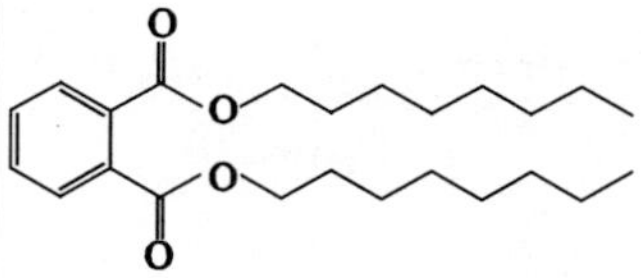	di-*n*-octyl phthalate(GB), dioctyl phthalate (IUPAC)	117-84-0
DIOP	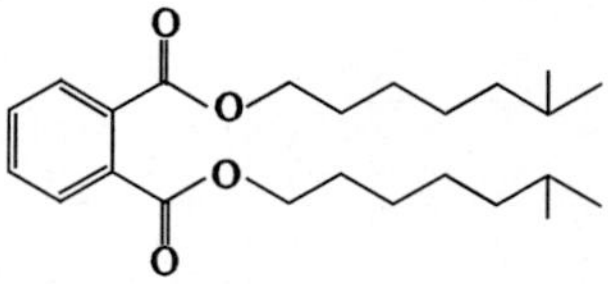	diisooctyl phthalate(GB)	27,554-26-3
DBP	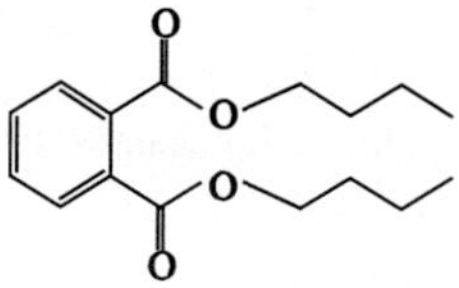	dibutyl phthalate(GB)	84-74-2
DIBP	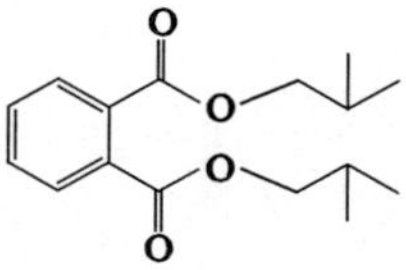	diisobutyl phthalate(GB)	84-86-9

(continued)

Table 1.2 (continued)

Abbreviations	Molecular formula	English name	CAS#
BBP	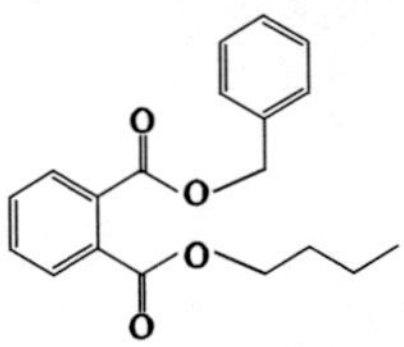	benzyl butyl phthalate(GB)	85-68-7
DMP	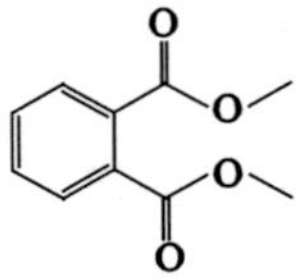	dimethyl phthalate(GB)	131-11-3
DCHP	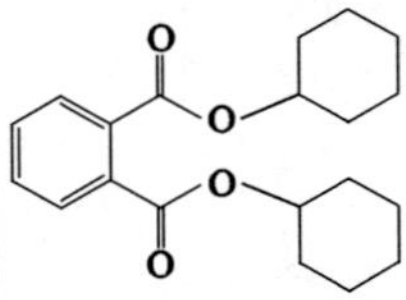	dicyclohexyl phthalate(GB)	84-61-7
DPP	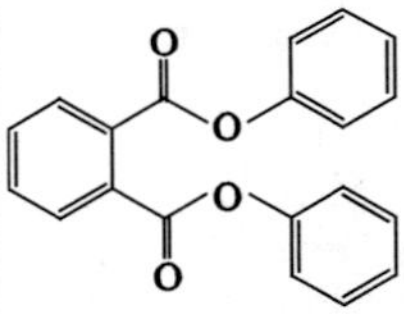	diphenyl phthalate(GB)	84-62-8
DEP	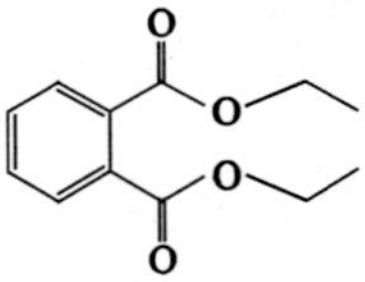	diethyl phthalate(GB)	84-66-2
DHXP	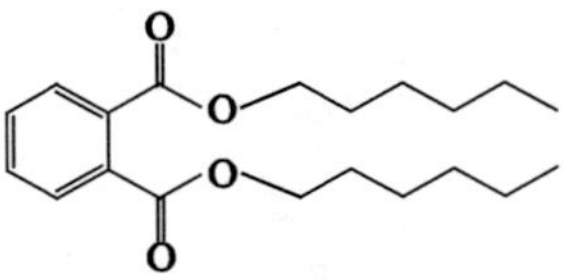	dihexyl phthalate(GB)	84-75-3
DNP	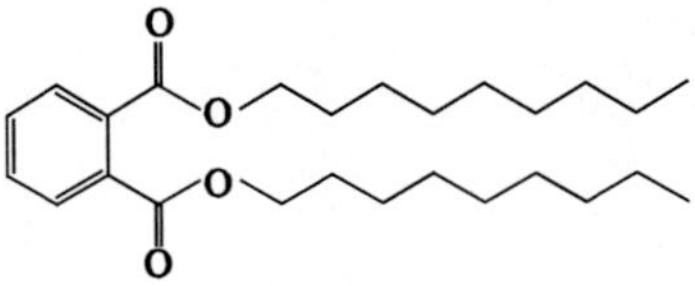	dinonyl phthalate(GB)	14,103-61-8

The benzene ring provides phthalates with superior compatibility with the polymer and meanwhile makes the polymer chain highly flexible. Longer alkyl chains in phthalates result in less miscibility with polymer. Shorter chain phthalates diffuse faster in the polymer enabling more space for polymer chain mobility, but are more volatile.

The origin of the toxicity of benzoates relies on the benzene ring with conjugated π bonds, leading to slow natural degradation. These plasticizers can enter human and animal bodies via migration, extraction, or evaporation, and the residues can accumulate in our environments. Therefore, phthalates have been strictly banned by EU's REACH and RoHS 2.0, and the USA's CPSIA, but without DOTP, trioctyl trimellitate (TOTM), ethylene glycyl dibenzoate, alkyl benzenesulfonates, and di (2-propylheptyl) phthalate (DPHP) in the banned list.

2. **Non-benzoate plasticizers**

Non-benzoate plasticizers mainly include aliphatic acid esters, citrates, epoxy plant oils, polyesters, cyclohexane-1,2-dicarboxylates, etc.

Aliphatic acid esters, such as dialkyl adipates, dialkyl azelates, dialkyl sebacates, and triglycerides, can only be used as secondary plasticizers because of their low molecular weights, inferior PVC compatibility, and risk of leaching (typically in small amounts as cold-tolerant plasticizers).

In comparison, citrates, such as acetyl tributyl citrate (ATBC), acetyl trioctyl citrate, and n-Butyryl tri-n-hexyl citrate, have better PVC compatibility, are non-toxic, are odorless, are flame retarding, and are degradable. But citrate plasticizers can only be used in PVC articles because of their low boiling points, high volatility, inferior heat tolerance, ease of aging, and high cost.

Epoxy plant oil plasticizers, including epoxy soybean oil, epoxy linseed oil, etc., can improve the anti-aging and anti-shock properties of PVC and also have good light and heat stability [26–29]. In the recent 2 years, the use of epoxy plant oils has increased due to their heat tolerance and performance improvement, but are used only as secondary plasticizers due to the technical hurdle of leaching.

Polyester plasticizers, such as polyethylene glycol (PEG), are produced via condensation polymerization between polyacids and polyols. They have molecular weight similar to PVC, are non-volatile and resistant to leaching out plastics. Thus, they have remarkable plasticization effects and are also called permanent plasticizers with heat tolerance, anti-aging, anti-extraction, and anti-migration properties. Different raw materials lead to different viscosities and performance. But currently, polyesters are not manufacturable in China due to production difficulty and narrow use range [30, 31].

Cyclohexane-1,2-dicarboxylates are a type of non-benzoate plasticizers specially developed for sensitive uses of PVC. They have similar plasticization effects to phthalates and also have anti-ultraviolet and anti-aging advantages as well as non-toxicity demonstrated in animal tests [32]. Albeit, the molecular weights are similar to phthalates; they also have the leaching issue due to poor PVC compatibility resulting from molecular polarity.

In addition, other types of plasticizers include ionic liquids, such as [$Bmim^+$][PF_6^-], [$Hmim^+$][$doss^-$], and [$ThtdPh^+$][Cl^-], chlorinated paraffin, phosphates, etc. [32–37].

1.3 Cyclohexane-1, 2-Dicarboxylate Plasticizers

Cyclohexane 1,2-dicarboxylate plasticizers are produced via catalytic hydrogenation of related phthalates. In 2002, BASF in Germany achieved industrial production of DINCH (di-isononyl-cyclohexane-1,2-dicarboxylate) by hydrogenation of DINP. The product DINCH is considered as a new type of non-phthalate plasticizer. With no benzene ring in its structure, DINCH is low-toxic and can be used as a manufacturing additive for PVC articles that have high health requirements. Therefore, it was recommended by the German Federal Institute for Risk Assessment (BfR) and was thought highly of by the European Food Safety Authority (EFSA). Now it has a continuously growing market in PVC articles such as toys and medical devices. The migration of certain elements can be controlled below 60 mg per kg of food, which is within the international standards. Therefore, DINCH is widely used in PVC articles such as food wraps, drinking straws, and food seals [38].

According to public reports, cyclohexane 1, 2-dicarboxylates not only afford better plasticization of PVC than phthalates, but also have no environmental toxicity, no carcinogen, no peroxisome proliferation, no biological accumulation, and no reproductive toxicity. Cyclohexane 1,2-dicarboxylates plasticizers have been considered as the most promising phthalate alternative for use in food packages, medical supplies, and children's toys. A number of countries have approved their use in food packages, medical devices, and children's toys without harm to human health [39–41].

China has started the investigations of the hydrogenation of phthalates to produce cyclohexane-based plasticizers [42, 43], but the commercialization has not been achieved. Currently, the main channel for using cyclohexane-based plasticizers relies on the importation of DINCH products from BASF.

It is reported that there are two methods to produce cyclohexane-based plasticizers: esterification and hydrogenation.

Esterification: using cyclohexane 1, 2-dicarboxylic acid or anhydride as the raw materials to undergo esterification reaction with alcohols to produce the esters, as shown in Fig. 1.1. This method is simple but the raw materials are very expensive.

Hydrogenation: using maleic anhydride and 1,3-butadiene as the raw materials to synthesize 1,2-cyclohex-diene dicarboxylate, followed by hydrogenation; or hydrogenating cyclohexene dicarboxylic anhydride to make cyclohexane dicarboxylic anhydride, followed by esterification reaction with alcohols to produce cyclohexane 1,2-dicarboxylate esters. This method is shown in Fig. 1.2. This method involves multiple reactions and is difficult to develop. Meanwhile, this method has problems such as side products and poor recyclability.

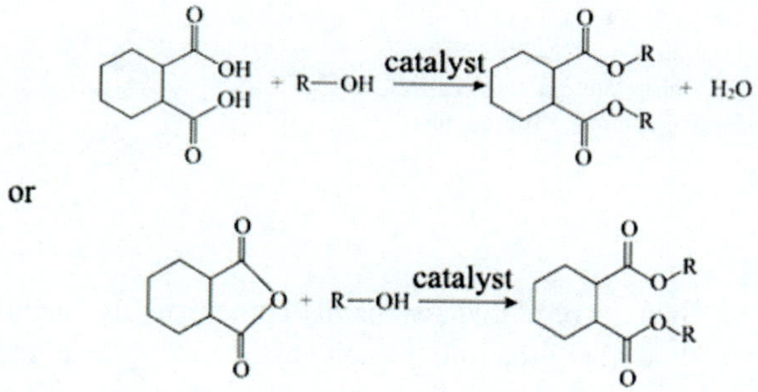

Fig. 1.1 Preparation of cyclohexane-1,2-dicarboxylate plasticizers via esterification

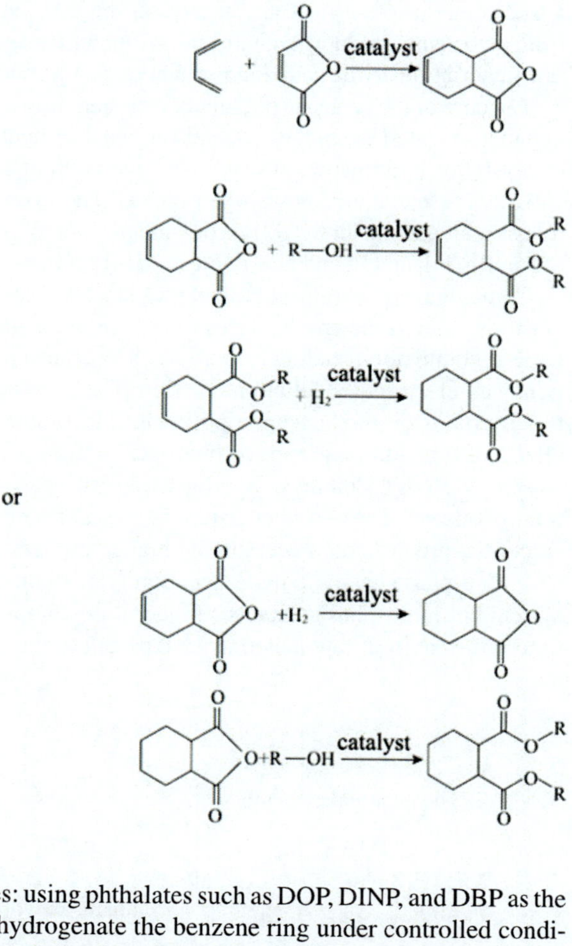

Fig. 1.2 Preparation of cyclohexane-1,2-dicarboxylate plasticizers via hydrogenation

Hydrogenation of phthalates: using phthalates such as DOP, DINP, and DBP as the raw materials to catalytically hydrogenate the benzene ring under controlled conditions to produce cyclohexane 1,2-dicarboxylates [44, 4546], as shown in Fig. 1.3. With the only need for a hydrogenation facility, China has been in an advantageous position because industrial facilities for phthalate production have already been established in China.

Fig. 1.3 Preparation of cyclohexane-1,2-dicarboxylate plasticizers via hydrogenation of phthalates

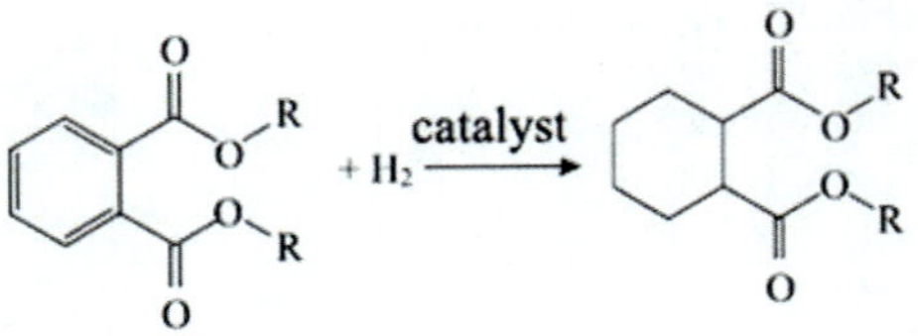

Sources of phthalates mainly come from the naphthalene method or the o-xylene method via oxidation of naphthalene or o-xylene, respectively, to produce phthalic anhydride, followed by esterification reaction with alcohols, as shown in Fig. 1.4. Because naphthalene comes from coal tar, the naphthalene method product contains a trace amount of sulfur that can poison the catalyst. But o-xylene that originates from petroleum products contains no sulfur contaminations and thus is the optimum raw material for hydrogenation production of cyclohexane-based plasticizers.

The core of the catalytic phthalate hydrogenation technology is the catalyst. Since Sabatier invented the nickel catalyst for benzene hydrogenation in 1926 [47], metal catalysts for hydrogenation reactions have continuously been developed, which is still a hot research area nowadays [48–55]. The hydrogenation catalysts are usually transition metals, with the VIII group metals mostly common including noble metals of Ru, Rh, Pd, and Pt and non-noble metals of Raney Co, Raney Ni, and Ni–Al alloy.

Compared with unsaturated aldehyde, ketone, alkene, nitro compounds, the hydrogenation of benzene and its derivatives is more challenging, mainly because of the large π-conjugation leading to more stable chemical properties. Benzene hydrogenation is an electrophilic addition reaction. The presence of electron-donating substitution groups on the benzene ring is more favorable for the reaction [56–61]. The effects of substitution groups on hydrogenation activity follow the trend: tert-butyl > iso-propyl > ethyl > methyl > hydrogen; phenol > para-chloroaniline > chlorobenzene > nitrobenzene. The two ester groups in phthalates make the catalytic hydrogenation reaction more difficult, necessitating higher temperatures and pressures [62–64].

Liquid-phase catalytic hydrogenation typically uses a solid–liquid–gas triphasic system but multi-phasic catalysis kinetics are complicated. To simplify the separation of the catalyst, raw materials, and products, the catalysts are mostly supported.

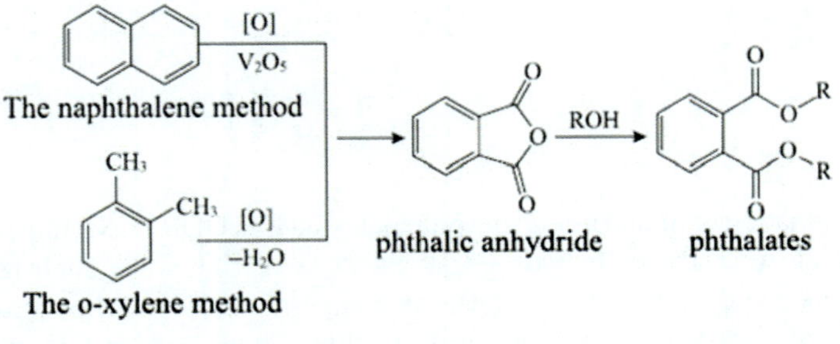

Fig. 1.4 Preparation of phthalate plasticizers via the naphthalene method and the o-xylene method

The support can improve the dispersion and utilization of the catalyst, and different substrates will have different effects on the hydrogenation process [65, 66].

Because of the high boiling points, hydrogenation of phthalates after gasification will cause high energy consumption and high equipment and facility requirements. Thus, liquid-phase hydrogenation will be adopted. Co and Ni have relatively low catalytic activity and require higher reaction temperatures. So, the major research focus is on the noble metals of Pd, Rh, Ru, Pt, etc.

Hydrogenation of phthalates will convert the benzene ring to cyclohexane while maintaining the plasticizing esters. If the benzene ring is not fully converted, separation of unreacted phthalates from the final product to reach a $< 0.1\%$ content is a must. Physical property tests must be carried out to determine whether the products can replace phthalates. Mechanisms of the toxicity and disease and human exposure limit need to be tested to determine whether the articles can be used in medical devices. Therefore, the following three major technical bottlenecks need to be overcome for the development and application of phthalate hydrogenation technology:

(1) Synthesis of catalyst: using high-activity, high-selectivity catalysts to convert phthalates under mild conditions while suppressing side products.
(2) Separation of product: the products have very similar boiling points to unreacted phthalates and are difficult to separate out via conventional distillation; phthalates may undergo reverse or isomerization reactions upon heating to produce acids, anhydrides, alcohols, monoesters, or even low-boiling-point alkenes/alkanes via further decarboxylation, making product separation very difficult.
(3) Use of product: the raw materials, particulate materials, and articles must follow the national standards (GB/T 16,881–2011) and industrial standards, and must be certified, in terms of physical properties, biological properties, drug compatibility, pharmacology, toxicology, and human exposure limits, by widely accepted authorization organizations. The physico-chemical properties and safety characteristics, especially the literature and documents of safety history, formula design, technical protocols, quality control, etc., must be provided.

1.4 Environmentally Friendly Plasticizers

Environmentally friendly plasticizers mainly refer to benzene-free non-phthalate plasticizers, as well as benzoate plasticizers such as terephthalates, trimellitates, and trimesates that are not in the banned list of EU's REACH, RoHS 2.0, and the USA's CPSIA [67–70]. However, with the implementation of these regulations, more and more strict limitations on plasticizers have been established by the EU's PVC market that even require no acceptable use of benzene-containing plasticizers. "Benzene-free" structure has become an important future direction in global chemical development. According to this viewpoint, DOTP is only a temporary plasticizer product despite its relatively high market acceptance. Its toxicity tests are already under way.

In China, more than 80% of the overall production capacity is on phthalate plasticizers and only less than 20% on non-phthalate plasticizers. With the implementation of EU and USA regulations, the production, sale, and consumption of plasticizers have all dropped [71, 72].

According to environmental protection and the increasing concerns of food safety, the development of environmental plasticizers has become a general and necessary trend. This relies on (1) the development of aliphatic acid ester, citrate, epoxy plant oil, and polyester plasticizers; and (2) catalytic hydrogenation of phthalates.

China has a solid industrial foundation for catalytic hydrogenation of phthalates with the use of currently available resources and technologies. Meanwhile, the production chain can be extended for upgrading and green development.

1.5 Developmental Direction of Plasticizer Industry in China

The national "Thirteenth Five-Year Plans" have proposed the economic concept of "innovative, coordinated, green, open, and inclusive development" and established the fundamental strategies of resource conservation and environmental protection. The plastics industry needs to set the developmental direction of environmental plasticizers.

We should actively participate in international policy making to increase our voices. The developed countries are trying to set barriers for the importation of Chinese products and protect their own market via increasing the inspection standards and items. We should fight back with our strong capability and respond actively.

We should promote the sci-tech innovation projects for green plasticizers and create a market-oriented, industry-centered, research instituted-supported, research-education-production-coordinated innovation mechanism. The purposed is to achieve critical technical breakthroughs and provide a strong intellectual foundation for advancing the green plasticizer industry, via the establishment of key laboratories, technological allies, inspection standards and systems, technical centers, R&D bases, industrial incubators, intellectual property service platforms, and press release centers for green plasticizer industry. We should strengthen the industrial integration and informationalization–industrialization integration to increase the green plasticizer developmental capability. The way of doing it is to form new profit chains, develop high-end products, and achieve continuous innovations via mutual union, penetration, and promotion of raw material manufacturing and end use. Meanwhile, we will strive to build our advantages in commercial scale-up, environmental protection, quality control, and added values of products and gain space in international market competitions [73].

1.6 Main Contents of This Book

The following important books have greatly promoted the plasticizer industry in China: "Plasticizers and Their Applications" edited by Wancong Shi, Zhibo Shi, and Pingping Jiang [74]; "Handbook of Plasticizers" edited by Wancong Shi, Junjie Si, and Wenguo Liu [75]; "Environmental Plasticizers" edited by Pingping Jiang and Yongfang Zhou [76]; "Examples of Plasticizer Formulations and Applications" edited by Duoren Wang [38]; etc.

Since 2001, in order to break through the technological monopoly by foreign countries, our research team has carried out the basic and applied sciences of selective hydrogenation of benzene to prepare cyclohexene [77–101102]. The selective hydrogenation catalyst technology we developed is resource-saving and environmentally friendly. Its application has been promoted since 2010 leading to exceptional social and economic benefits. In recent years, to tackle the technical and trade barriers of our plasticizer industry, our team has carried out the research on hydrogenation of phthalates. With similar research and scholarly concepts, these two technological problems need urgent solutions to advance industrial transformation and green development.

The main contents of this book cover the fundamentals of plasticizers, domestic and foreign laws and regulations, research status and invention patents, and the R&D, pilot, and industrialization of catalytic phthalate hydrogenation technology. The goal is to develop the catalytic phthalate hydrogenation technology with our independent intellectual property, offer technical and product supports for industrial transformation and green development of plasticizers, and provide useful references for relevant professions.

References

1. Staples CA, Peterson DR, Parkerton TF, et al. The environmental fate of phthalate esters: a literature review [J]. Chemosphere. 1997;35:667–749.
2. Onorato TM, Brown PW, Morris P. Mono-(2-ethylhexyl)phthalate increase spermatocyte mitochondrial peroxiredoxin and cyclooxygenase [J]. J Androl. 2008;29:293–303.
3. Erkekoglu P, Rachidi W, Favier A, et al. Evaluation of cytotoxicity and oxidative DNA damaging effects of di(2-ethylhexyl)-phthalate (DEHP) and mono (2-ethylhexyl)-phthalate (MEHP) on MA-10 leyding cells and protection by selenium [J]. Toxicol Appl Pharmacol. 2010;248(1):52–62.
4. Hao YL, Zheng GY, Li QZ et al. Combined toxic effects of DBP and DEHP on sperm in male mice. Du W J. Informatics and Management Science I[M]. London: Springer, 2013, 204: 729–734.
5. Luo ZH, Huang XL, Ye DZ. Advances in research of biodegradation of environmental endocrine disruptors-phthalate esters' [J]. Chin J Appl Environ Biol. 2008;14(6):890–7.
6. Piersma AH, Verhoefa A, Biesebeeka J. Developmental toxicity of butyl benzyl phthalate in the rat using a multiple dose study design [J]. Reprod Toxicol. 2000;14:417–25.
7. Singh S, Li SS. Phthalate: toxicogenomics and inferred human diseases [J]. Genomics. 2011;97(3):148–57.

8. Murphy J. Additives for Plastics Handbook [M]. 2nd ed. New York: Elsevier; 2001.
9. Peter K. How dangerous are phthalate plasticizers? Integrated approach to toxicity based on metabolism, electron transfer, reactive oxygen species and cell signaling [J]. Med Hypotheses. 2010;74(4):626–8.
10. Nass LI, Heiberger CA. Encyclopedia of PVC [M], vol 2, 2nd ed. New York: Marcel Dekker, 1988: 143–151.
11. Edenbaum J. Plastics additives and modifiers handbook [M]. New York: Van Nostrand Reinhold; 1992.
12. Haim J, Hyatt D. International plastics handbook [M]. 3rd ed. New York: Carl Hanser; 1995.
13. Wypych G. Plasticizer types. Wypych G. Handbook of Plasticizers [M]. Toronto: ChemTec Publishing, 2004.
14. Bonilla E, del Mazo J. Deregulation of the Sod1 and Nd1 genes in mouse fetal oocytes exposed to mono-(2-ethylhexyl) phthalate (MEHP) [J]. Reprod Toxicol. 2010;30(3):389–92.
15. Kricheva YI, Kostov VD, Chichovska M. *In vitro* and *in vivo* studies of the concentration of plasticizer di (2-ethylhexyl) phthalate on the blood compatibility of plasticized poly(vinyl chloride) drain tubes [J]. Biomaterials. 1995;16(7):575–9.
16. Li X, Nie XP, Pan DB, et al. Content and distribution characteristics of phthalates in cultured fish [J]. J Environ Health. 2008;25(3):202–5.
17. Xiao KE, Mo CH, Cai QY. Concentrations of PAEs in vegetable fidds of pearl river delta [J]. Sichuan Environ. 2012;34(3):49–55.
18. Kirkpatrick A. Some relations between molecular structure and plasticizing effect [J]. J Appl Phys. 1940;11:255–61.
19. Aiken W, Alftey T, Janssen A, et al. Creep behavior of plasticized vinylite VYNW [J]. J Polym Sci. 1947;2:178–95.
20. Wadey BL. An innovative plasticizer for sensitive applications [J]. J Vinyl Addit Technol. 2003;9(4):172–6.
21. Daniels PH. A brief overview of theories of PVC plasticization and methods used to evaluate PVC-plasticizer interaction [J]. J Vinyl Addit Technol. 2009;15(4):219–23.
22. Moreno R. The role of slip additives in tape casting technology. II: binders and plasticizers [J]. Am Ceram Soc Bull. 1992; 71(11):1647–57.
23. Rahman M, Brazel CS. The plasticizer market: An assessment of traditional plasticizers and research trends to meet new challenges [J]. Prog Polym Sci. 2004;29(12):1223–48.
24. Jiang PP, Cong MB, Zhang DP et al. Accelerate the research and application of citrate ester, a new environmental friendly plasticizer in China [J]. Plast Add. 2003; 4(5):1–8.
25. Jin D. Production and application of green plasticizer tributyl citrate [J]. Chem Technol Mark. 2009;32(4):18–21.
26. Zhang L. Market status and prospect of citric acid ester plasticizer [J]. Plast Add. 2008; (1):10–1.
27. Adhvaryu A, Erhan SZ. Epoxidized soybean oil as a potential source of high-temperature lubricants [J]. Ind Crop Prod. 2002;15(3):247–54.
28. Lu H, Bi YL, Liu YL. Development status and prospect of epoxy plasticizer [J]. Grain oil processing. 2010;6:26–30.
29. He SF. Discussion on quality and control of epoxy soybean oil [J]. Guangzhou Chem Ind. 2005;33(3):81–2.
30. Tang YF. Development and application of green environmental protection polyester plasticizer, [C]. Nanjing: Nanjing Publishing media Group; 2016.
31. Gosse C, Larson TM, Legrand PJP et al. Plasticized polyvinyl chloride: USA, US 7297738 [P]. 2007–11–20.
32. He YC. Discussion on plasticizer in food packaging materials [J]. Grand View Weekly. 2012;51(5):141–2.
33. Frados J. Plastics engineering handbook of the society of the plastics industry [M]. New York: Van Nostrand Reinhold; 1976.
34. Harper CA. Modern plastics handbook [M]. New York: McGraw-Hill Professional; 2000.

35. Hansjürgen S. International plastics handbook: for the technologist, engineer and user [M]. Munchen: Hans Publishers; 1987.
36. Földes E. Study of the effects influencing additive migration in polymers [J]. Angew Makromol Chem. 1998;261(1):65–76.
37. Rahman M, Brazel CS. Ionic liquids: new generation stable plasticizers for poly (vinyl chloride) [J]. Polym Degrad Stab. 2006;91(12):3371–82.
38. Jiang PP, Zhou YF. Environmental protection plasticizer [M]. Beijing: National Defence Industry Press; 2009.
39. Shi WC. New plasticizer cyclohexane dicarboxylic acid dialkyl ester [J]. Plastic Add. 2010;1:17–22.
40. Shi WC. Toxicity of plasticizers and related restrictions [J]. Plastic Add. 2010;3:43–7.
41. Brian LW. An innovative plasticizer for sensitive applications [J]. J Vinyl Addit Technol. 2003;9(4):172–6.
42. Ding YJ, Diao YJ, Ma LX. The invention relates to a catalyst for the preparation of 1, 2-cyclohexane diformic acid ester: China, CN 200810224900. 9 [P]. 2008–10–24.
43. Ding Y J, Diao Y J, Ma L X. The invention discloses a method for preparing 1, 2-cyclohexane diformic acid ester: China, CN 200810224912. 1 [P]. 2008–10–27.
44. Li CC, Jiang PP, Dong YM. Synthesis of environmental protection plasticizer cyclohexane 1, 2-diisooctyl dicarboxylate [J]. Fine Chem. 2011;28(11):1124–8.
45. Jiang PP, Shen ML, Dong YM. Homogeneous catalytic synthesis of polyhexahydrophthalic anhydride diethylene glycol ester by titanium complex [J]. Chem Prog. 2011;30(3):634–7.
46. Zhao J, Xue M, Huang Y, et al. Hydrogenation of dioctyl phthalate over supported Ni catalysts [J]. Catal Commun. 2011;16(1):30–4.
47. Sabatier P. How I have been led to the direct hydrogenation method by metallic catalysts [J]. Ind Eng Chem. 1926;18(10):1005–8.
48. Hu SC, Chen YW. Partial hydrogenation of benzene: a review [J]. J Chin Inst Chem Eng. 1998;29(6):387–96.
49. Rothwell IP. A new generation of homogeneous arene hydrogenation catalysts [J]. Chem Commun. 1997;15:1331–8.
50. Harman WD. The activation of aromatic molecules with pentaammine-osmium (II) [J]. Chem Rev. 1997;97(6):1953–78.
51. Stanislaus A, Cooper BH. Aromatic hydrogenation catalysis: A review [J]. Catal Rev: Sci Eng. 1994;36(1):75–123.
52. Muetterties EL, Bleeke JR. Catalytic hydrogenation of aromatic hydrocarbons [J]. Acc Chem Res. 1979;12(9):324–31.
53. Maitlis PM. (Pentamethylcyclopentadienyl) rhodium and-iridium complexes: approaches to new types of homogeneous catalysts [J]. Acc Chem Res. 1978;11(8):301–7.
54. Moyes RB, Wells PB. The chemisorption of benzene [J]. Adv Catal. 1973;23:121–56.
55. Garnett JL. π-Complex intermediates in homogeneous and heterogeneous catalytic exchange reactions of hydrocarbons and derivatives with metals [J]. Catal Rev. 1972;5(1):229–67.
56. Mccusker-Orth JE. Process for a cyclohexanedimethanol using raney metal catalysts: USA, US 6919489 [P]. 2005–07–19.
57. Sumner J, Gustafson BL. Hydrogenation of phthalic acids to cyclohexane-dicarboxylic acid: USA, US 6291706 [P]. 2001–09–18.
58. Gustafson B L, Kuo Y J, Ptice T W, et al. Low pressure process for the hydrogenation of dimethyl benzenedicarboxylates to the corresponding dimethyl cyclohexanedicarboxlates: USA, US 5286898 [P]. 1994–02–15.
59. Bohnen H, Klein T, Bergrath K. Process for preparing cyclohexanedicarboxylic esters: USA, US 6740773[P]. 2004–05–25.
60. Du X, Shi Y, Hu JY. Comparative study on liquid-phase hydrogenation activities of different substituted benzenes over Ru/AC catalysts [J]. J Sichuan Univ. 2004;41(2):384–8.
61. Chaeles ESJ, Gustafson BL. Hydrogenation of phthalic acids to cyclohexane-dicarboxylic acid: USA, US 6291706 [P]. 2001–09–18.

62. Toppinen S, Rantakyla TK, Salmi T, et al. Kinetics of the liquid phase hydrogenation of di- and trisubstituted alkylbenzenes over a nickel catalyst [J]. Ind Eng Chem Res. 1996;35(12):4424–33.
63. Lu F, Liu J, Xu J. Fast aqueous/organic hydrogenation of arenes, olefins and carbonyl compounds by poly (*N*-vinylpyrrolidone)-Ru as amphiphilic microreactor system [J]. Adv Syn Catal. 2006; 348(7–8):857–61.
64. Lu F, Liu J, Xu J. Synthesis of PVP-Ru amphiphilic microreactors with Ru nanocatalysts and their application in the fast hydrogenation of unsaturated compounds in aqueous media [J]. J Mol Catal A: Chem. 2007;271(1):6–13.
65. Hu SC, Chen YW. Effect of preparation on Ru-Zn ultrafine catalysts in partial hydrogenation of benzene [J]. Ind Eng Chem Res. 2001;40(14):3127–32.
66. Li TX, Zhang XF. Study on amorphous Ultrafine Nickel alloy catalyst ii. Kinetics of Hydrogenation of benzene over Ni-B catalyst [J]. Chin J Catal. 1995; 16(5):399–402.
67. Hileman B. Panel concludes phthalate is safe [J]. Chem Eng News. 2001;79(26):12–5.
68. Hileman B. FDA suggests replacing DEHP in plastics [J]. Chem Eng News. 2002;80(37):6–9.
69. Wang B, Wang KZ, Gong YL. Research progress of environmental protection plasticizer [J]. China Plast Ind. 2013;41(5):12–5.
70. Ventrice P, Ventrice D, Russo E, et al. Phthalates: European regulation, chemistry, pharmacokinetic and related toxicity [J]. Environ Toxicol Pharmacol. 2013;36(1):88–96.
71. Sang L. Current situation and development of plasticizer industry [J]. Sci Technol Inform Devel Econ. 2009;19(27):124–5.
72. Shi WC, Zhao CY. Toxicity of plasticizers and related regulations [J]. Plastic Addit. 2007;2:46–51.
73. Feng SJ. Green plasticizer additives boost the green transformation of plastic artificial leather, plastic floor leather and plastic carpet [C]. Nanjing: Nanjing Publishing media Group; 2016.
74. Shi WC, Shi ZB, Jiang PP. Plasticizer and its application [M]. Beijing: Chemical Industrial Press; 2002.
75. Shi WC, Si JJ, Liu WG. Practical handbook for plasticizers [M]. Beijing: Chemical Industrial Press; 2009.
76. Jiang PP, Zhou YF. Environmental protection plasticizer [M]. Beijing: National Defence Industry Press, 2009.
77. Wang DR. Formulation process and application of plasticizer [M]. Beijing: Printing Industry Press; 2013.
78. Liu SC, Luo G, Han ML, et al. Characterization of the catalyst prepared by impregnation for selective hydrogenation of benzene to cyclohexene [J]. Chin J Catal. 2001;22:559–62.
79. Liu SC, Luo G, Xie YL. Study on Ru-Zn catalyst prepared by precipitation method for selective hydrogenation of benzene to cyclohexene [J]. Chin J Molec Catal. 2002;16(5):349–54.
80. Liu SC, Zhu BZ, Luo G, et al. Characterization of amorphous Ru-M-B/ZrO2 catalysts for partial hydrogenation of benzene to cyclohexene [J]. Chin J Molec Catal. 2002;16(3):217–22.
81. Liu SC, Guo YQ, Yang XL, et al. Kinetic equations for liquid-phase selective hydrogenation of benzene to cyclohexene [J]. Chin J Chem. 2003;24:41–6.
82. Liu SC, Wu YM, Wang Z, et al. Study of a Ru-La/ZrO2 catalyst prepared by precipitation method for selective hydrogenation of benzene to cyclohexene [J]. J Natur Gas Chem. 2005;14:226–32.
83. Si RJ, Liu SC, Liu ZY, et al. Study on Ru-M1-M2 catalyst prepared by precipitation method for selective hydrogenation of benzene to cyclohexene [J]. Chin J Molec Catal. 2005;19:141–5.
84. Liu SC, Liu ZY, Zhao SH, et al. Study on the nanosized amorphous Ru-Fe-B/ZrO2 alloy catalyst for benzene selective hydrogenation to cyclohexene [J]. J Natur Gas Chem. 2006;15:319–26.
85. Huang ZX, Liu ZY, Wu YM, et al. Selective hydrogenation of benzene to cyclohexene by a novel Ru-Zn catalyst [J]. Chin J Molec Catal. 2006;20(3):226–9.
86. Liu SC, Liu ZY, Liu YL, et al. Effect of lanthanum on performance of rub amorphous alloy catalyst for benzene selective hydrogenation [J]. J Rare Earths. 2006;24:456–60.

87. Liu SC, Liu ZY, Wang Z, et al. A novel amorphous alloy Ru-La-B/ZrO2 catalyst with high activity and selectivity for benzene selective hydrogenation [J]. Appl Catal A: Gen. 2006;313:49–57.
88. Liu SC, Liu ZY, Zhao SH, et al. Study on the amorphous alloy Ru-Fe-B/ZrO2 nano-sized catalyst for benzene selective hydrogenation to cyclohexene [J]. J Natur Gas Chem. 2006;150:1–9.
89. Liu SC, Liu ZY, Wang Z, et al. Characterization and study on performance of the Ru-La-B/ZrO2 amorphous alloy catalysts for benzene selective hydrogenation to cyclohexene under pilot conditions [J]. Chem Eng J. 2008;139:157–64.
90. Liu ZY, Sun HJ, Wang DB, et al. The modifiable character of a novel Ru-Fe-B/ZrO2 catalyst for benzene selective hydrogenation to cyclohexene [J]. Chin J Chem. 2010;28:1927–34.
91. Liu ZY, Sun HJ, Wang DB, et al. Selective hydrogenation of benzene to cyclohexene over Ru-Zn catalyst with nanosized zirconia as dispersant [J]. Chin J Catal. 2010;21:150–2.
92. Sun HJ, Guo W, Zhou XL, et al. Process in Ru-based amorphous alloy catalysts for benzene selective hydrogenation to cyclohexene [J]. Chin J Catal. 2011;32:1–16.
93. Sun HJ, Zhang XD, Chen ZH, et al. Monolayer dispersed Ru-Zn catalyst and its performance in the selective hydrogenation of benzene to cyclohexene [J]. Chin J Catal. 2011;32:224–30.
94. Sun HJ, Chen ZH, Guo W, et al. Effect of organic additives on the performance of nano-sized Ru-Zn catalyst [J]. Chin J Chem. 2011;29:369–73.
95. Zhou XL, Sun HJ, Guo W, et al. Selective hydrogenation of benzene to m cyclohexene on Ru-based catalysts promoted with Mn and Zn [J]. J Natur Gas Chem. 2011;20:53–9.
96. Sun HJ, Jiang HB, Li SH, et al. Selective hydrogenation of benzene to cyclohexene over nanocomposite Ru-Mn/ZrO2 catalysts [J]. Chin J Catal. 2013;34:684–94.
97. Sun HJ, Pan YJ, Li SH, et al. Selective hydrogenation of benzene to cyclohexene over Ce-promoted Ru catalysts [J]. J Eng Chem. 2013;22:710–6.
98. Sun HJ, Pany J, Jiang HB, et al. Effect of transition metals (Cr, Mn, Fe Co, Ni, Cu and Zn) on the hydrogenation properties of benzene over Ru-based catalyst [J]. Appl Catal A: Gen. 2013;464–465:1–9.
99. Sun HJ, Dong YY, Li SH, et al. The role of La in improving the selectivity to cyclohexene of Ru catalyst for hydrogenation of benzene [J]. J Molec Catal A: Chem. 2013;368–369:119–24.
100. Sun HJ, Jiang HB, Li SH et al. Effect of alcohols as additives on the performance of a nano-sized Ru-Zn(2.8%) catalyst for selective hydrogenation of benzene to cyclohexene[J]. Chem Eng J. 2013; 218:415–24.
101. Sun HJ, Wang HX, Jiang HB, et al. Effect of (Zn(OH)2)3(ZnSO4)(H2O)5 on the performance of Ru-Zn catalyst for benzene selective hydrogenation to cyclohexene [J]. Appl Catal A: Gen. 2013;450:160–8.
102. Liu Z Y, Liu SC, Li ZJ. Catalytic technology of selective hydrogenation of benzene to cyclohexene [M]. Beijing: science press.

Chapter 2
Foreign and Domestic Plasticizer Laws

2.1 Foreign Plasticizer Laws, Evolution

Since 1990, the plasticizer regulations and their evolutions in EU, USA, Japan, Canada, etc., are listed in Table 2.1.

2.2 Domestic Plasticizer Laws and Evolution

Since 2008, the plasticizer regulations and their evolutions in China are listed in Table 2.2.

In 2008, based on the Food Hygiene Law of the People's Republic of China, and referencing the Code of Federal Regulation and Commission Directive 2002/72/EC, China issued the Hygienic Standards for Uses of Additives in Food Containers and Packaging Materials (GB 9685-2008) to inspire more attention to the amount of phthalate plasticizer migration in model food. The Toys and Children's Articles-Determination of Phthalate Plasticizers in Polyvinyl Chloride Plastics (GB/T 22048-2008) indicated 6 phthalate plasticizers whose contents should be restricted. According to Determination of Phthalate Esters in Foods (GB/T 21911-2008), 16 types of phthalate plasticizers need to be controlled. In 2009, Limit of Harmful Substances of Coatings for Toys (GB 24613-2009) required the total content of DEHP, BBP, and DBP at no more than 0.1% and that of DIDP, DINP, and DNOP at no more than 0.1% in toys. On June 1, 2011, the Ministry of Health published put 17 types of phthalates into the 6th set of the List of Potential Illegal Inedible Substances and Abusive Additives in Food. In June 2011, the 551st Document signed by the Ministry of Health, i.e., Letter to Report the Maximum Phthalate Residue in Food and Additives, declared that the maximum residue of DBP was 0.3 mg/kg. In 2012, Determination of Phthalate Esters in Cosmetics (GB/T 28599-2012) indicated the 22

Z. Liu, *Green Catalytic Hydrogenation of Phthalate Plasticizers*,
https://doi.org/10.1007/978-981-16-9789-0_2

Table 2.1 Foreign laws and their evolutions for phthalate plasticizers

Time	Country and region	Law	Contents
12/1999	EU	1999/815/EC	The contents of 6 types of plasticizers in PVC articles that may contact the mouths of children below 3 years of age must not exceed 0.1%
2002	EU	2002/72/EC	The law lists the additives that can be used in manufacturing food contact plastic materials and articles as well as their restrictions and regulations
12/2005	EU	2005/84/EC	The contents of DEHP, DBP, and BBP in toys and children products much not exceed 0.1%
01/2007	EU	–	DBP, DEHP, and BBP are prohibited and DINP, DIDP, and DNOP are restricted for use in toys and children's products
06/2007	EU	REACH	Approximately 30000 chemicals must be registered to reinforce chemical product control
06/2008	EU	2007/19/EC	EU will ban the production and import of food contact plastic materials and articles that fail to meet the plasticizer content limits defined in 2007/19/EC
2009	EU	2009/48/EC	This law particularly prohibits the use of CMR (carcinogenic, mutagenic, reproductively toxic) materials in human contact parts of toys
08/2008	USA	CPSIA	Since February 10, 2009, the consumption, sale, and import of children's toys and care products having DEHP, DBP, and BBP contents exceeding 0.1% are prohibited. The consumption, sale, and import of edible children products having DINP, DIDP, and DNOP contents exceeding 0.1% are temporarily prohibited

(continued)

Table 2.1 (continued)

Time	Country and region	Law	Contents
01/2009	USA	AB1108	The sale, shipping, and production of children's toys and care products having DEHP, DBP, BBP, DINP, DIDP, and DNOP contents exceeding 0.1% are prohibited in California
08/2009	USA	ASTM F963-08	Children's pacifiers, rattles, and teethers must not contain DEHP
08/2009	Japan	Specifications and standards for food, additives, containers, packaging, toys, detergents	Toys must not use synthetic PVC resin with DEHP as the raw material
08/2009	Japan	Ordinance for Enforcement of the Food Sanitation Act	Toys must not use synthetic PVC resin with DINP as the raw material
08/2009	Japan	Toy Safety Standard 8th Edition (ST2002)	Synthetic PVC resin with DEHP, DBP or BBP as the raw material must not be used in toys. Synthetic PVC resin with DINP, DIDP or DNOP as the raw material must not be used in edible toys and toys for children under 3 years of age. Children's pacifiers and teethers must not use synthetic PVC resins
08/2009	Japan	Food Sanitation Act	DEHA must not be detected in PVC articles. The content of tricresyl phosphate must not exceed 0.1%
06/2009	Canada	–	The contents of DEHP, DBP, and BBP must not exceed 0.1% in all soft PVC-based children's toys and care products. The contents of DINP, DIDP, and DNOP must not exceed 0.1% in all edible children's toys and care products for children under 4 years of age
09/2008	Argentina	–	Since September 9, 2008, the production, import, export, sale or donation of children's toys and care products that contain >0.1% contents of the 6 phthalates must be prohibited

(continued)

Table 2.1 (continued)

Time	Country and region	Law	Contents
2009	Denmark	–	Besides the same content limits of the 6 plasticizers defined in EU laws, the phthalate contents in toys and care products for children below 3 years of age must not exceed 0.05%
02/2010	Malaysia	–	All toys for children below 14 years of age must pass the inspections and tests by Malaysian Specification and Inspection Co., Ltd. for sale

Table 2.2 Chinese laws and their evolutions for phthalate plasticizers

Time	Law	Contents
2008	Hygienic Standards for Uses of Additives in Food Containers and Packaging Materials (GB 9685-2008)	The standards closely followed the international standards. The concept of "specific migration limit" was introduced to raise attention to certain plasticizers with limited migration into food mimics
2008	Toys and Children's Articles-Determination of Phthalate Plasticizers in Polyvinyl Chloride Plastics (GB/T 22048-2008)	This law listed the 6 phthalates whose contents should be restricted
2009	Determination of Phthalate Esters in Foods (GB/T 21911-2008)	This law defined the 16 types of phthalates as controlled substances
2009	Limit of Harmful Substances of Coatings for Toys (GB 24613-2009)	This law required the total content of DEHP, BBP, and DBP not exceeding 0.1% and that of DIDP, DINP, and DNOP not exceeding 0.1% in toys
05/2011	Food Announcement #1001301729 (Taiwan Health Authorities)	This law showed the 6 controlled plasticizers possibly contained in food
06/2011	6th List of Potential Illegal Inedible Substances and Abusive Additives in Food (Ministry of Health)	This law listed 17 phthalates and pointed clearly out that "phthalates are not food raw materials or additives and are strictly prohibited to be added to food"
06/2011	Letter to Report the Maximum Phthalate Residue in Food and Additives (Ministry of Health #551 Document)	This law declared that the maximum residue of DBP was 0.3 mg/kg
2012	Determination of Phthalate Esters in Cosmetics (GB/T 28599-2012)	This law indicated the 22 controlled phthalate substances in cosmetics

(continued)

Table 2.2 (continued)

Time	Law	Contents
01/2016	Safety of Toys (GB 6675-2014)	This law listed 6 plasticizers including DBP as controlled substances with the same maximum content requirements as those of the EU
12/2016	Index of Government Encouraged Substitutes for Toxic, Harmful Raw Materials (Products) (2016)	This law supported substitutes of phthalate plasticizers with citrates, dioctyl terephthalate, and diacetyl epoxy vegetable glycerin oleate
02/2017	Technical Requirement for Environmental Labeling Products-Furniture (HJ 2547-2016)	The new standards added the content limits for migratable elements and phthalates in products

regulated phthalate substances in cosmetics. Safety of Toys (GB 6675-2014), effective on January 1, 2016, listed 6 plasticizers including DBP as regulated substances with the same maximum content requirements as those of the EU.

Therefore, the use of hexahydrophthalate esters as substitutes to the traditional phthalate ester plasticizers well agrees with the national development strategies.

2.3 Interpretations of Foreign and Domestic Plasticizer Laws

2.3.1 EU Bans

On December 7, 1999, EU approved the temporary ban 1999/815/EC throughout the EU member countries, which sets a <0.1% content limit of the six phthalate plasticizers (Table 2.3) in PVC articles with frequent mouth contact in children less than 3 years old. Now, this temporary ban has become a permanent ban.

In Table 2.3, the six phthalate plasticizers are abbreviated as 6P. In addition, there is also a 16P term to name the following 15 phthalate plasticizers, inlcuding diisononyl phthalate (DINP), di(2-ethyl hexyl) phthalate (DEHP), dibutyl phthalate

Table 2.3 The six phthalate plasticizers limitted in the temporary ban 1999/815/EC

#	English name	Abbreviations	CAS#
1	Di(2-ethylhexyl) phthalate	DEHP	117-81-7
2	Di-*n*-octyl phthalate	DNOP	117-84-0
3	Butyl benzyl phthalate	BBP	85-68-7
4	Dibutyl phthalate	DBP	84-74-2
5	Diisononyl phthalate	DINP	28553-12-0
6	Diisodecyl phthalate	DIDP	26761-40-0

(DBP), diisodecyl phthalate (DIDP), diisobutyl phthalate (DIBP), butyl benzyl phthalate (BBP), di(*n*-octyl) phthalate (DNOP), diisooctyl phthalate (DIOP), dimethyl phthalate (DMP), di(*n*-nonyl) phthalate (DNNP), diheptyl phthalate (DHP), dipentyl phthalate (DPP), diethyl phthalate (DEP), dicyclohexyl phthalate (DCHP), dipropyl phthalate (DPRP), and diisononyl adipate (DINA).

On December 14, 2005, EU's 2005/84/EC policy set limits for phthalates: (1) the contents of 3 phthalates (DEHP, DBP, and BBP) in toys and children care supplies must be lower than 0.1%; (2) the contents of 3 phthalates (DINP, DIDP, and DNOP) in mouth-contacting toys and children care supplies must be lower than 0.1%. Those containing such phthalates at >0.1% must not be sold in EU markets. This policy took effect on December 14, 2005. All EU countries have made this policy into a law before July 16, 2007, and have implemented it since January 16, 2008.

The 2007/19/EC policy requires no production and sale of food-contacting plastic materials and articles in the EU that contain plasticizers beyond the set content limits shown in Table 2.4.

The 2009/48/EC policy newly released by the EU sets even stricter requirements in concepts and scopes, safety limits, warning labels, responsibilities of sellers, and market supervision. CMR (carcinogenic, mutational, reproductive toxic) materials are particularly banned for use in contactable areas of toys.

Table 2.4 Phthalate-related limit index defined in EU's 2007/19/EC

Name	limit and/or index
Di(2-ethylhexyl) phthalate (DEHP) Dibutyl phthalate (DBP)	Only used for: (i) plasticizers in non-fat food contact materials and articles of repeated uses; (ii) the content of DEHP in final products not exceeding 0.1% and of DBP not exceeding 0.05%. $SML_{DEHP} = 1.5$ mg/kg; $SML_{DBP} = 0.3$ mg/kg
Butyl benzyl phthalate (BBP) phthalates with primary saturated C_8–C_{10} alkyl ($C_9 > 60\%$) phthalates with primary saturated C_6–C_{11} alkyl ($C_{10} > 90\%$)	Only used for: (i) plasticizers in materials and articles of repeated uses; (ii) plasticizers in disposable non-fat food contact materials and articles (except children products and products defined in 91/321/EEC and 1999/815/EC); (iii) not exceeding 0.1% in final products. $SML_{DBP} = 0.3$ mg/kg; $SML_{BBP} = 30$ mg/kg

Note SML is the specific migration limit

2.3.2 *EU's REACH Policy [1, 2]*

1. General Information of REACH

The full name of REACH is Registration, Evaluation, Authorization and Restriction of Chemicals. It is a management regulation and system for current and new chemicals, a law to strengthen and standardize the production and use rules of chemicals inside EU, and has been in effect since June 1, 2007. All raw materials and chemicals produced and imported in EU are under the regulations of REACH. Non-registered chemicals must not be used in EU market.

2. Contents of REACH

(1) Registration system: all chemicals produced or imported in EU at ≥1 t/a
(2) Evaluation system: evaluated collectively by EU Chemical Center
(3) Authorization: only for substances of very high concern (SVHC)
(4) Safety assessment: all known uses of chemicals (≥10 t/a)
(5) Risk management: chemicals with unacceptable risks will be banned.
(6) Information transparency: full information circulation of all chemicals in the whole supply chain
(7) Substances of very high concern (SVHC): i.e., those substances that may have serious or irreversible effects on human health and environment, such as carcinogen, gene mutation, reproductive toxicity, long environmental damages, bio-accumulation, etc. Currently, 168 substances are closely supervised in the SCHC list, which is also updated annually. A number of small-molecule phthalates are listed as SVHC by REACH.

The results of risk evaluation for DBP, BBP, DEHP, DINP, and DIDP by the European Food Safety Authority (EFSA) are shown in Table 2.5.

Table 2.6 lists the EU categories and labels of phthalates.

Table 2.5 Risk assessment results by the EFSA

Name	Notes
Dibutyl phthalate (DBP)	Male and female reproduction
Butyl benzyl phthalate (BBP)	Multi-generation study
Di(2-ethylhexyl) phthalate (DEHP)	Testicular toxicity
Phthalates with primary saturated C_8–C_{10} alkyl (C_9 > 60%)	No observable effects on chronic liver and kidney diseases related to peroxisome hyperplasia
Phthalates with primary saturated C_6–C_{11} alkyl (C_{10} > 90%)	Based on the effects on dog's livers (dogs are considered insensitive to peroxisome hyperplasia)

Table 2.6 EU categories and labels of phthalates

Small-molecule phthalates
Large-molecule phthalates
18 categories, reproductive toxicity

3. Effects of REACH

The REACH is apparently a regulation policy for the chemical management among EU members, as well as a law on the production, sale, and safety of chemicals. On the other hand, REACH is also a new trade barrier including technological, environmental, and social factors. Therefore, while the global economy is highly integrated nowadays, the REACH barrier will result in a huge negative impact on the economy and trade of China. It will not only greatly affect the production and commerce chemical industry and downstream industries, but also indicate the large gap existing for the Chinese laws and inspection/quarantine policy away from developed countries.

Since January 2007, the EU has banned the use of DBP, DEHP, and BBP and also has limited the use of DINP, DIDP, and DNOP in toys and other childcare articles. Meanwhile, many developed countries and regions have made policies and regulations to restrict or limit the use of certain plasticizers in toys, medical devices, food and drug packing materials, etc.

On July 3, 2009, the official website of RAPEX, EU's rapid alert system for dangerous non-food articles, announced that three sets of articles, part of which were made in China, failed to comply with the REACH. The children's toys among these articles had chemical hazards due to the existence of DEHP and DINP; thus, they were restricted for market sale and those already sold were recalled.

The REACH includes the requirements for the performance of chemicals, as well as strict regulations of their toxicity, mutagenic carcinogenicity, heredity, nervous system, allergy, immunity, etc. Therefore, all the sellers must establish the philosophy of green manufacturing and marketing of their products.

The fundamental pathway to breaking barriers such as REACH lies in the increase in product quality. This can be achieved via following the international standards, using high-tech and sustainable manufacturing, and researching the related laws and regulations.

2.3.3 The EU's RoHS [1, 3]

The Restriction of Hazardous Substances (RoHS) is an EU's law-reinforced mandatory standard, effective from July 1, 2006, which is mainly to regulate the materials and manufacturing protocols of electronics and electrical articles for better protection of human health and the environment.

In 2011, the EU issued the 2011/65/EU, i.e., RoHS 2.0, as part of CE (Conformite Europeanne) certifications, which is a must-have for all articles entering Europe. The new RoHS has the requirements of RoHS inspection and RoHS certification (i.e., the certificate and inspection report) for electronics and electrical articles. With the CE label, the manufacturers must ensure the articles to comply with RoHS and have the related disclaimer and technical documents ready.

All European members have translated the 2011/65/EU into their national laws before January 2, 2013. On June 4, 2015, the EU officially announced (EU)2015/863, a revision to RoHS 2.0, which puts DEHP, BBP, DBP, and DIBP in the list of limited substances increasing the number of regulated substances from 6 to 10. The EU members must translate this revision into their immediately effective laws before December 31, 2016. Starting on July 22, 2019, all electronics and electrical articles, except medical and monitoring devices, must comply with the new laws; and this will be the case for all the medical devices (including in vitro diagnostic devices) and the monitoring and control equipment (including those for industrial use) from July 22, 2021. The (EU)2015/863 revision is the law enacted by the EU to protect the environment and human health. The electronics and electrical equipment covered by it are shown in Table 2.7.

Table 2.7 RoHS 2.0 applicable product types

Numbering	Category	Product examples
1	Large appliances	Refrigerators, washers, microwaves, air conditioners, etc.
2	Small appliances	Vacuum cleaners, irons, blowers, clocks, etc.
3	IT and telecommunication devices	Computers, faxes, telephones, cell phones, etc.
4	Consumables	TV, radio, VCR, camera, etc.
5	Lighting (excluding tungsten bulbs and household lamps)	Non-household fluorescent lamps, neon lights, lighting control devices, etc.
6	Electrical tools (excluding large static industrial tools)	Driller, chainsaw, welder, mower, etc.
7	toys, sports	Electrical train toys, game console, large entertainment facilities such as race cars, treadmill, etc.
8	Medical devices	X-ray tester, ultrasonic, MRI, etc.
9	Product monitoring devices	Industrial monitoring devices, etc.
10	Vending machines	Various automatic machines for selling products such as drinks, candy, tokens, etc.
11	Other electrical and electronic devices not included in 1–10	Cables, other types of parts

2.3.4 USA's CPSIA

The EPA, in the 1990s, selected the plasticizers that have acute toxicity (AT), chronic toxicity (CT), mutagenicity (M), teratogenicity (T), and ecotoxicity (E). The toxicity data are shown in Table 2.8.

In 2002, the National Toxicity Program (NTP) of the United States reported the toxicity of phthalates and their potential targets, as shown in Table 2.9.

On August 14, 2008, the USA signed the Consumer Product Safety Improvement Act (CPSIA) and the Toy Safety Standards ASTM F963.0. Starting on February

Table 2.8 The toxicity data of 15 plasticizers

CAS#	Plasticizer	The toxicity data				
		Acute toxicity (AT)	Chronic toxicity (CT)	Teratogenicity (T)	Mutagenicity (M)	Ecological toxicity (E)
84-61-7	Dicyclohexyl phthalate	Y	N	Y	N	N
84-66-2	Diethyl phthalate	Y	N	Y	Y	Y
84-69-5	Diisobutyl phthalate	Y	N	Y	Y	Y
84-74-2	Dibutyl phthalate	Y	N	Y	Y	Y
84-77-5	Didecyl phthalate	Y	N	N	N	N
85-68-7	Butyl benzyl phthalate	Y	Y	Y	Y	Y
85-69-8	Butyl octyl phthalate	N	N	N	N	N
117-817	Di(2-ethylhexyl) phthalate	Y	Y	Y	Y	Y
117-84-0	Dioctyl phthalate	Y	N	Y	Y	Y
119-06-2	Ditridecyl phthalate	Y	N	N	Y	Y
131-11-3	Dimethyl phthalate	Y	Y	Y	Y	Y
131-17-9	Diallyl phthalate	Y	N	N	Y	Y
6422-86-2	Di(2-ethylhexyl) terephthalate	Y	Y	Y	Y	Y
26761-40-0	Diisodecyl phthalate	Y	Y	Y	Y	Y
28553-12-0	Diisononyl phthalate	N	N	N	N	N

Table 2.9 Toxicity and affected groups for some phthalates

Phthalates	Type of toxicity	Affected groups
Butyl benzyl phthalate (BBP)	Reproductive toxicity	Adults/children
Diisodecyl phthalate (DIDP)	Developmental toxicity	Fetuses/children
Diisononyl phthalate (DINP)	Reproductive/developmental toxicity	Children
	Reproductive toxicity	Adults
Dibutyl phthalate (DBP)	Reproductive toxicity	Adults
	Reproductive/developmental toxicity	Children
Di(*n*-octyl) phthalate (DNOP)	Reproductive toxicity	Fetuses/children/adults
Di(2-ethylhexyl) phthalate (DEHP)	Reproductive/developmental toxicity	Adults (healthy males)/children (critically ill patients)
	Developmental toxicity	Fetuses
Di(*n*-hexyl) phthalate (DNHP)	Insufficient data, further research needed	

10, 2009, the CPSIA prohibits the sale, distribution, and importation of toys and childcare articles that contain >0.1% DEHP, DBP or BBP. In addition, the CPSIA also temporarily prohibits the sale, distribution, and importation of bitable children's articles that contain >0.1% DINP, DIDP or DNOP, until the research report from the Chronic Hazard Advisory Panel (CHAP) determines whether to abolish the prohibition or to permanently prohibit their use. After a transition period of 180 days from the effective date (February 10, 2009), the manufacturing and sale of any toy or childcare product with >0.1% DEHP, DBP or BBP is illegal. During this time, the manufacturing and sale of any bitable toy or childcare product with >0.1% DINP, DIDP or DNOP is also considered illegal. The Consumer Product Safety Committee (CPSC) of the USA is asked to investigate whether to continue this action or to extend the prohibition to other phthalates or their substitutes. Since August 17, 2009, the ASTM F963-2008 became a mandatory toy standard, which requires that no DEHP be used in baby pacifiers, rattles, and biterings.

The AB1108 Act of California puts stricter restrictions on phthalates than the CPSIA. According to AB1108, the sale, shipping, and manufacturing of toys and childcare articles containing the above 6 phthalates at >0.1% are prohibited in California after January 1, 2009. The CPSIA has a one-year transition time for the prohibition of DINP, DIDP, and DNOP, but the AB1108 has a permanent prohibition of them.

2.3.5 Related Phthalate Laws in Other Countries

The Japanese Ordinance for Enforcement of the Food Sanitation Act, Toy Safety Standard 8th Edition (ST2002), and Food Sanitation Act, all effective in August 2003, require that no DEHP, DBP or BBP be used in toys. DINP, DIDP, and DNOP cannot be used in bitable toys, as well as toys, pacifiers, and biterings for children under 3 years old. In PVC-based plastic food packaging articles, di(2-ethylhexyl) adipate (DEHA) should not be detected and the content of tricresyl phosphate (TCP) plasticizers should not exceed 0.1%.

The Argentinean Government announced the prohibition of the manufacturing, importation, exportation, sale or free offer of toys and childcare articles that contain the above 6 phthalates at >0.1%.

According to two Canadian law drafts, the contents of DEHP, DBP or BBP in all soft PVC-based toys and childcare articles should not exceed 0.1%; and the contents of DINP, DIDP or DNOP in soft PVC-based bitable toys and childcare articles for children under 4 years old should not exceed 0.1%.

The Malaysian Government determined that, after February 2010, the sales of domestic and imported toys for kids under 14 years old be permitted only after they pass the inspection and testing performed by the Malaysian Certification and Inspection Co., Ltd.

Since March 2008, Brazil has executed the policy that the contents of DEHP, DBP or BBP in all olefin-based toys must be below 0.1% and that the contents of DEHP, DBP, BBP, DINP, DIDP or DNOP in all olefin-based toys for children under 3 years old must be below 0.1%.

2.3.6 Domestic Laws [4–6]

In 2008, based on the Food Hygiene Law of the People's Republic of China, the Chinese Ministry of Health enacted the Hygienic Standards for Uses of Additives in Food Containers and Packaging Materials (GB 9685-2008) and the Toys and Children's Articles-Determination of Phthalate Plasticizers in Polyvinyl Chloride Plastics (GB/T 22048-2008), after using USA's Code of Federal Regulation and EU's Commission Directive 2002/72/EC as references. Table 2.10 shows the plasticizer-related regulations in GB 9685-2008.

In 2009, the Chinese Ministry of Health released the Determination of Phthalate Esters in Foods (GB/T 21911-2008) which includes 16 regulated phthalates in food. The Limit of Harmful Substances of Coatings for Toys (GB 24613-2009), released by the Chinese Ministry of Health in the same year, requires the total content of DEHP, BBP, and DBP at no more than 0.1% and that of DIDP, DINP, and DNOP at no more than 0.1% in toys.

On June 1, 2011, the Ministry of Health published the 16th Public Proclamation that puts phthalates into the 6th set of the List of Potential Illegal Inedible Substances

Table 2.10 Phthalate-related rules in GB9685-2008

Plasticizers	Amount of use	Limit
Di(2-ethylhexyl) phthalate (DEHP)	PE, PP, PS, AS, ABS, PA, PET, PC, PVC: in suitable amounts	SML = 1.5 mg/kg (only for non-fat food contact containers)
Dimethyl phthalate (DMP)	PP, PE, PS: 3.0%	–
Diisobutyl phthalate (DIBP)	PVC: 10%	–
Diisononyl phthalate (DINP)	PVC: 43%	–
Diisooctyl phthalate (DIOP)	Bottle cushion plastics: 50%; PE, PA, ABS, PP, PS, AS, PET, PC, PVC, PVDC: 40%; rubber: 40%	This material cannot contact fatty article for long
Dibutyl phthalate (DBP)	PE, PP, PS, AS, ABS, PA, PET, PC, PVC, PVDC: 10%; rubber: 10%	–
Di(C_9–C_{11} alkyl) phthalates (C_{10} rich)	PVC: 43%	–
Di(C_6–C_{10} alkyl) phthalates (C_9 rich)	PVC: 43%	–

Note PE polyethylene; PP polypropylene; PS polystyrene; AS acrylonitrile-styrene copolymer; ABS acrylonitrile-butadiene-styrene copolymer; PA polyamide; PET polyethylene terephthalate; PC polycarbonate; PVDF poly(vinylidene fluoride)

and Abusive Additives in Food. The Letter to Report the Maximum Phthalate Residue in Food and Additives, the 551st Document signed by the Ministry of Health, clearly points out that phthalates are neither foodstuff nor food additive and should not be added in food and additives. The use of phthalates in food containers and packaging materials should strictly follow the Hygienic Standards for Uses of Additives in Food Containers and Packaging Materials (GB 9685-2008). This document also sets the maximum residues of DEHP, DINP, and DBP in food and additives.

In 2012, the Ministry of Health released the Determination of Phthalate Esters in Cosmetics (GB/T 28599-2012), which indicates the 22 regulated phthalate substances in cosmetics.

In 2014, China made the first ISO international toy standard. China is the biggest toy manufacturer and exporter, producing 70% of toys in the world. According to the Chinese Ministry of Commerce, there were 6000 large-scale toy production companies in China in 2013. The gross toy exports reached 12.38 billion US dollars with an increased rate of 8%. At the beginning of 2009, the Chinese Toys Standardization Administration Technical Committee proposed a bill of making the ISO plasticizer standards to the International Toys Standardization Committee (ISO/TC181). In November 2009, the bill was approved at the Annual Meeting of ISO-TC181 in New York, USA, and China was authorized to take the lead in making the toy standards. After five years' effort, the first international toy standard, Certain Phthalate Esters in Toys and Children's Products (ISO 8124-6), was officially published by the International Organization for Standardization (ISO) on August 15, 2014. The Standardization Administration of China published it on August 18. This is a breakthrough for China in the international toy standardization, which was achieved via a collaborative effort. The Chinese Toys Standardization Administration Technical Committee organized domain institutes and experts to form a Domestic Drafting Committee, which then formed the ISO/TC181/WG6 working team with international experts. In ISO 8124-6, plasticizers are defined as chemicals that can soften plastics and are widely used in toys, food packaging materials, detergents, lubricants, and personal care articles. Among the many types, phthalate plasticizers are the most widely used, have the best performance, and have the lowest cost. However, certain phthalate plasticizers have been identified as a type of estrogens and can adversely affect human growth. Before the ISO 8124-6, many countries had issued restrictions to the use of phthalate plasticizers but lacked international standards for plasticizers. The ISO 8124-6 have promoted breakthrough progress in identifying the types of toys, restricting the types of plasticizers in toys, controlling the inspection cost, ensuring the inspection accuracy, and facilitating the operation methods. This has offered practical guidelines for toy manufacturers and inspection laboratories.

The mandatory national standards, Safety of Toys (GB 6675-2014), co-published by the General Administration of Quality Supervision, Inspection and Quarantine of China and the Standardization Administration of China, became effective on January 1, 2016. Besides some revisions to certain technical requirements and inspection methods, the new standards also list 6 plasticizers including DBP as regulated substances with the same maximum content requirements as those of the EU.

In order to achieve the goals described in the "Made in China 2025" and the "Green Industry Development Plan (2016–2020)", China has a great need to guide sustainable industrial development, utilize low-toxicity, low-hazard raw materials, decrease the contents of hazardous substances in articles, and decrease or even avoid the generation of pollutions. For this purpose, the Ministry of Industry and Information Technology, the Ministry of Science and Technology, and the Ministry of Environmental Protection co-organized the edition of "Recommended Substances for Toxic and Hazardous Raw Materials/Products" and published it in December 2016. According to it, if the content of a substance in an article exceeds 1 ton/year and 0.1 wt%, the manufacturer and exporter must report it to the European Chemicals Agency (ECHA) within 6 months after the substance is categorized in the candidate list of SVHC.

On February 1, 2017, the Ministry of Health published a newer version of Technical Requirement for Environmental Labeling Products: Furniture (HJ 2547-2016), which added the maximum content requirements for migratable elements and phthalates in furniture products.

2.4 Comparison and Trend of Plasticizer Laws for Food Contact Applications

2.4.1 Current Status

Criticizers continue to criticize soft PVC. Especially, phthalate plasticizers are still the focus of attention. The regulatory authority continues to investigate phthalate plasticizers with an obviously increasing trend on restrictive requirements.

A *Science* article "*Link to Children's Health Scientific Studies Raise Concern*" points out that the commonly used phthalate plasticizers can cause symptoms such as decreased newborn weights, child obesity, autistic disorder, etc. The child health-associated research results published in *Science* have attracted great attention.

2.4.2 Chinese Laws and Regulations

The related clauses in "Food Safety Law of the People's Republic of China":

No. 2: Production, sale, and use of food-related articles.

No. 17 and 18: Risk monitoring and evaluation.

No. 26: Food safety standards.

No. 34: Prohibition of production and sale.

No. 37: License for new food-related articles.

No. 41: Regulatory compliance for food-related articles. Production of food-related articles must comply with the laws, regulations, and national standards. Licensing for the packaging materials with high risks of food contacting must follow the Measures for the Administration of Manufacturing License for Industrial Products. The Quality Supervision Authority must strengthen the quality monitoring to the productions of food-related articles.

No. 66: The food additives including preservatives and packaging materials used in the packaging, preservation, storage, and shipping of edible agricultural produces much comply with the national food safety standards.

No. 92: Imported food, food additives, and food-related articles much meet the national food safety standards.

No. 152: The detailed measures for the productions of food-related articles will be made by the State Council Quality Supervision Department, following this Law.

2.4.3 EU's Laws and Regulations

(1) EC 178/2002: General Principles of Food Law.
(2) EC 882/2004: Official Controls of Foodstuffs.

(3) EC 1935/2004: Food Contact Materials and The European Framework Regulation.
(4) EC 10/2011: Commission Regulation on plastics.
(5) EU CFR regulations on phthalate esters.

2.4.4 USA's Laws and Regulations

(1) Federal Food, Drug, and Cosmetic Act
(2) Food and Drug Administration Modernization Act
(3) Code of Federal Regulations Title 21: (i) previously approved substances, 21CFR 170-189, generally recognized as safe (GRAS); (ii) plasticizers in polymers (22 types), 21CFR178-374; (iii) food contact notification (FCN); (iv) phthalate plasticizers: DBP, BBP, DEHP, DINP, DIDP.

2.5 Conclusion

Using hexahydrophthalate esters to replace the traditional phthalate ester, plasticizers has become a general trend with domestic and foreign laws, regulations, and standards.

1. Foreign Laws and Regulations

On December 7, 1999, the 1999/815/EC regulation of EU required that the contents of 6 phthalate plasticizers in PVC-based toys and childcare articles intended to be placed in the mouth by children under three years of age be no higher than 0.1%. On December 14, 2005, the 2005/84/EC regulation of EU required that the contents of DEHP, DBP, and BBP in all toys and childcare articles be no higher than 0.1%. The REACH of EU, effective since 2007, categorized phthalate plasticizers as "highly concerned substances" that may cause serious, irreversible damages to human health and the environment. In June 2008, the 2007/19/EC regulation of EU declared that EU will prohibit the production and importation of food contact materials and articles in which the contents of plasticizers fail to comply with the 2007/19/EC. On August 14, 2008, the CPSIA of USA declared that the consumption, distribution, and importation of toys and childcare articles that contain >0.1% DEHP, DBP, and BBP be prohibited since February 10, 2009. On January 1, 2009, the AB1108 regulation of the USA prohibited the sale, distribution, and manufacturing of toys and childcare articles with 6 phthalate esters of >0.1% in California. The Toy Safety Standard 8th Edition (ST2002) of Japan declared that no DEHP, DBP or BBP be used in toys, and that DINP, DIDP, and DNOP not be used in bitable toys, as well as toys, pacifiers, and biterings for children under 3 years old. From September 9, 2008, Argentine prohibited the manufacturing, importation, exportation, sale or free offer of toys and childcare articles that contain 6 phthalates above 0.1%. In June 2009, Canada required that the contents of DEHP, DBP or BBP in all soft PVC-based toys and childcare articles do not exceed 0.1% and that the contents of DINP, DIDP or DNOP in soft PVC-based bitable toys and childcare articles for children under 4 years old not exceed 0.1%. In 2009, Danish regulations required that, besides the same content requirement as EU, the contents of phthalate esters in toys and childcare articles for children under 3 years of age do not exceed 0.05%. The (EU)2015/863 regulation, an amendment to RoHS 2.0, added DEHP, BBP, DBP and DINP to the list of toxic or hazardous substances used in electronics and electrical articles.

2. Domestic Laws and Regulations

In 2008, based on the Food Hygiene Law of the People's Republic of China and referencing the Code of Federal Regulation and Commission Directive 2002/72/EC, China issued the Hygienic Standards for Uses of Additives in Food Containers and Packaging Materials (GB 9685-2008) to inspire more attention to the amount of phthalate plasticizer migration in model food. The Toys and Children's Articles-Determination of Phthalate Plasticizers in Polyvinyl Chloride Plastics (GB/T 22048-2008) indicated 6 phthalate plasticizers whose contents should be restricted. According to Determination of Phthalate Esters in Foods (GB/T 21911-2008), 16 types of phthalate plasticizers need to be controlled. In 2009, Limit of Harmful Substances of Coatings for Toys (GB 24613-2009) required the total content of DEHP, BBP, and DBP at no more than 0.1% and that of DIDP, DINP, and DNOP at no more than 0.1% in toys. On June 1, 2011, the Ministry of Health published put 17 types of phthalates into the 6th set of the List of Potential Illegal Inedible Substances and Abusive Additives in Food. In June 2011, the 551st Document signed by the Ministry of Health, i.e., Letter to Report the Maximum Phthalate Residue in Food and Additives, declared that the maximum residue of DBP was 0.3 mg/kg. In 2012, Determination of Phthalate Esters in Cosmetics (GB/T 28599-2012) indicated the 22 regulated phthalate substances in cosmetics. Safety of Toys (GB 6675-2014), effective on January 1, 2016, listed 6 plasticizers including DBP as regulated substances with the same maximum content requirements as those of the EU.

Therefore, the use of hexahydrophthalate esters as substitutes to the traditional phthalate ester plasticizers well agrees with the national development strategies.

References

1. Wang HT. How to interpret the latest major international laws and regulations on plasticizers. Nanjing: Nanjing Publishing media Group; 2016.
2. Shi WC, Jia CX. Introduction of the REACH. Plast Addit. 2008;2:53–5.
3. Huai P. How does the cable industry respond to the latest EU environmental law. Cables Cords. 2016;2:7–10.
4. Han P. The DOP market has both opportunities and challenges in 2017. Nanjing: the 9th Jinlianchuang Plasticizer Industry Chain Market Summit Forum and intra-circle fair in 2017; 2017.

5. Jiang PP. Technological change and market trend of plasticizer industry in the process of regulation. Nanjing: the 9th Jinlianchuang Plasticizer Industry Chain Market Summit Forum and intra-circle fair in 2017; 2017.
6. Chen FR. Comparison and development trend of plasticizer laws and regulations for food contact materials at home and abroad. Nanjing: Nanjing Publishing media Group; 2016.

Chapter 3
Current Status of Plasticizer Research

This chapter reviews the progress of plasticizer research since the 1990s. The references used in this review include other reviews, introduction of plasticizer performance, interpretation of laws, regulations, policy studies, lab research results published by domain experts. These references are important resources for understanding the developmental dynamics, trend of plasticizers, innovating green plasticizer preparation technologies, promoting the transformation, sustainability of the plasticizer industry.

This chapter basically follows the chronical sequence of plasticizer research, but the research accomplishments of the same authors or team even at different times will be discussed together. It ends with a brief conclusion.

3.1 Foreign Research Status

3.1.1 Phthalate Hydrogenation Technology

In the 1990s, phthalate esters including DBP, DOP, DNOP, DIOP, and DINP were used as plasticizers in plastic materials such as PVC. Because of the benzene ring involved in their structures, these phthalates were considered hazardous and thus were criticized when used in toy-manufacturing plastics and even were banned in some countries. To address this problem, the USA and Germany started research on the hydrogenation of phthalate ring, catalytically converting the benzene ring to cyclohexane, and filed patents in the USA, Germany, Europe, China, etc. For example, in 1998 BASF (Germany) filed an invention patent in China titled "A method for hydrogenation of phenyl multi-carboxylic acids and their derivatives via catalysts with macropores" (CN1962599A) [1]. This method uses substrate-supported Ru catalysts to hydrogenate phthalates to hexahydrophthalates at a yield of 99.5% and a selectivity of 99.2% in steel vertical pressurized tubular fixed-bed reactors under an average temperature of 125 °C, an average pressure of 20 MPa, a phthalate flux

Z. Liu, *Green Catalytic Hydrogenation of Phthalate Plasticizers*,
https://doi.org/10.1007/978-981-16-9789-0_3

of 0.45 kg/h, and a 20% excess of H_2. After the second hydrogenation of the 99.5% hexahydrophthalate at a flux of 1 kg/h, no phthalate esters were detected and the selectivity for hexahydrophthalate was 99%. The 1% low-boiling-point side products were removed by distillation at 170 °C and 5 kPa, producing 99.7% hexahydrophthalate.

Compared with phthalate esters, hexahydrophthalate esters have lower densities and viscosities for use as plasticizers, which can improve the low-temperature softness of plastics, while other properties such as Shore hardness and mechanical strength. Therefore, hexahydrophthalate esters can enhance the processibility and increase the production efficiency of blended plastics.

In 2002, BASF achieved the ring hydrogenation of DINP. The product Hexamoll® DINCH (a trademark for diisononyl cyclohexane-1,2-dicarboxylate) was recommended by the German Federal Institute for Risk Assessment (German: Bundesinstitut für Risikobewertung, BfR) and was also advocated by the European Food Safety Administration (EFSA) in 2006 to export to the EU, the USA, Japan, China, etc. In China, this product is used to substitute phthalates for use in PVC articles intended to have close human contact such as medical devices. Here is the information about it:

Trademark: Hexamoll® DINCH.

Note: this is a plasticizer used in PVC and other polar polymers, especially for toxicologically sensitive articles.

Chemical name: diisononyl cyclohexane-1,2-dicarboxylate.

Molecular formula: $C_{26}H_{48}O_4$.

CAS number: 166412-78-8; 474919-59-0 in Europe and Asia.

EC number: 431-890-2.

Specifications: See Table 3.1.

Properties: Hexamoll® DINCH is a colorless, transparent, anhydrous, odorless liquid that is soluble in common organic solvents, is insoluble in water, and is miscible with all the low-molecular-weight plasticizers in PVC.

Physico-chemical data: molecular weight, 424.7 g/mol; pour point, −54 °C (DIN ISO 3016); vapor pressure, see Table 3.2; density and viscosity, see Table 3.3; cloud point, 151 °C (5% S-PVC, K value 71, DIN 53408); surface tension at 20 °C (DIN EN 14370), 30.7 mN/m; saponification value (DIN EN ISO 3681), 264 mg KOH/g.

Storage: this product can be stored for >1 year in suitable containers under <40 °C and moisture-free conditions.

GC analysis conditions: CP-Wax 52 CB (a trademark of Varian) capillary column (stationary phase: polyethylene glycol), length, 25 m; ID, 0.25 mm; membrane thickness 0.2 μm; injection port temperature (with a split inlet), 265 °C; oven temperature, 60 °C, elevated to 250 °C at a rate of 3 °C/min, kept for 52 min; detector temperature, 300 °C; carrier gas, high-purity nitrogen or helium (1 mL/min, 110 kPa), via hydrogen flame ionization detection (H_2/syngas 1/10 by volume), calculated by using the area percentage method.

It is reported that Hexamoll® DINCH is compatible with a number of polymers, including PVC, polyurethane, acrylic resin, phenol-formaldehyde resin, etc., and has mechanical properties comparable to phthalate esters, superior processibility,

Table 3.1 Product specifications

Property	Index	Test method
Dynamic viscosity at 20 °C*/(mPa s)	44–60**	DIN 51562/D445
Density at 20 °C*/(g/cm³)	0.944–0.954	DIN 51757/D4052
Pt–Co chromaticity/Hazen	<40	DIN EN ISO 6271-2/D5386
Refractive index* n_D^{20}	1.460–1.466	DIN 51423/D1045
Acid value/(mg KOH/g)	<0.07	DIN EN ISO 2114/D1045
Ester content/%	>99.5	GC analysis based on area***
Water content/(%)	<0.1	DIN 51777/E203
Phthalate content/%	<0.01	UV-BASF
Sb, As, Ba, Sn contents/ppm	<1	ICP-MS
Cr, Pb, Hg, Se contents/ppm	<1	ICP-MS
Cd content/ppm	<0.6	ICP-MS

Note * these properties are not items with conventional measurements; ** Obtained via the measured dynamic viscosity multiplied by density according to DIN 51562; *** See below for GC conditions

low volatility, good cold tolerance, odorlessness, environmental benignness, non-toxicity, resistivity to mildew/fungi, and excellent toxicological properties. During granulation or plasticization, it has remarkable PVC compatibility and low viscosity, making it suitable for special injection molds even with delicate designs. In the final plasticized articles, it has very low migration and can ensure long durability and calendar life.

Applications of Hexamoll® DINCH in PVC: sensitive soft PVC-based articles including food packaging, food wraps, food sealing gaskets, food contact gloves, environmentally friendly wallpapers, decorative sheets, sports and entertainment products, shower curtains, shoes, anti-slip mats, water pipes, high-end electronics, electrical cables and wires, medical supplies, toys for children under 3 years of age, glues, spices, inks, nail oils, etc.

The market demand for this product, mainly for use in PVC-based toys and medical devices, has continuously increased since its launching in 2002. The production facilities in Port Ludwig, Germany, have been scaled up from 100 to 200 k tons per year. Because of the increasing importance in food contact, PVC articles such as food wraps and sealing gaskets, hexahydrophthalate esters are considered as a promising new type of phthalate substitute plasticizers.

Table 3.2 The vapor pressure of Hexamoll® DINCH at different temperatures

Temperature/°C	Vapor pressure/Pa
50	1.3×10^4
60	5.5×10^4
70	2.2×10^3
80	7.5×10^3
90	2.3×10^2
100	6.7×10^2
120	0.44
140	2.2
160	0.09
180	0.31
200	0.95
220	2.6
240	6.2
260	13.9
270	20.2

Note Antoine Equation $\ln p = A + B/(C + T)$ (unit of p is 10^5 Pa; $A = 11.6057$; unit of T is °C; $B = -6601.25$; $C = 155.61$). The Antoine constant was obtained based on the vapor pressure measured with the dynamic method under N_2 atmosphere in the temperature range of 190–270 °C. The data in the table were obtained using the Antoine Equation, which can only be used as approximate estimates

Table 3.3 The density and viscosity of Hexamoll® DINCH at different temperatures

Temperature/°C	Density*/(g/cm^3)	Dynamic viscosity**/(mPa s)
5	0.9597	135
10	0.9560	96
20	0.9486	52
30	0.9415	30
40	0.9344	19
50	0.9273	13

Note * Based on BASF data, obtained via the equation: $\rho = -0.00072T + 0.96205$ (ρ is density (g/cm^3); T is temperature (°C)); ** Obtained via the measured dynamic viscosity multiplied by density according to DIN 51562

3.1.2 Research on Toxicity of Phthalate Compounds

The recent research in Germany indicates that the decrease in the number of sperms may be related to the wide use of phthalate plasticizers in the light industry. Due to their wide use in cosmetics, toys, and food wraps, they can cause serious damages to human health if their contents exceed the standards. Research has suggested that phthalates have similar effects to estrogens for humans or animals, possibly causing reproductive issues to males such as interference of the endocrine system, reduction of semen and sperms, decrease in sperm mobility, abnormal sperm morphology, and even testicular cancer. Among the many PVC articles, cosmetics and nail oils have the most phthalates, which can enter the female body via the respiratory system or skin and, if overused, can increase the possibility of breast cancer and even damage the reproductive system of their future male infants. Long-term contact with phthalate-containing toys or childcare articles may cause damages, and even cancer, to children's liver and kidney [2–8]. Therefore, many countries have made rules and regulations in order to minimize any further damages phthalate esters may cause to humans.

3.2 Domestic Research Status

3.2.1 Synthesis of DIBCH (1999) [9]

Diisobutyl cyclohexane-1,2-dicarboxylic acid ester (DIBCH) is also called diisobutyl hexahydrophthalate. Its molecular structure is:

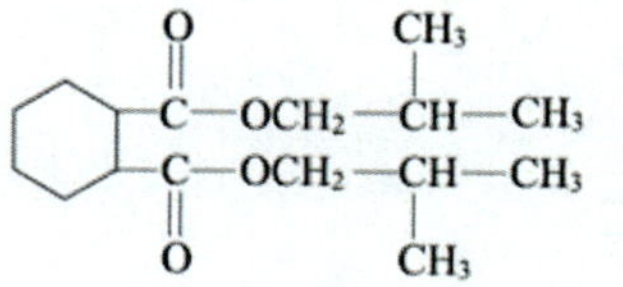

DIBCH is an important chemical industry product as a plasticizer and also as an indispensable gas-phase absorbent solvent in maleic anhydride production. In the 1990s, China depended on importation to obtain it and had a high need for localization. The key to the synthesis of DIBCH is the formation of the cyclohexane ring. Catalytic hydrogenation of the DIBP is difficult to form the cyclohexane ring due to the stability of the benzene ring.

The synthesis of DIBCH from 1,3-butadiene proceeds via the following stepwise reactions. First, 1,3-butadiene reacts with maleic anhydride to produce cis-1,2,3,6-tetrahydrophthalic anhydride, followed by a hydration reaction to produce cis-1,2,3,6-tetrahydrophthalic acid, shown as follows:

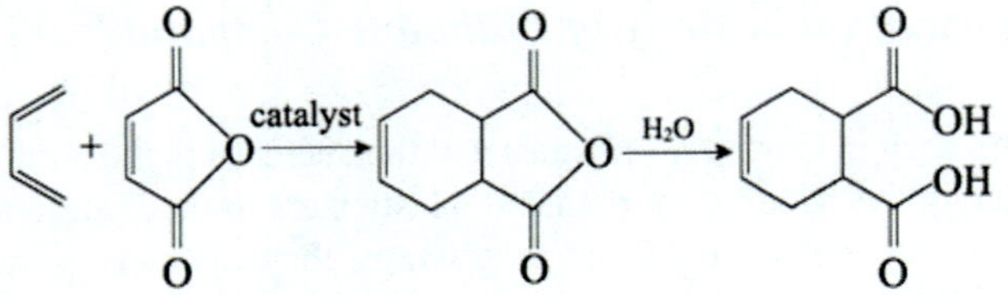

There are two routes for synthesizing DIBCH from cis-1,2,3,6-tetrahydrophthalic acid. The first route is the catalytic hydrogenation with esterification followed, while the second one is esterification with catalytic hydrogenation followed. This research was focused on catalytic hydrogenation reactions with an emphasis on the effects of reaction factors, including the catalysts, pressure, temperature, and medium (pH), on the hydrogenation results for cis-1,2,3,6-tetrahydrophthalic acid and diisobutyl cis-1,2,3,6-tetrahydrophthalate ester. The route of esterification then hydrogenation was finally chosen with the determination of optimal reaction conditions and selective hydrogenation catalysts. The main conclusions are as follows:

(1) With the same temperature, hydrogen partial pressure, and catalyst (e.g., 5% Pd–C or Raney Ni), the C=C double bond in diisobutyl cis-1,2,3,6-tetrahydrophthalate ester is easily reduced, which is not the case for cis-1,2,3,6-tetrahydrophthalic acid under either acidic or alkaline conditions. Thus, the structure of the raw material is an important factor to determine the hydrogenation results.

(2) The optimal synthetic route for DIBCH is esterification then hydrogenation, as shown in Eqs. (3.1) and (3.2).

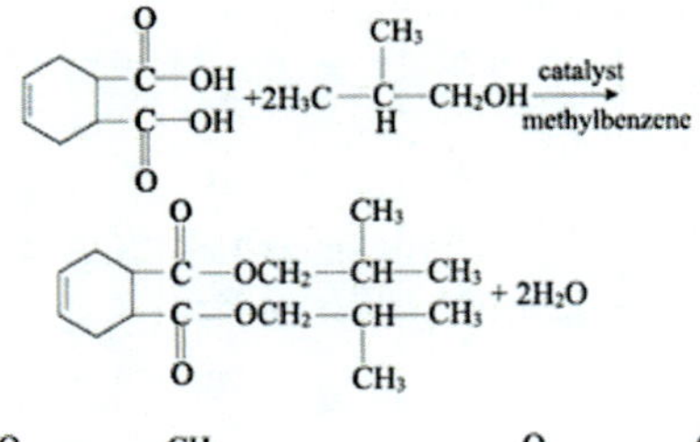

(3.1)

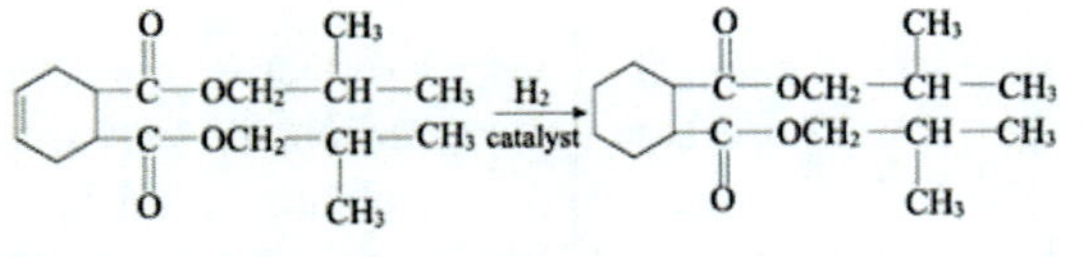

(3.2)

(3) Both 5% Pd–C and Raney Ni are catalysts with high activity and high selectivity to the C=C bond with a conversion ratio of 100% at relatively low temperature and pressure. Even 40 °C can result in good hydrogenation results and 80 °C yields a 100% conversion. A hydrogen partial pressure of 0.56 MPa with 5% Pd–C can produce good hydrogenation and a 100% conversion is achieved under 1.01 MPa. The ester group remains intact.

(4) The synthetic route designed by this research achieves high yields under mild conditions with a weight percentage of >97% after separation and purification.

3.2.2 *Synthesis of HHPA (2000) [10]*

Hexahydrophthalic anhydride (HHPA), also called cyclohexane-1,2-dicarboxylic anhydride, has a molecular formula of $C_8H_{10}O_5$, CAS # 13-149-00-3, and chemical structure:

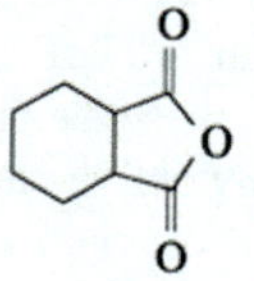

HHPA can undergo esterification reactions with alcohols to produce cyclohexane-1,2-dicarboxylate plasticizers that is also an indispensable raw material for polyester-based high-end coatings. In the production of alkyd amino paints, amino polyester paints, polyester paints, polyurethane paints and powder coatings, and substitution of aromatic multi-carboxylic acids with HHPA can significantly improve the distinctness of image and glossiness, decrease the viscosity, increase the solid content, and particularly enhance the weather tolerance of these paints. For example, polyester paints made from HHPA retain 78.2% of their glossiness after natural light exposure for 2 years, while those made from aromatic multi-carboxylic acids retain only 28%.

Besides its use in plasticizers and polyester paints, HHPA is also the raw material for the production of pesticides, rust inhibitors, herbicides, etc. The rapid development of information technology catalyzes the higher quality and quantity requirements to semiconductor molding and potting compounds. HHPA is widely used in this field as a curing agent for epoxy resins. In the beginning of the 1990s, China established several tens of production lines for semiconductor molding compounds and dry-type transformers, but the curing agents needed for these production lines relied heavily on importation.

The catalytic hydrogenation reaction of tetrahydrophthalic anhydride (THPA) to hexahydrophthalic anhydride (HHPA) is shown here:

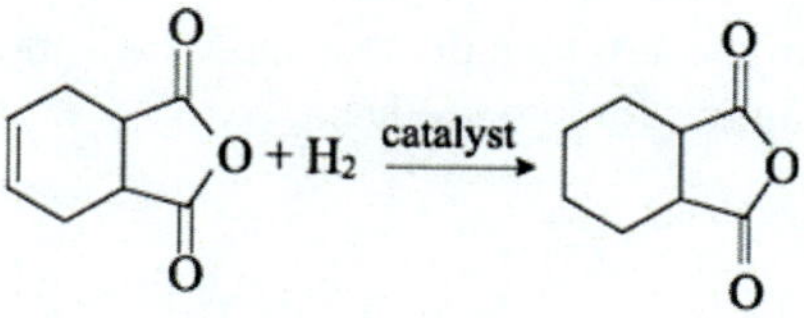

A double bond activation step occurs during the hydrogenation reaction, but the transfer of the double bond shown below is also very possible.

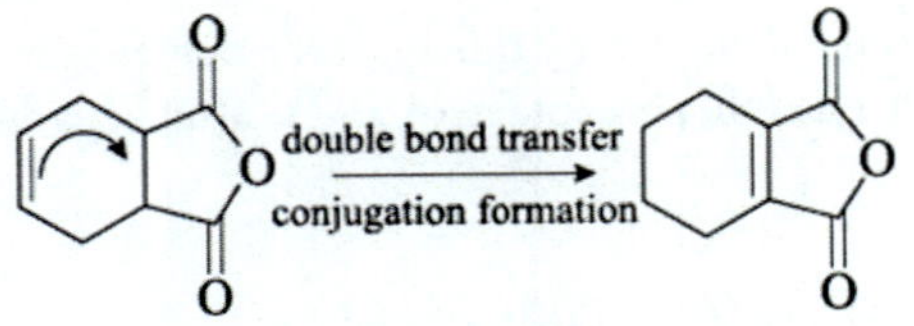

The new double bond conjugates with the two carbonyl groups, causing incomplete hydrogenation. Meanwhile, the anhydride can have condensation, hydrogenation, or crosslinking side reactions with the presence of hydrogen to form inseparable side products. These factors adversely affect the conversion ratio, selectivity, and yield. The high-boiling-point impurities are easy to carbonize under the reaction temperature and attach to the catalyst surface, resulting in poisoning of the catalysts. THPA has a high boiling point similar to HHPA, making it difficult to separate out via distillation. This will seriously decrease the product quality. Therefore, the development of highly selective and highly active catalysts is the key to the production of quality HHPA. The research team's invention of a type of Pd-based composite catalysts has successfully solved the problems with a selectivity of 100%.

The authors investigated the effects of temperature, pressure, and catalysts on the hydrogenation reaction. Temperature plays an important role in controlling product quality. Too low temperatures cause incomplete hydrogenation leading to high-iodine-number products. Too high temperatures yield satisfying iodine numbers but expedite the side reactions resulting in lower anhydride purity, more free acids, inferior product yield, and quality. Here are the optimal technical conditions. Under a temperature of 170–180 °C, hydrogen partial pressure of 1.4–1.6 MPa and 4% catalyst [11], the solvent, THPA, and catalyst are added to a 1 L autoclave. The atmosphere is exchanged to nitrogen, then to hydrogen. After heating to a certain temperature, hydrogen is added under stirring. When the reaction is done, the solvent is removed to give the crude product that undergoes reduced pressure distillation to afford the final product HHPA.

The melting point, acidity, iodine number, and color of the obtained HHPA are measured according to "Handbook of Chemical Commodity Inspections" [12]. The purity of HHPA is analyzed using a Shimadzu GC-14B spectrometer with a hydrogen flame ionization detector, XE-60 on ChromaSorb W-AW-DMCS (60–80 mesh) stainless steel column, and C-R6A data processer. Acetone is used as the eluent solvent.

Compared to the traditional Ni catalysts, the Pd composite catalysts used in this technique lead to significant advantages including higher catalytic efficiency, shorter reaction time, lower temperature, an HHPA selectivity of 100%, and constant product quality. This innovation lays a foundation for the localization of this product with comparable quality to similar foreign products.

3.3 Di(2-Propylheptyl) Phthalate (2006) [13]

BASF (Germany) has launched a new plasticizer product Palatinol® 10-P, i.e., di(2-propylheptyl) phthalate (DPHP), with a molecular formula of $C_{28}H_{46}O_4$, molecular weight of 446.7, CAS # 53306-54-0, and the chemical structure is as follows:

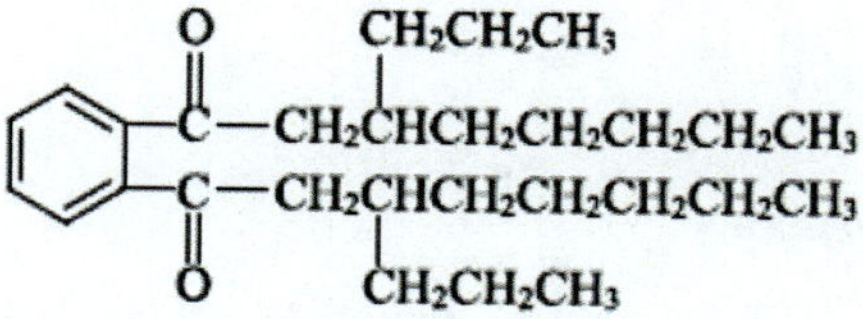

The physical properties and performance (mechanical properties, processibility, volatility, extractivity, electrical properties, fogging, and weather tolerance) are compared between DPHP and common phthalates including DOP, DINP, and DIDP.

3.3.1 *Physical Properties*

DPHP is a viscous, colorless liquid at room temperature and its typical physical properties are shown in Table 3.4. For comparison, the properties of common plasticizers

Table 3.4 Comparative physical properties of DEHP and a few common plasticizers

Index	DPHP	DOP	DINP	DIDP
Density (25 °C)/(g/cm^3)	0.960	0.983	0.971	0.965
Boiling point/°C	254	231	252	255
Refractive index (25 °C)	1.484	1.485	1.484	1.484
Pour point/°C	48	46	45	48
Flash point/°C	232	218	224	232
Viscosity/(Pa s)				
0 °C	0.59	0.381	0.50	0.617
20 °C	0.122	0.081	0.103	0.119
40 °C	0.087	0.022	0.032	0.037
Vapor pressure/kPa				
50 °C	5.7×10^8	1.1×10^6	4.0×10^8	1.7×10^8
80 °C	3.2×10^6	2.7×10^5	2.6×10^6	1.4×10^6
160 °C	4.6×10^3	0.017	4.7×10^3	3.0×10^3
240 °C	0.4	1.4	0.411	0.27
Specific heat capacity (20–200 °C)/[J/(g K)]	1.63–1.90	1.69–1.91	1.75–2.00	1.901
Surface tension (20 °C)/(mN/m)	30.4	31.1	31.0	32. 2

are also shown.

From Table 3.4, DPHP is obviously less volatile than DOP with a 20-fold lower vapor pressure at 50 °C, indicating that DPHP can improve the temperature tolerance of plastics and decrease the contact damages to the human body. Furthermore, DPHP has the lowest density potentially leading to enhanced volumetric efficiency.

3.3.2 *Performance*

1. Mechanical performance

Based on the 100% modulus data in Table 3.5, DPHP is better than DIDP and is comparable to DOP. DPHP has the highest breaking elongation, indicating its better plasticization performance. DPHP plasticization grants PVC with a better low-temperature tenacity than DOP and with a lower brittle temperature than DIDP, indicating the better low-temperature performance of DPHP. Based on its high mechanical performance, DPHP can be used in PVC as a primary plasticizer.

2. Processibility

Under the same formulation and extrusion conditions, the extrusion test results for PVCs using DPHP and DIDP as plasticizers, respectively, are shown in Table 3.6.

With the formulation in Table 3.6, the packing density of the dry DPHP blend is higher than the DIDP blend, leading to higher product rates for DPHP. Moreover, the smaller torques needed in the case of DPHP suggest the lower melt density, in good agreement with its lower density. In total, compared to DIDP, DPHP has potential advantages for melt processing.

3. Volatility and extractivity

The volatility and extractivity are important properties of plasticizers. Lower volatility indicates the product can survive higher temperatures. The extractivity suggests the scope of application of the product. In addition, these two properties

Table 3.5 Comparative mechanical properties of DEHP and a few common plasticizers

Properties	DPHP	DOP	DINP	DIDP
Shore hardness A (15 s)	86	79	87	92
Brittleness temperature/°C	28	26	30	27
cold flex temperature/°C	20	18	23	24
tensile strength/MPa	17.03	1.45	19.86	21.03
elongation at break/%	370	275	342	340
100% modulus/MPa	10.89	10.76	12.07	13.03

Note 100% modulus is the tensile modulus of elasticity of a material at 100% elongation

Table 3.6 Extrusion test results of dry PVC blends

Screw speed/(r/min)	feeder speed/(r/min)	Torque/(N m)		Production/(kg/h)	
		DPHP blend (ρ = 0.636 g/cm^3)	DIDP blend (ρ = 0.656 g/cm^3)	DPHP blend (ρ = 0.636 g/cm^3)	DIDP blend (ρ = 0.656 g/cm^3)
40	35	11.2	12	2426	2371
40	36.5	12.2	–	2522	–
50	44	12.1	–	3155	–
60	50	11.7	12	3500	3246

Note ρ is the packing density of dry blend. Formulation: PVC resin 100 parts, lead sulfate tribasic 3 parts, lead stearate tribasic 0.8 parts, $CaCO_3$ 80 parts, and plasticizer 55 parts

are worth special attention because they are also used as one of the main safety standards. DIDP is difficult to be extracted. The volatilities of DPHP and other common plasticizers tested under standard conditions are shown in Table 3.7.

According to Table 3.7, PVC thin films plasticized by DPHP have the lowest volatility suggesting their ability to sustain higher temperatures as well as their high safety. The weight losses of DPHP and DIDP under the same conditions are shown in Table 3.8.

Compared to DIDP, DPHP is easier to be extracted by oily media, but is more difficult to be extracted under 100% humidity.

Table 3.7 Volatility of PVC films with different plasticizers

Plasticizer	Content/(parts/100 parts PVC)	Volatility/(%)	
		70 °C, 24 h	90°C, 24 h
DPHP	40	0.2	–
	50	0.2	–
	70	0.3	–
DOP	40	0.7	–
	50	0.8	–
	70	1.2	–
DINP	43	0.9	1.4
	54	1.1	1.6
	67	1.3	1.8
DIDP	43	0.8	1.3
	54	0.8	1.4
	67	0.9	1.6

Note Formulation for DPHP and DOP: PVC resin 100 parts, lead stearate 1 part, lead phosphite 2 parts; formulation for DINP and DIDP: PVC resin 100 parts, liquid Ba/Cd stabilizer 4 parts. PVC film thickness: 0.5 mm

Table 3.8 Extractive weight loss of PVC films with different plasticizers

Tests	DPHP	DIDP
Oil soaking weight loss (7 days, 23 °C)/%	6.0	4.5
100% humidity weight loss (12 weeks, 120 °C)/%	0.28	0.39

Note Plasticizer content: 50 parts

4. Electrical properties

When used for cables and wires, plasticizers need to pass strict insulation tests, including those after accelerated aging. Many countries have made corresponding standards to ensure safety. DIDP has been considered as an excellent candidate in earlier applications. BASF performed comparative tests between DPHP and DIDP, which were completed by the USA's Underwriters Laboratories Inc. (UL) following the UL1581 standard. The results are shown in Table 3.9.

The results in Table 3.9 indicate similar electrical insulations for DPHP and DIDP. The DPHP/DUP blends can pass the 90C standard tests.

5. Atomization

The atomization is related to volatility and is a must-do test for the decorative articles in cars. DPHP has the same molecular weight as DIDP, but the volatilities of the pure DPHP and PVC articles are obviously better than DOP. The atomization-resistant performance tests carried out by BASF followed the SA EJ1756 standard method: after co-existence with the test subject under specified conditions, the changes of a transparent material in its glossiness and the deposited test subject weight were measured. The results are shown in Table 3.10.

Table 3.9 Electrical insulation test results of PVC films with different plasticizers

Tests	DIDP	DPHP	DPHP/DUP
60C standard (7 day, 100 °C)			
0.5 mm	Pass	Pass	–
1.0 mm	Pass	Pass	–
75C standard (10 day, 100 °C)			
0.5 mm	Pass	Pass	–
1.0 mm	Pass	Pass	–
80C standard (7 day, 113 °C)			
0.5 mm	Pass	Fail	Pass
1.0 mm	Pass	Pass	Pass
90C standard (7 day, 121 °C)			
0.5 mm	Fail	Fail	Pass
1.0 mm	Fail	Fail	Pass

Note DUP diheneicosyl phthalate

Table 3.10 Atomization resistance test results

Tests	DIDP	DPHP
Pure plasticizer		
Gloss (3 h, 100 °C)/%	68	70
Mass of condensed substances (16 h, 100 °C)/mg	0.74	0.72
PVC test sample		
Gloss (3 h, 100 °C)/%		
67 parts	95	91
54 parts	93	91
43 parts	93	99
Mass of condensed substances (16 h, 100 °C)/mg		
67 parts	0.67	0.75
54 parts	0.68	0.68
43 parts	0.63	0.67

The atomization test results indicate that DPHP and DIDP have almost identical atomization-resistant performance and thus can be used in decorative articles in cars.

6. Weather resistance

The weather resistance of a plasticizer describes the relationship between its aging and time under simulated or real natural conditions, typically expressed as weight loss. Thus, the weather resistance reflects the stability to light, heat, oxidation, humidity, and even pollutant. The comparative weather resistance tests performed by BASF under simulated conditions (light exposure for 8 h at 70 °C, in dark for 5 h at 50 °C, alternatingly repeated) are summarized in Table 3.11. And the comparative test results under natural conditions are shown in Table 3.12.

The test results reveal the significantly better weather resistance of DPHP, indicating that DPHP is more suitable than DIDP for use in outdoor PVC articles.

Table 3.11 Weight loss in mimic weather resistance tests for DIDP- or DPHP-containing PVC films

Time/h	Weight loss of DIDP sample/%	Weight loss of DPHP sample/%
118	0.48	0.39
256	1.53	1.07
612	4.93	3.26
940	8.08	5.60
1226	10.74	7.6
1483	12.77	9.42

Note Formulation: DIDP/DPHP 64 parts, PVC resin 100 parts, Ba/Zn stabilizer 3 parts, light stabilizer 0.6 parts

Table 3.12 Weight loss in natural weather resistance tests for DIDP- or DPHP-containing PVC films

Time/months	Weight loss of DIDP sample/%	Weight loss of DPHP sample/%
6	0.32	0.23
12	0.81	0.46
18	1.51	0.74
24	1.35	0.73
36	2.90	1.07
48	11.90	2.70

Note Formulation: DIDP/DPHP 64 parts, PVC resin 100 parts, Ba/Zn stabilizer 3 parts, light stabilizer 0.6 parts. Test environment is outdoor in middle Europe

Conclusions: phthalate esters will still dominate the plasticizer market for a rather long time because of their advantages in cost and combinatorial performance. But the global trend is the substitution of DOP with safer C9 and C10 alcohol-based ester products. Among them, DPHP, produced from butene and 2-propylheptanol, is being rapidly developed because of its low-cost raw materials, abundant resources, and superior performance. DOP occupies almost 80% of plasticizer consumption in China. Due to the lack of domestic production of C_9 and C_{10} alcohols, the development of high-carbon plasticizers is very slow. The priority is recommended to be given to the development and production of DPHP when planning the plasticizer industry.

3.4 Cyclohexane Dicarboxylate Plasticizers (2010) [14]

Cyclohexane-1,2-dicarboxylate esters, also called ring-hydrogenated phthalates, have a general chemical structure shown as follows:

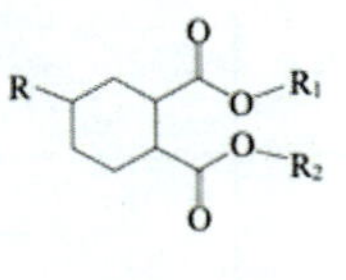

where R is H or alkyl groups; R_1 and R_2 are identical or different alkyl groups, typically C_5–C_{13}.

This article investigated the properties, preparation, and plasticization performance for PVC articles of this new cyclohexane-1,2-dicarboxylate plasticizer. This plasticizer features a good combination of performance, is biodegradable, and is not eco-toxic, therefore suitable for use in food packaging, medical devices, and toys.

Because of their unique properties, phthalate plasticizers are important in the preparation and application of PVC articles, but suffer from serious drawbacks including low molecular weights, poor resistance to heat, UV and cold weather, and ease of evaporation, migration, and extraction, causing inferior the quality and lifetime of the final products. In addition, the burning of phthalate esters generates a large amount of smoke, which will bring fatal damages to humans in the case of fire accidents.

Numerous animal tests indicate that long exposure to phthalates causes peroxisome proliferation and induces liver tumors. Many countries have restricted or even prohibited the use of phthalate plasticizers in medical devices, food packaging, and toys, and instead are seeking their substitutes. According to the research of Brian [15], Gosse [16], Melanie [17], etc., ring-hydrogenated phthalate esters not only have better plasticization performance than related phthalates but also demonstrate no toxicity. Therefore, they are considered as the most promising phthalate substitutes for use in PVC articles such as food packaging, medical devices, and toys, and were approved by the European Food Safety Administration (EFSA) in 2006. The production capacity of 25 kt/y of BASF facilities in Ludwig, Germany, could not satisfy the market need and therefore the second set of 100 kt/y was launched in 2007.

3.4.1 Preparation of Cyclohexane Dicarboxylate Plasticizers

1. Hydrogenation

Example 1: To a 300 mL pressurized vessel were added THF (100 g), trimethyltrimesate (25.2 g, 0.1 mol) and Ru catalyst (10 g). Pure H_2 was introduced and maintained at 20 MPa for about 4 h until no more hydrogen was consumed. The conversion ratio of trimethyltrimesate was 97% and the yield of trimethyl hexahydrotrimesate was 93%.

Example 2: To a 1.2 L pressurized vessel were added DIDP (760 g) and substrate-supported Ru catalyst (53 g), with pure H_2 maintained at 20 MPa and 80 °C. The hydrogenation reaction was completed in about 10 h with 100% DIDP conversion and a hexahydro-DIDP yield of 99.5%.

Example 3: To a 1.2 L pressurized vessel were added tri(2-ethylhexyl) trimellitate (800 g) and substrate-supported Ru catalyst (53 g), with pure H_2 maintained at 20 MPa and 100 °C. The hydrogenation reaction was completed in about 20 h with 95% tri(2-ethylhexyl) trimellitate conversion and a tri(2-ethylhexyl) hexahydrotrimellitate yield of 94%.

2. Olefin Process

The olefin process is started with a Diels–Alder reaction between C_4-diene and maleic anhydride to produce cis-4-cyclohexene-1,2-dicarboxylic anhydride, followed by hydrogenation to produce cyclohexane-1,2-dicarboxylic anhydride and then by an esterification reaction with an alcohol to produce the cyclohexane-1,2-dicarboxylate esters.

If butadiene is used to react with maleic anhydride, the final product is a cyclohexane-1,2-dicarboxylate ester without any functional groups. If pentadiene is used, the final product is a methyl-substituted cyclohexane-1,2-dicarboxylate ester.

Example 1: The reaction pathway using butadiene is shown as follows:

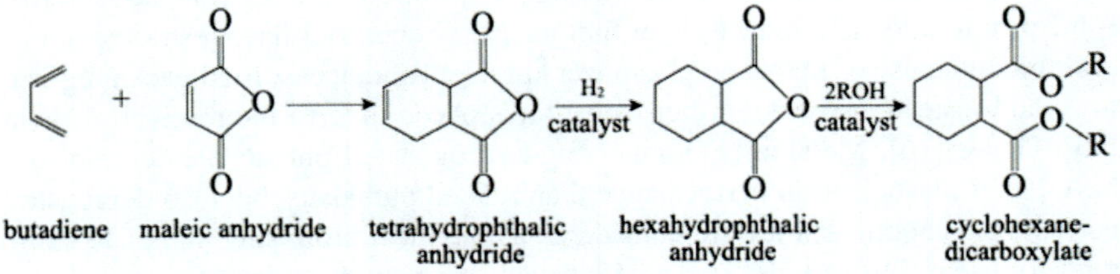

Example 2: If pentadiene is used, the products are tetrahydrophthalic anhydride, hexahydrophthalic anhydride, and hexahydrophthalate ester, successively.

The reaction pathway can also follow an order of cyclization, esterification, and hydrogenation, with almost no sacrifice of yield.

3.4.2 Properties and Assessment of Cyclohexane Dicarboxylate Plasticizers

1. Sample Preparation

A PVC suspension was added to a high-speed mixer under vigorous stirring and was heated to 50 °C. The stir rate was maintained at 1500 rpm, and then was dropped to 750 rpm when the plasticizer was added, and then was reset to 1500 rpm with dry blending. The stabilizer was added and the stir rate was elevated to 2500 rpm. The temperature was raised to 120 °C and cooled back to 30 °C to discharge the mixture to obtain soft PVC. The formulation is shown in Table 3.13.

A formulated mixture of plasticizer, stabilizer, and PVC was added to the high-speed mixer and heated to 35 °C until no clumps were observed in the dry blend. The rest plasticizer was added to the mixture that was then stirred to homogeneity

Table 3.13 Formulations of soft PVC and PVC plastisol

Raw materials	Soft PVC	PVC plastisol
PVC (suspension method)/parts	100	100
Plasticizer/parts	40–70	75
Epoxidized soybean oil (ESO)/parts	–	5
Liquid Ba–Zn stabilizer/parts	2	3

Note The plasticizer was acetyl tributyl citrate, diisononyl cyclohexane-1,2-dicarboxylate (DINCH), or diisononyl phthalate (DINP)

to obtain the PVC plastisol. Bubbles needed to be removed from the plastisol before measuring its mechanical properties.

2. Performance Comparison

(1) The comparative physical properties of DINCH with DINP and ATBC are shown in Table 3.14.

DINCH has higher PVC plasticization efficiency because of its lower relative density, e.g., 2.5% lower than DINP and 10% lower than ATBC. And the viscosity of DINCH is significantly lower than that of DINP, affording its better processibility and the lower plastisol viscosity.

(2) Comparative PVC plasticization performance of DINCN, DINP, and ATBC is shown in Table 3.15.

The brittle temperature and thermal stability of DINCH-plasticized PVC are improved to certain extents. Especially, the thermal stability is much better than the ATBC-plasticized PVC. The Clash–Berg temperature of cyclohexane-1,2-dicarboxylate esters exhibits a shift with the number of C atoms, which is also about 3.3 °C lower than that of the phthalate with the same C number, shown in Table 3.16.

Table 3.14 Comparative physical properties of DINCH, DINP, and ATBC

Items	DINCH	DINP	ATBC	Test method (ASTM)
Relative density (20 °C)	0.947	0.972	1.052	D-4052
Ester content/%	99.6	99.6	99.5	D-3465
Acid value/(mg KOH/g)	0.04	0.04	0.04	D-1045
Water content/%	0.02	0.03	0.01	E-203
Chromaticity (APHA)	5	5	10	D-5386
Viscosity (20 °C)/(mPa s)	48	72	42	D-445

Table 3.15 Comparative physical properties of DINCH-, DINP-, and ATBC-plasticized PVC

Items	DINCH	DINP	ATBC	Test method (ASTM)
Shore hardness A	76	74	69	D-2240
Shore hardness D	25	24	22	D-2240
Tensile strength/MPa	16.2	16.4	17.3	D-638
Elongation at break/%	355	375	340	D-638
100% constant elongation modulus/MPa	8.6	8.1	7.5	D-638
Brittleness temperature/°C	41	38	29	D-746
Volatility (24 h, 130 °C)/%	2.4	1.2	13.1	D-1023
Thermal stability (180 °C)/min	19.3	17.5	10.0	D-2115
Compatibility (28 d)/%	0.6	0.5	1.4	D-2383

Table 3.16 Evolution of Clash and Berg temperature with the number of alkyl carbon atoms in cyclohexane-dicarboxylates

Alcohols	Clash and Berg temperature/°C	
	Phthalate	Cyclohexane-1,2-dicarboxylate esters
$C_8H_{17}OH$	6.0	9.0
$C_9H_{19}OH$	4.4	8.1
$C_{10}H_{21}OH$	4.9	7.7

(3) Comparative mechanical properties of PVC plasticized by di(2-ethylhexyl)-1,2-cyclohexane dicarboxylate (DENCH), DINCH, diisodecyl-1,2-cyclohexane dicarboxylate (DIDCH), and related phthalates are shown in Table 3.17.

DENCH is relatively too volatile to be used in cables and wires, while DINCH and DIDCH exhibit low areal weight loss, making them suitable for such uses.

(4) Comparative light stability of DENCH, DINCH, DIDCH, and related phthalates are shown in Table 3.18.

(5) Comparative viscosity inhibition effects of different plasticizers are shown in Table 3.19.

Table 3.17 Mechanical properties of DEHCH-, DINCH-, DIDCH-, and phthalate-plasticized PVC

Items	Dioctyl ester		Diisononyl ester		Diisodecyl ester	
	DOP (50 parts)	DEHCH (50 parts)	DINP (53 parts)	DINCH (54.5 parts)	DIDP (55 parts)	DIDCH (57.5 parts)
Mechanical properties						
100% constant elongation modulus/MPa	10.9	11.4	10.7	10.5	11.2	10.4
Stress at break/MPa	15.6	14.5	15.1	14.7	16.0	13.5
Elongation at break/%	301	223	289	290	299	235
Aging properties (100 °C, 7 day, natural venting)						
100% residual modulus/MPa	98.6	95.1	100.7	100.4	107.0	99.0
Residual stress at break/MPa	95.5	96.2	94.1	94.2	98.5	96.8
Residual elongation/%	90.8	105.6	87.5	84.3	87.8	100.5
Mass loss/(mg/cm^2)	0.60	1.92	0.26	0.53	0.21	0.27
Volatility/%	0.79	2.49	0.35	0.72	0.27	0.38

Note PVC 100 parts

Table 3.18 Comparative light stability of DEHCH, DINCH, DIDCH, and corresponding phthalates

UV radiation time/h	Dioctyl ester		Diisononyl ester		Diisodecyl ester	
	DOP	DEHCH	DINP	DINCH	DIDP	DIDCH
0	White	White	White	White	White	White
220	Light yellow	White	Light yellow	White	Light yellow	White
456	Yellow	White	Yellow	White	Yellow	White
626	Yellow	White	Yellow	White	Yellow	White
794	Yellow	Light yellow	Yellow	White	Yellowish brown	Yellowish brown
1 056	Dark yellow	Yellow	Brown	Yellow	Brown	Yellowish brown

Table 3.19 Comparative viscosity inhibition effects of different plasticizers

Items	1#	2#	3#	4#	5#	6#
PVC/parts	100	100	100	100	100	100
Butyl benzyl phthalate (BBP)/parts	30					
Diisoheptyl phthalate (DIHXP)/parts		40	40	30	25	
Diisohexyl cyclohexane-1,2-dicarboxylate (DIHCH)/parts			8	10	15	20
Dioctyl phthalate (DOP)/parts	10					
Diisononyl phthalate (DINP)/parts						20
Dodecyl phthalate (DDP)/parts	1	1	1	1	1	1
Total liquid/parts	48	48	48	40	40	40
Brookfield viscosity/(mPa s)						
2 h	7600	3400	5300	23,250	19,000	12,500
4 h	8300	3500	5550	23,550	18,750	12,800
1 d	11,400	4450	7650	31,500	25,500	17,250
5 d	14,600	5600	10,400	44,500	37,000	23,500
Hot stage gelling temperature/°C	58	66	59	59	60	66
Volatility (pre-gelation*)/%	1.87	1.56	0.96	0.97	0.97	0.92

* Pre-gelation: 160 °C, 20 s, wind speed 1000 rpm; evaporative weight loss: 2 min, 180 °C, wind speed 2300 rpm

(6) The eco-toxicity investigation results of DINCH-based soft sheets are shown in Table 3.20.

The test results in Table 3.20 suggest the following conclusions about DINCH: (i) biodegradability, no acute toxicity, no toxicity to aquatic, bacteria, and algae (the bioconcentration toxicity data being collected); (ii) no acute toxicity in oral and skin tests; (iii) no skin allergy or eye irritation in animal tests; (iv) no genotoxicity; (v) no

Table 3.20 Toxicological test results of DINCH-based soft sheets

Items	Test standard (OECD)	test method	Items	Test standard (OECD)	Test method
Acute toxicity (oral)	423	Mice	Repeated toxicity		
Acute toxicity (skin)	402	Mice	28 d	407 + MITI	Mice
Skin irritation			90 d	408 + FDA Red Book 11	Mice
DINCH	404	Rabbit	Repeated toxicity		
DINCH foil	404	Rabbit	Ames test	471	*S. typhimurium* 和*E. coli*
DINCH/PVC plastisol	404	Rabbit	Chromosome aberration test	473	V79
Eye irritation	405	Rabbit	Hypoxanthine-guanine phosphoribosyltransferase	476	CHO (Chinese hamster ovary cells)
Sensitization	406	Cavia porcellus	Micronucleus test (in vivo)	474	Mice
Chronic carcinogenicity	453	Mice	Acute aquatic toxicity	202	*Daphnia magna*
Biological kinetics	417	Mice		201	Algae
Reproductive toxicity			Chronic Daphnia toxicity	211	In observation
Fetus toxicity	414	Mice	Bacterial toxicity	209	Bacterial
Two-generation reproduction toxicity	416	Mice	Soil parasite toxicit	207	Soil
Ecological toxicity			Biodegradation	301B	*S. typhimurium* 和*E. coli*
Acute aquatic toxicity	203	Zebra fish	Adsorption coefficient K_{oc}	121	HPLC

Note OECD: Organization for Economic Cooperation and Development; MITI: Malaysian Ministry of International Trade and Industry; FDA: US Food and Drug Administration; Ames test: bacterial reverse mutation test peroxisome proliferation; (vi) no reproductive toxicity after 28 days of feed or ovary tests.

3.4.3 *Application Examples of Formulation*

1. Cables and Wires

The comparative hardness and low-temperature softness of PVC plasticized by DOP, DENCH, DINP, DINCH, DIDP, and DIDCH, respectively, are shown in Table 3.21.

From Table 3.21, cyclohexane-1,2-dicarboxylate esters (DENCH, DINCH, DIDCH) obviously improve the low-temperature softness and brittleness compared to related phthalate esters (DOP, DINP, DIDP).

2. Foils and Sheets

The formulation for foils and sheets is PVC 100, plasticizer 35, epoxidized soybean oil 0.5, Ba/Zn thermal stabilizer 2. The hardness of the articles is shown in Table 3.22.

Table 3.21 Mechanical properties of wires and cables with different plasticizers

Mechanical properties	Test method (ASTM)	Plasticizers					
		DOP (50 parts)	DENCH (50 parts)	DINP (53 parts)	DINCH (54.5 parts)	DIDP (55 parts)	DIDCH (57.5 parts)
Shore harness A (HA)	D-2240	90.3	90.9	91.2	91	90.4	91.7
Shore harness D (HD)	D-2240	38.2	38 3	8.7	38	38	36.4
Clash and Berg temperature/°C	D-1043-84	17.0	20.7	18.1	26.0	21.7	29.7
Brittleness	D-746	14.0	16.2	16.8	22.0	19.6	26.5

Note PVC 100 parts

Table 3.22 Hardness of PVC with different plasticizers

PVC hardness	Dioctyl ester		Diisononyl ester		Diisodecyl ester	
	DOP	DEHCH	DINP	DINCH	DIDP	DIDCH
Shore Hardness D (HD)	47	51.4	52.9	55.2	48.3	52.7

3. Floor Tiles

An example formulation for floor tiles is PVC 100, fast-melting plasticizer (i.e., one of the common plasticizers: phthalate, benzoate, phosphate, aliphatic dicarboxylate, citrate, etc.) 10–30, diisononyl cycloalkane-1,2-dicarboxylate 20–30, epoxidized soybean oil 0–6, $CaCO_3$ 500–800, and others (pigment, stabilizer, or other fillers) 0–10.

4. Synthetic Leather, Wallpapers, Conveyer Belts

An example formulation for synthetic leather, wallpapers, and conveyer belts is PVC (suspension) 100, fast-melting plasticizer 5–25, DINCH 30–50, filler 0–50, stabilizer 1–4, and pigment 0–10.

5. Automobile Chassis Sealant

An example formulation for automobile chassis sealant is PVC 100, fast-melting plasticizer 10–30, DINCH 20–30, filler 500–800, and epoxidized soybean oil 0–6.

3.4.4 Conclusion

Cyclohexane-1,2-dicarboxylate esters are featured with inexpensive, abundant raw materials (alkenes), great combinatorial performance, and compatibility with a variety of polymers such as PVC, polyurethane, polyacrylic acid, and phenol-formaldehyde resin. Cyclohexane-1,2-dicarboxylate plasticizers can decrease the amount of PVC by 2.5–3.0% compared with related phthalates. When used with other plasticizers, the content of cyclohexane-1,2-dicarboxylates can reach up to 90%.

The blends of cyclohexane-1,2-dicarboxylate ester and PVC have excellent processibility, superior resistance to heat, cold, UV, and flame, and produce no suffocating smoke when burned. Short-chain alkyl esters of cyclohexane-1,2-dicarboxylic acid can serve as plasticizers and also as solvents with good affinity with many active ingredients and strong dissolution uniformity.

Animal tests indicate that cyclohexane-1,2-dicarboxylate esters do not induce peroxisome proliferation or liver tumor. Many countries have approved their use in human contact articles including food packaging, medical devices, and toys. Cyclohexane-1,2-dicarboxylate esters are biodegradable in nature, do not cause environmental pollutions, and are safe for human health.

Because of the above advantages, cyclohexane-1,2-dicarboxylate esters have been considered as a promising type of phthalate substitute plasticizers.

3.5 Development Status of Plasticizers (2010, 2011) [18, 19]

The recent progress of plasticizer R&D is reviewed here, with an introduction of new environmentally friendly plasticizers and suggestions for a sustainable path of plasticizer development in China.

3.5.1 *Status of Plasticizers in Foreign Countries*

Recently, the focus of international plasticizer research has been the improvement of their safety and specialty. Production of the dominant DOP has been switched from traditional sulfuric acid catalysts to the current clean solid acid catalysts. Non-toxic, green plasticizers for use in electrical insulation, food packaging, and medical devices, as well as new plasticizers with superior resistance to oil, extraction, and migration, have been developed and employed continuously. Environmentally friendly plasticizers have attracted wide attention.

With the increasing awareness of environmental protection, plastic articles for drug and food packaging, regular supplies, and toys have put forward more strict requirements, resulting in a general trend of using green plasticizers to replace traditional plasticizers.

For a long time, the dominant plasticizers are phthalates such as DOP, but their toxicity has received increasing attention following their wide use in the food and medicine industry. When the blood stored in PVC bags is transfused into the human body, DOP materials are found in the lung. In 1982, the National Cancer Institute (NCI) of the USA carried out a bioassay of the carcinogenicity of DOP and concluded that DOP is a liver carcinogen to laboratory rats and mice. Since then, the toxicity of DOP attracted wide attention. Although debates still exist so far, many countries have restricted the use of DOP due to its potential carcinogenic effect. Based on NCI's research, the Environmental Protection Agency of USA has terminated the production of 6 phthalates.

3.5.2 *Status of Plasticizers in China*

Currently, the performance of primary plasticizers produced in China cannot fully satisfy the requirements for environmental protection, especially poor cleanliness and high toxicity.

DOP and dioctyl adipate (DOA), especially the former, are the plasticizers widely used in PVC-based food wrap and packaging. Due to the high cost of DOA, most PVC food wrap film producers in China use DOP and DBP as the primary plasticizer.

Such films can release the plasticizer into food when wrapping high-fat meats or co-heating with food in a microwave, which may cause potential damages to human health.

The technology backwardness has limited the development of plasticizers. After several tens of years, the plasticizer industry in China has grown to a relatively large scale. Currently, China has the largest plasticizer production and consumption in Asia, with a capacity of 2.2 million tons, an actual volume of 1.3 million tons, and an apparent consumption of 1.7 million tons. Although some plasticizer products are in a leading position globally, in general obvious gaps still exist compared with developed countries. First, the industry is big but not strong with relatively small production scales. Second, self-reliant technologies are lacking and the R&D levels are low. Although the research and production systems have been established, the R&D in China still heavily relies on imitation. Finally, the technologies are relatively old with low efficiency, making it difficult to win international competitions on plasticizers.

Due to the potential carcinogenic risks of plasticizers, especially DOP, many countries have taken measures to restrict their use and also have increased the R&D investment in non-toxic plasticizers. However, the plasticizers such as DOP that have been prohibited or restricted in foreign countries still have large market in China. The producers have not paid sufficient attention to the development and promotion of non-toxic plasticizers. Because of the low prices, the plasticizers sold in the Chinese market are mainly DOP and DBP. Currently, a large number of new types of plasticizers have been developed abroad with increasingly delicate functions, while high-end products are rarely seen in China. The technical level of the whole plasticizer industry in China has fallen 20 years behind than that in developed countries. For example, DBP has been banned in many countries, but China still has large-scale productions. The national standards still list DOP as one of the plasticizers that can be used for food packaging. In contrast, non-toxic, long-lasting, high-performance plasticizers, such as citrates, 810 ester, polyester, trimellitates, and epoxy ester, are facing slow development.

Considering the global trend of large-scale manufacturing, product specialization, multifunctionality, and environmental friendliness for plasticizers, Chinese producers should stick to the sustainability concept, increase the R&D on green plasticizers, and establish the leading positions in production scale, environmental benefits, product quality, and added value, with the purpose of occupying a significant part of the international market. To achieve this goal, research institutes and industry should collaborate more closely to solve the current problems of small scale, limited variety, technology backwardness, high consumption quota, and inconsistent product quality. Large consortiums should be formed to decrease the production cost and energy consumption. Technological R&D should be strengthened to innovate new products and satisfy personalized needs. The plasticizer industry should be restructured with emphasis on environmental protection and non-toxicity to broaden the performance and applications. These can serve as a powerful support to push forward the plasticizer industry in China.

3.5.3 New Environmentally Friendly Plasticizers

Plasticizer producers should increase the R&D investment to develop new products and circumvent the technical barriers to win the international market. Currently, the research on environmentally friendly plasticizers obviously falls behind the practical needs. The development of non-toxic, cost-effective, high-performance plasticizers is a key to the plastic article industry. Diisopropyl cyclohexane-1,2-dicarboxylate is a colorless, transparent, slightly odorous liquid with a slow migration property. It is insoluble in water but is soluble in many organic solvents and is compatible with common PVC plasticizers. It can be used in a variety of soft medical articles such as stomach tubes and infusion bags, and can also provide safe solutions to toy producers. The quality of the resulting PVC articles meets the requirements of the new EU regulations.

3.5.4 Sustainable Development of Plasticizers in China

1. Updating the Lagging Plasticizer Standards

Currently, the export trade of China is facing a more challenging international environment. Especially, more and more countries have increased the inspection standards and items as means to limit the export of Chinese products and protect their own industry. Countries and regions including the EU, the USA, Japan, South Korea, etc. have made new standards and regulations prohibiting DOA and DOP in human contact articles such as food packaging, medical supplies, and toys. But the Chinese standards still permit their use. Although debates over their carcinogenicity still exist, the lagging Chinese standards have limited the export of a number of plastic articles and also have encouraged the import of products containing outdated plasticizers. Therefore, it's very urgent to solve the problem of lagging plasticizer standards.

2. Promoting Risk Assessment

The European Commission on Plasticizers and Intermediates (ECPI) states that the determining factor for whether a plasticizer is useful should be the risk assessment results instead of toxicity test results. The China Plasticizer Industry Association also thinks that any publicity stunts without scientific evidence will be detrimental to the healthy progress of the plasticizer and plastics industry of China. Therefore, an increase in the efforts of toxicity tests and risk assessment for plasticizers is certainly necessary.

In determining what applications a plasticizer is useful for, the key lies not in the toxicity tests in animals, but in an in-depth investigation of the relationships between dose and toxicity, between exposure and damage, and between migration level and rate and exposure. These scientific analyses will help make related risk control measures and minimize the damages to the public and environment. It is too hasty if restrictive measures are taken simply based on hazard identification.

The management of chemicals including plasticizers will be increasingly strict and scientific internationally.

The EU regulations on the registration, evaluation, authorization, and restriction of chemicals were implemented before 2008. The risk assessment is being recognized by the public and has become the foundation for related laws and policies. Currently, the animal tests and risk assessment of plasticizers are still rather weak in China, which is not helpful to maintain the healthy development of this industry. Attention should be paid to the risk assessment.

3. Expediting Plasticizer Industry Restructuring

Plasticizers are indispensable in high-quality plastic articles. There are continuous changes in their types and structures with a faster pace of new technology development, advanced inspection methods, and reasonable assessment approaches. Progress is being gradually made for plasticizers in the loop of "innovation-application-toxicity testing-risk assessment-new innovations". Therefore, proper use of plasticizers is key to the healthy development of this industry. To achieve this, plasticizer producers should strive for structural tailoring. Non-toxic, eco-friendly, biodegradable plasticizers such as DINCH are being greatly promoted in developed countries. Currently, the production of non-toxic, eco-friendly, biodegradable plasticizers in China is facing two critical challenges: (1) low yields with numerous side reactions; (2) low production volume and high cost. The costs of these plasticizers such as citrates are rather high due to the use of expensive catalysts, forcing the users to choose low-cost, traditional options. Accelerated elimination of toxic plasticizers and the invention of non-toxic candidates have become an inevitable trend. Chinese producers should largely increase the R&D of new catalyst and separation technologies to push next-generation cost-effective, safer, eco-friendly plasticizers to the market.

4. Application-Driven Market Specialization

Market specialization driven by the application fields is an important measure to avoid plasticizers "accidents". In addition to increasing the production volume, Chinese plasticizer producers should also investigate the compatibility of different plasticizers in plastic articles. Choose the lesser of two evils and the better of two goods. They should offer constructive suggestions to end users and show them with practice and evidence that it is safe to use plasticizers in a reasonable way.

3.6 Diisononyl Cyclohexane-1,2-Dicarboxylate (2010, 2014) [20, 21]

Diisononyl cyclohexane-1,2-dicarboxylate (DINCH), or ring-hydrogenated DINP: molecular formula $C_{26}H_{48}O_4$, CAS # 166412-78-8, a colorless, transparent, oily liquid, density 0.955 g/cm^3, viscosity 45.5 mPa s, acidity <0.003 mg KOH/g, molecular structure is as follows:

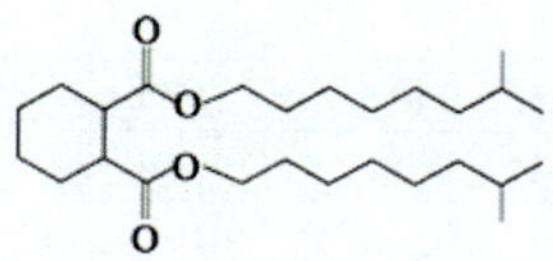

DINCH can be prepared via esterification of hexahydrophthalic anhydride (HHPA) or catalytic hydrogenation of DINP. The reactions are schemed as follows:

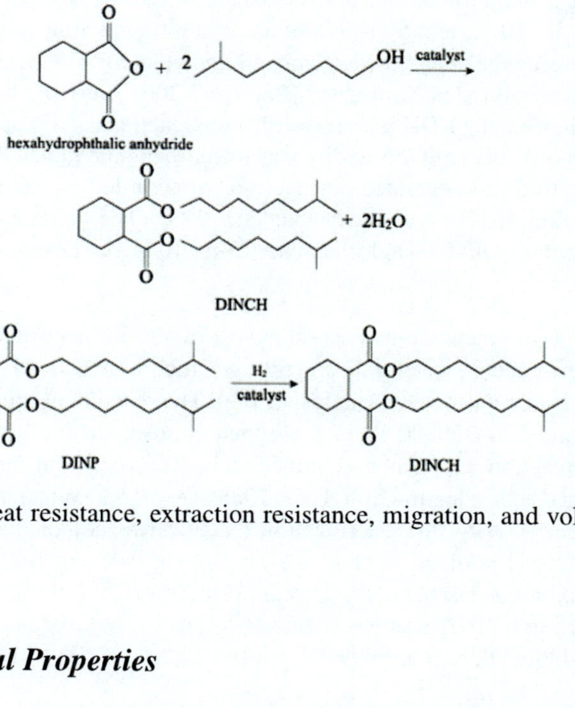

DINCH has heat resistance, extraction resistance, migration, and volatility very similar to DINP.

3.6.1 *Physical Properties*

Some physical properties of dialkyl cyclohexane-1,2-dicarboxylates are listed in Table 3.23.

3.6.2 *Production Routes of DINCH*

1. Route via Esterification of HHPA

Example 1 Using *p*-toluenesulfonic acid catalyst: To a 1 L four-necked flask equipped with a stirrer, thermometer, water trap, and nitrogen tube were added HHPA

Table 3.23 Selected physical properties of dialkyl cyclohexane-dicarboxylates

Items	Test method (ASTM)	DIHCH	DEHCH	DINCH	DIDCH
Viscosity (25 °C)/(mPa s)	D445	34.0	45.5	64.7	87.0
Density (20 °C)/(g/mL)	D4052	0.961	0.955	0.946	0.943
W (H_2O)/(μg/g)	D1064	154	64	111	69
Acid value (25 °C)/(mg KOH/g)	D1045	0.11	0.03	0.03	0.04

(154 g), 5-methyl-1,2-hexahydrophthalic anhydride (MHHPA, 7.7 g), isononanol (INA, 370 g), and *p*-toluenesulfonic acid catalyst (2.6 g). The reaction mixture was maintained at 250 °C and 500–760000 Pa in a nitrogen atmosphere for 8 h. Water generated during the reaction was removed immediately. The acidity of the reaction mixture was measured according to GB/T 1668-2008 every 0.5 h. When the acidity dropped below 3.0 mg KOH/g, the esterification reaction was completed, followed by the addition of KOH until the acidity was lower than 0.08 mg KOH/g. The resulting reaction mixture was separated via steam distillation to the extent that the residual weight fraction of INA was lower than 3×10^{-4}. The crude ester was dried and filtered to obtain DINCH and diisononyl 5-methyl-cyclohexene-1,2-dicarboxylate ester.

Example 2 Using stannous octoate catalyst: To a 1 L four-necked flask equipped with a stirrer, thermometer, water trap, and nitrogen tube were added HHPA (154 g), INA (370 g), and stannous octoate catalyst (2.6 g). The reaction mixture was maintained at 230 °C and 500–760000 Pa in a nitrogen atmosphere for 7 h. Water generated during the reaction was removed immediately. The acidity of the reaction mixture was measured according to GB/T 1668-2008 every 0.5 h. When the acidity dropped below 3.0 mg KOH/g, the esterification reaction was completed, followed by the addition of KOH until the acidity was lower than 0.08 mg KOH/g. The resulting reaction mixture was separated via steam distillation to the extent that the residual weight fraction of INA was lower than 3×10^{-4}. The crude ester was dried and filtered to obtain DINCH with 99.5% purity.

Example 3 Using titanium tetraisononanolate catalyst: To a distillation flask were added cis-HHPA (462 g), INA (1296 g), toluene (1.2 L), and titanium tetraisononanolate catalyst (1.94 g). The reaction mixture was refluxed at 180 °C for 7.5 h. The measured acidity was ≤0.1 mg KOH/g and the HHPA conversion ratio was >99.9%. Then, toluene was removed via distillation at 120 °C and 50 Pa, and the INA residue was removed via distillation at 180 °C and 5 Pa. NaOH solution was added to the reaction mixture for neutralization and then was stirred at 80 °C for 30 min. The temperature was raised to 180 °C, and water was added slowly to the reaction mixture to purify the ester at 180 °C and 5–20 Pa. Then the product was cooled to room temperature under reduced pressure and was finally filtered.

2. Route via Catalytic Hydrogenation of DINP

Example 1: To a 1.2 L high-pressure reaction vessel was added a hanging basket containing substrate-supported Ru catalyst (53 g) and DINP (800 g). The reaction proceeded with pure hydrogen at 80 °C and 20 MPa until no more hydrogen was consumed (~6 h). The DINP conversion ratio was 100% and the yield of DINCH was 99.5%.

Example 2: To a 0.3 L high-pressure reaction vessel was added a hanging basket containing substrate-supported Ru catalyst (10 g) and DINP (150 g). The reaction proceeded with pure hydrogen at 120 °C and 20 MPa until no more hydrogen was consumed (~62 h). The DINP conversion ratio was 100% and the yield of DINCH was 99.4%.

Example 3: To a 1.2 L high-pressure reaction vessel was added a hanging basket containing substrate-supported Ru catalyst (53 g) and DINP (780 g). The reaction proceeded with pure hydrogen at 120 °C and 20 MPa until no more hydrogen was consumed (~4 h). The DINP conversion ratio was 100% and the yield of DINCH was 99.4%.

Example 4: To an autoclave was added DINP (137.4 g) and substrate-supported Ru catalyst (MCM-41, 10 g). The autoclave was sealed and heated to 80 °C with a pressure of 20 MPa for 3.0 h to produce DINCH with a yield of 99.5%.

3.6.3 Applications of DINCH

1. Cables and Wires

DINCH has great potential to replace DINP and DIDP for use in cables and wires.

2. Printing Inks

DINCH can significantly improve the printing performance of inks. With a needed amount of only 5% of common plasticizers, DINCH can largely drop the cost of printing inks to harvest high economic benefits. The reduction in the amount of needed plasticizer does not sacrifice any other ink properties including adhesion, glossiness, abrasion resistance, etc.

3. PVC Plasticization

Cyclic aliphatic carboxylates including DINCH have comparable and even better use value and combined performance with lower plastisol viscosity and less thermal aging weight loss than traditional polycarboxylate blends. DINCH is a superior new safe, environmentally friendly plasticizer because of its extremely low toxicity, biodegradability, good cold tolerance and anti-brittleness performance, water extraction resistance, and good thermal stability.

As a representative of cyclic aliphatic carboxylates, DINCH can be used in PVC articles such as electrical devices, kitchen equipment, computer chassis, casing and parts, TV parts, pipelines, cables, wire sheath, insulation belts, door-window profiles, indoor decorations, cars, furniture, architecture constructions, plastic cements, flooring, medical supplies, toys, food and beverage packaging, gaskets, composite thin films, disks, synthetic leather, adhesives, clothing, coating, and fiber fabrics. Moreover, DINCH can also be used as additives for metallurgic processing, and as adhesives and sealants in paints/coatings.

4. Market Outlook

At present, phthalate plasticizers are dominating >80% of plasticizer consumption in China. DINCH is an environmentally friendly plasticizer with comparable plasticization performance to phthalates and is an ideal substitute for phthalate plasticizers. But the lagging of related policy and the lack of related standards have seriously hampered the development of the plasticizer industry.

Plasticizers are used mainly in PVC articles. In 2013, the total production volume of PVC articles reached as high as 5.5 million tons including 3 million tons of soft articles, which consumed 1.1 million tons of plasticizers. Plasticizers are necessary additives for synthetic resin processing to improve the processibility.

The current production volume of synthetic resins reaches 9.6 million tons per year, making China one of the major plastics production and consumption countries in the world. But this still fails to satisfy the market need necessitating an annual import of 10 million tons. Therefore, the plastics industry of China has huge market potential and so is the plasticizer industry. After China joined the WTO, foreign plasticizer producers have strategically shifted their focus to China and Asia–Pacific region, resulting in high competition. Facing the chances and challenges, the plastics additive industries of China have been self-adjusting the product structure. The production of DINCH only needs to add the hydrogenation facilities after the esterification, which is not a big change. This new plasticizer can restore the whole plasticizer industry. Therefore, China should largely increase the R&D of DINCH series products to achieve a new high peak in the plasticizer industry.

Seriated, diversified plasticizers have advantages of strong ability to adapt to the market and high production flexibility and thus can satisfy the varied requirements of different plastic articles for specialty plasticizers.

Production of DINCH can improve the asset use of manufacturing facilities. Simply modified facilities can be used to produce DOP substitutes with even better combined plasticization performance. This can decrease energy consumption for further cost reduction.

Production of DINCH uses DINP as the raw material with a highly consistent purity of almost 100%. The DIOP production facilities can be used to produce the DINCH series products. This makes it easy to improve the grade, export exchange, and market of PVC articles, potentially leading to huge economic and social benefits.

3.7 Environmentally Friendly Plasticizers

3.7.1 Di(2-Ethylhexyl) Cyclohexane-1,2-Dicarboxylate Plasticizer (2011) [22]

Plasticizers are the most consumed plastics additive. Phthalate plasticizers are the dominant type in this field and occupy >80% of the total consumption in China, of which 70% is di(2-ethylhexyl) phthalate (DEHP) [23–25]. Since the 1990s, based on the investigations of their pseudo-estrogen pollution, many countries have called for the restriction of phthalate plasticizers. Currently, the EU, the USA, Switzerland, and Japan have legislated laws to restrict the use of phthalate plasticizers in food packaging, cosmetics, and toys [26–29]. The research and development of non-toxic, environmentally friendly plasticizers have become the focus of attention.

Cyclohexane-1,2-dicarboxylate plasticizers exhibit comparable performance to phthalates additionally with advantages of non-toxicity and environmental friendliness, and therefore are ideal substitute plasticizers for food packaging, medical devices, and toys. The major synthetic methods include the hydrogenation method and the olefin method [30–34], both using HHPA and isooctanol as raw materials with the activated carbon-supported methanesulfonic acid catalyst.

Di(2-ethylhexyl) cyclohexane-1,2-dicarboxylate (DENCH) was synthesized using HHPA and isooctanol as raw materials with the activated carbon-supported methanesulfonic acid catalyst. The optimal conditions include an isooctanol:HHPA molar ratio of 2.4:1.0, catalyst equal to 2.1% of the total weight of HHPA and isooctanol, hydrated cyclohexane 6 mL, temperature 160–180 °C, reaction time 2.5 h, and esterification ratio 99.82% (still higher than 90% after 4 repeated use of catalyst). The product is a light yellow oily liquid with an HPLC purity of 99.15%. The measured acidity, density, viscosity, and thermogravimetric curve indicated that DENCH meets the performance requirements for a plasticizer.

1. Preparation of Catalyst

Dry activated carbon (30 g) was soaked in a 10 wt% methanesulfonic acid solution (150 g) for 48 h and then was filtered. The solid was dried at 130 °C for 24 h to obtain the catalyst (34.41 g) with a catalyst loading of 14.70 wt%.

2. Synthetic Method of DENCH

To a four-necked flask equipped with a stirrer, thermometer, water trap, and nitrogen tube were added HHPA and isooctanol. The mixture was heated to reflux under stirring, followed by the addition of the catalyst. The reaction was maintained under reflux for specified periods of time with well-controlled temperature. The acidity of the reaction mixture was measured according to GB/T 1668-2008 every 0.5 h until no change was observed. The reaction mixture was cooled and filtered to separate the catalyst and the solution phase. The excess isooctanol was removed under reduced pressure. The solution phase was neutralized, washed with water, and dried to obtain

DENCH. The catalyst was washed with a small amount of ethanol and dried at 130 °C for 12 h for reuse.

3. Synthetic Route of DENCH

The pathway of the esterification reaction between HHPA and isooctanol is shown as follows:

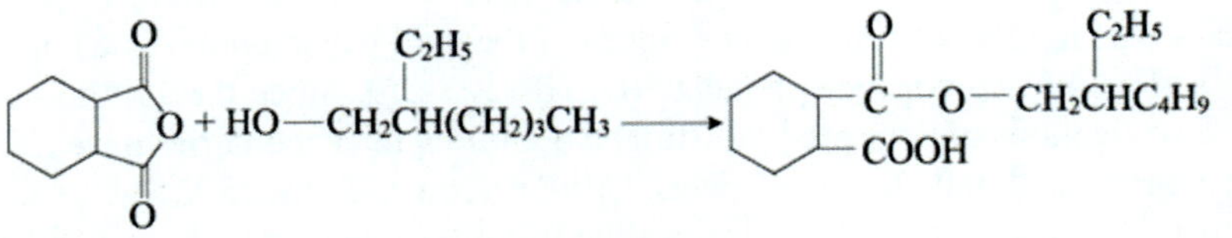

(3.3)

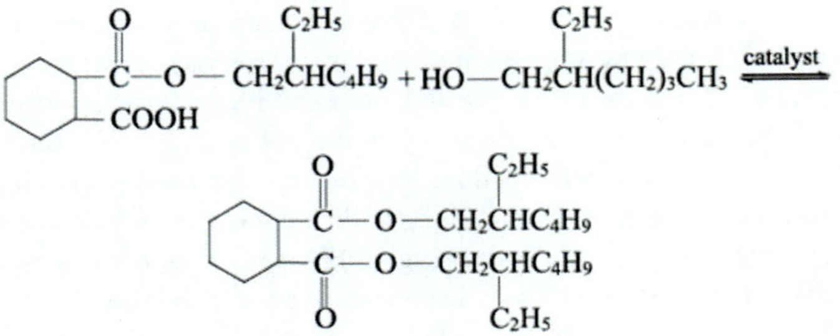

(3.4)

Equation (3.3) is an irreversible, fast reaction without the catalyst involved. Equation (3.4) is a reversible reaction that can only proceed with the catalyst.

4. Property Measurements of the Product

The properties and performance of DENCH are measured and compared with those of common plasticizers including DEHP and DBP. The results are shown in Table 3.24. DENCH has advantages of light color, low viscosity, and good heat resistance and satisfies the requirements for a plasticizer.

5. Conclusion

The optimal conditions for the synthesis of DENCH, using HHPA and isooctanol as raw materials with the activated carbon-supported methanesulfonic acid catalyst, include a HHPA:isooctanol molar ratio of 1.0:2.4, hydrated cyclohexane 6 mL, catalyst equal to 2.1% of the total weight of HHPA and isooctanol, temperature 160–180 °C, reaction time 2.5 h, and esterification ratio 99.82%.

The HPLC purity of DENCH was 99.15%. The thermogravimetric curve showed an extrapolated onset temperature of 265 °C, indicating high thermal stability. The measured acidity, density, viscosity, and thermal weight loss indicated that DENCH meets the requirements for a plasticizer and can be used as a high-performance environmentally friendly plasticizer.

Table 3.24 Comparative properties of DEHCH, DEHP, and DBP

Items	DEHCH	DEHP	DBP
Appearance	transparent oily liquid	transparent oily liquid	transparent oily liquid
Acid value/(mg KOH/g)	≤0.10	0.02	0.02
Chromaticity (Co–Pt)/Hazen	<50	40	25
Chromaticity after thermal treatment (Co–Pt)/Hazen	<100	<100	<100
Density (20 °C)/(g/mL)	0.955 ± 0.003	0.985 ± 0.003	1.046 ± 0.002
Viscosity (25 °C)/(mPa s)	45	80	16
Heated weight loss (125 °C, 3 h)/%	≤ 0.25	0.3	0.5

3.7.2 Research and Application Progress of Catalytic Synthesis of Environmentally Friendly Plasticizers (2012) [15]

Pingping Jiang et al. reviewed the catalytic synthesis of new plasticizers including the synthetic methods, catalytic reaction mechanisms, and catalyst types, and preparations. They focused on the properties, performance, synthetic methods, application, and research status of new environmentally friendly plasticizers, including plant oil, polyester, citrates, trimellitates, and dioctyl terephthalate. The new synthetic methods and catalysts were compared with the traditional ones. They pointed out the trend of catalytic plasticizer synthesis and the potential applications of plasticizers in the high-tech fields in China. Suggestions were given that Chinese plasticizer producers should emphasize the adjustment of product structures to promote the R&D of non-phthalate, plant oil-based plasticizers.

After the plasticizer incident in Taiwan, the Chinese government and the additive industry have paid high attention to the potential solutions to such a crisis.

In 2011, the production volume of plasticizers in China reached beyond 3 million tons. Among them, the traditional phthalate plasticizers, synthesized via esterification reactions between phthalic anhydride and aliphatic alcohols with proton acids as the catalysts, were the most used, ~80% of the total. However, their possible carcinogenicity and toxicity have raised wide concerns. The EU and the USA systematically investigated the performance and safety of traditional plasticizers and launched a series of laws and regulations to restrict their use in toys and medical plastic articles. Meanwhile, with the increasing depletion of petroleum resources and concerns of the environment and human health, heterogeneous synthesis of non-toxic, eco-friendly, biodegradable, high-performance, multifunctional plastics additives has

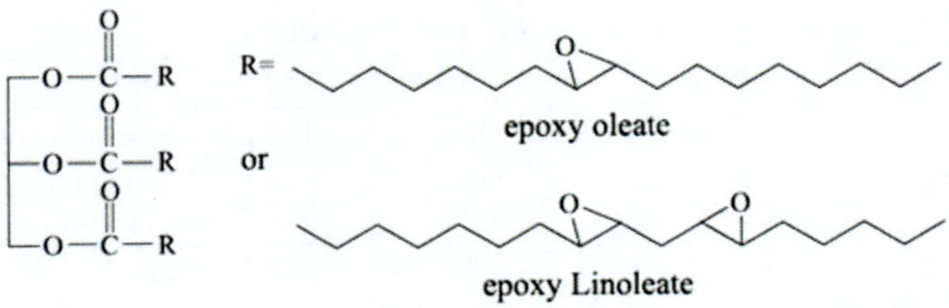

Fig. 3.1 Molecular structures of the major components in epoxidized soybean oil

become a hot spot. It has been an important research topic for domain scientists and industrial engineers to synthesize non-phthalate plasticizers with new catalysts and synthetic protocols [35–37]. The following are the major environmentally friendly non-phthalate plasticizers.

1. Plant Oil Plasticizers

Because of the absence of aromatic rings and renewable raw materials, epoxidized plant oil-based compounds have been widely applied in plastics, coatings, polymers, rubbers, etc., with a continuously increasing production volume. Due to the very low toxicity, epoxy plasticizers have been approved for use in food and drug packaging materials in many countries, are the only type of plasticizer approved by the USA's FDA for food packaging, and have grown to the third largest plasticizer. Major commercial epoxidized oil products include soybean oil, cottonseed oil, rice husk oil, and sunflower oil. Among them, epoxidized soybean oil is the most important one, ~70% of the total consumption. Figure 3.1 shows the molecular structure of the major component in epoxidized soybean oil.

In the USA, the consumption of epoxidized plant oil is at the third position, only lower than those of phthalates and aliphatic dicarboxylates. In China, epoxidized plant oils are already ~20% of the overall plasticizer consumption [38–40]. Epoxidized soybean oil contains a highly active epoxy group that can adsorb the HCl released upon heating PVC resins, and is a non-toxic, odorless PVC plasticizer and thermal stabilizer. It is stable to light and heat, is compatible with PVC resins, is non-volatile, has low migration, and can be used in almost any type of PVC articles such as non-toxic articles, transparent articles, food and drug packaging, medical supplies (e.g., transfusion bags), outdoor plastic supplies, waterproof membranes, plastic doors and windows, and wallpaper base membrane. The addition of epoxidized soybean oil to agricultural thin films, which require high weather tolerance, can largely improve their lifetime. Epoxidized soybean oil not only can reduce the material cost of PVC articles but also can improve their performance including good processibility, aging resistance, folding endurance, etc.

2. Polyester Plasticizers

Polyester plasticizers are a type of superior plastics additives, not only as the primary plasticizers but also as specialty additives in various PVC articles, polymeric materials, and rubber articles [41–43]. Polyester plasticizers are yellow or colorless viscous oily liquids that are odorless, non-toxic, and water-insoluble. Typically, they are prepared via condensation reactions between dicarboxylic acids (or anhydrides) and diols and have average molecular weights of 1000–8000. Based on the diacid monomers, major polyester plasticizers include adipates, azelates, sebacates, glutarates, phthalates, etc. The most commonly used diols include ethylene glycol, 1,2-propylene glycol, 1,3-propanediol, 1,4-butanediol, diethylene glycol, etc. The main synthetic methods are esterification and polymerization reactions in the presence of metal coordination catalysts. Here is an example using diacids and diols to make polyesters, as shown in Eq. (3.5):

$$\begin{aligned} & n\,\mathrm{HO{-}R{-}OH} + n\,\mathrm{HOOC{-}R'{-}COOH} \rightleftharpoons \\ & \mathrm{H{+}OROOCR'CO{+}_{n}OH} + (2n-1)\mathrm{H_2O} \end{aligned} \tag{3.5}$$

For an alcohol used as the end-capping agent, the reaction is shown in Eq. (3.6):

$$\begin{aligned} & n\,\mathrm{HO{-}R{-}OH} + (n+1)\,\mathrm{HOOC{-}R'{-}COOH} + 2\mathrm{R''OH} \rightleftharpoons \\ & \mathrm{R''O{+}OCR''COORO{+}_{n}OCR'COOR''} + (2n+2)\mathrm{H_2O} \end{aligned} \tag{3.6}$$

For an acid used as the end-capping agent, the reaction is shown in Eq. (3.7):

$$\begin{aligned} & (n+1)\,\mathrm{HO{-}R{-}OH} + n\mathrm{HOOC{-}R'{-}COOH} + 2\mathrm{R''OH} \rightleftharpoons \\ & \mathrm{R''O{+}OCR'COORO{+}_{n}OCR'COOR''} + (2n+2)\mathrm{H_2O} \end{aligned} \tag{3.7}$$

The major research progress of polyester synthesis is described below:

In 2000, Oriol-Hemmerlin et al. [44] synthesized poly(adipate-1,3-butylene glycol ester) with an average molecular weight of ~600 and investigated the effects of its addition on PVC performance using FTIR and NMR.

In 2001, Farahat [45] obtained several polyesters with specified acidities via reactions between sebacic acid and ethylene glycol, 1,2-propylene glycol, and 1,3-butanediol, respectively. Then the end carboxylic acids of these polyesters reacted with chromium and barium salts to form thermally stable polyester plasticizers. A comparative study indicated that the chromium-coordinated polyesters exhibited better thermal stability than the barium counterpart.

In 2006–2007, Lindström et al. [46, 47] introduced straight and branched poly(diol adipate esters) into PVC resins and investigated the effects of the side chain, molecular weight, and end-capping agent on the durability of the polyesters. The comparative study indicated that the side chains enhanced the compatibility with PVC, the mechanical properties, and water extraction resistance. The introduction of end-capping agents improved the hydrolysis resistance of the final articles. The presence of both side chains and end-capping agents increased the migration resistance of the plasticizers.

In China, Pingping Jiang's team carried out much research on the synthesis of polyesters. Their major accomplishments are introduced below.

In 2008, this team [48] prepared a series of adipate-based polyester plasticizers via esterification reactions between adipic acid and 1,2-propylene glycol, 1,3-butanediol, and 1,4-butanediol, respectively, with 2-ethylethanol as the end-capping agent. The effects of reaction time, temperature, alcohol/acid molar ratio, and amount of the end-capping agent were investigated to control the average molecular weight at ~1900. The decomposition temperatures of poly(1,2-propylene glycol adipate), poly(1,3-butanediol adipate), and poly(1,4-butanediol adipate) were 334 °C, 358 °C, and 349 °C, respectively, indicating remarkable thermal stability.

In 2011, this team [49] used HHPA, diethylene glycol, and isooctanol as raw materials, with methanesulfonic acid, titanium *n*-butanolate, activated carbon-supported methanesulfonic acid, respectively, as catalysts, to prepare poly(diethylene glycol cyclohexane-dicarboxylate) and DENCH. They also synthesized poly(triethylene glycol adipate) using adipic acid and triethylene glycol as raw materials with titanium *n*-butanolate as the catalyst [50].

Polyester plasticizers have low toxicity and can be used in plastic articles with high safety requirements. The USA's FDA has approved its use in food contact materials [51]. Polyester plasticizers have better extraction resistance than traditional phthalates and are highly compatible with PVC resins [52].

The catalysts for the synthesis of polyester plasticizers include strong acids such as sulfuric acid and *p*-toluenesulfonic acid; metal compounds, mainly antimony-, germanium-, titanium-, tin-, and aluminum-based catalysts; enzymes such as porcine pancreatic enzyme, pseudomonas, and candida rugosa lipase; rare earth metal catalysts such as compounds of La, Y, and Sc [53, 54]. Because of the high viscosity of polyester products, separation of the catalyst is a challenging and complicated step. Therefore, liquid-phase catalysts are typically chosen to simplify the workup procedures. The development of highly active, selective, non-toxic liquid catalysts is a general trend.

Polyester plasticizers are widely applied in various articles including oil-resistant cables, gas hoses, waterproof membranes, synthetic leather, toys, drinking straws, diary bottle gaskets, high-temperature wire coatings, oil-resistant hoses, indoor decorations, etc.

3. Citrate Plasticizers

Citrate plasticizers are widely applied in food, textile, leather, and cosmetics industries such as lubricants, emulsifiers, wetting agents, softeners, detergents, and conditioners.

Tributyl citrate (TBC) is synthesized using citric acid and butyric acid as raw materials with a catalyst. The reaction is shown in Eq. (3.8):

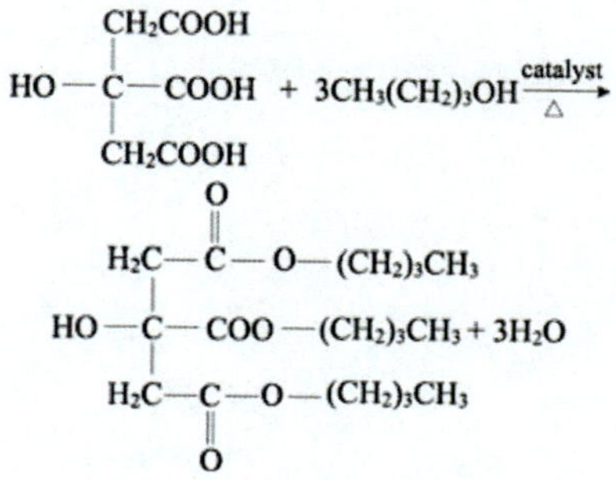

(3.8)

TBC has attracted wide attention as the first green substitute of phthalates because of its characteristics of good compatibility, high plasticization efficiency, non-toxicity, non-volatility, strong weather tolerance, high light stability, good water resistance, high thermal stability, no color change when heated, high deflectivity even in cold regions, etc. TBC is a safe, durable, long-lasting plasticizer suitable for use in food, drug packaging, plasma bags, disposable infusion tubes, etc. TBC can plasticize PVC, PP, and cellulosic resins with great compatibility. When co-used with other non-toxic plasticizers, TBC can improve the hardness of plasticized articles especially soft cellulose ethers. TBC is non-toxic, antibacterial, and flame-retarding. Therefore, it is especially needed as a sterilizing, flame-retarding plasticizer for membranes, drinking straws, food bottle gaskets, medical devices, hospital walls, sidings, ceilings, kitchens and bathrooms for houses, restaurants, and hotels, as well as plastic articles used in transportation compartments including defense aircraft, battleships, armors, etc. In addition, industrial tri(2-ethylhexyl) citrate and tri(2-propylheptyl) citrate products have been used in Europe.

Meng Wei et al. [55] investigated the synthetic protocols for tributyl citrate via reactions between citric acid and *n*-butanol with the $Ce(SO_4)_2{\cdot}4H_2O/NH_2SO_3H$ complex catalyst. NH_2SO_3H was introduced into $Ce(SO_4)_2{\cdot}4H_2O$ to solve its drawback of easy deactivation and thus increase the catalyst stability. The complex catalyst had a great promise of commercialization because of its advantages of cost reduction, light color, simple workup procedure, and no corrosion to facilities, which could also actively promote the use of rare earth metals as esterification reaction catalysts.

Acetyl tributyl citrate (ATBC) is synthesized using TBC and acetic anhydride as raw materials with a catalyst. The reaction is shown in Eq. (3.9):

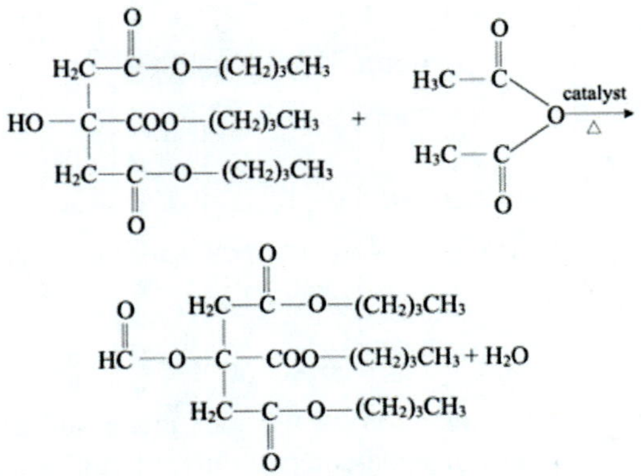

(3.9)

ATBC has been approved by many countries for use in food packaging because of its pharmacological safety, absence of odor, and high performance. Moreover, it is the best plasticizer for manufacturing children's toys and is the most widely used citrate ester. In foreign countries, ATBC is largely used in capsules for delayed drug release and gums. Due to its high plasticization performance and no environmental hazards, it has a great promise to replace the traditional phthalate plasticizers.

4. Tri(2-Ethylhexyl) Trimellitate

Trimellitic anhydride undergoes esterification reactions with alkyl alcohols in the presence of acid catalysts to produce trimellitate esters. For example, isooctanol (2-ethylhexanol) reacts with trimellitic anhydride to produce tri(2-ethylhexyl) trimellitate (TOTM). The reaction is shown in Eq. (3.10):

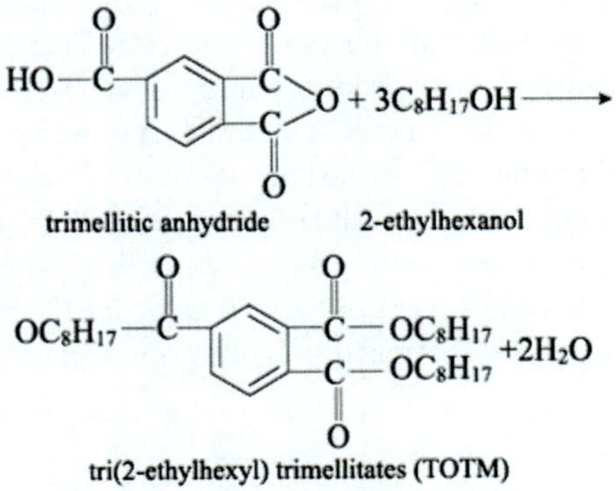

(3.10)

Due to the presence of anhydro rings and carbonyl groups of different activities, the reactions also produce isomeric side products of different ratios, as shown in Fig. 3.2.

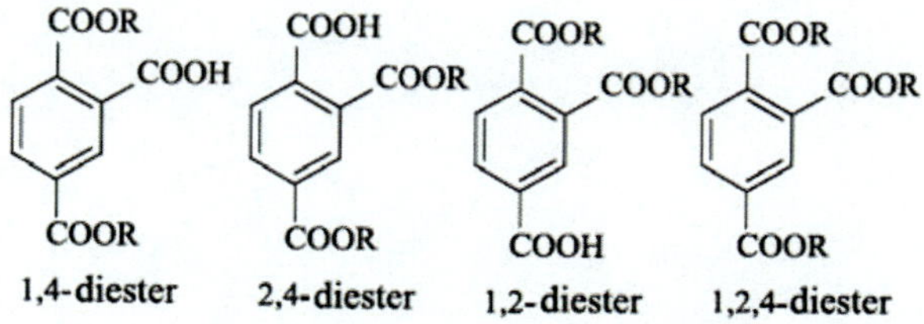

Fig. 3.2 The synthesis routes of DIBCH

Amoco Chemicals Corporation (USA) commercializes trimellitic anhydride with which to react with different alcohols to produce trimellitate esters of different molecular structures for use as plasticizers. The most important product is TOTM, an excellent plastics processing additive. Because of its low volatility and high resistance to extraction and migration, TOTM has advantages similar to polyester plasticizers. Meanwhile, its compatibility, processibility, and low-temperature performance are very similar to phthalate plasticizers. Therefore, TOTM possesses the advantages of both phthalate and polyester plasticizers, and is widely used in heat-resistant PVC and solution-resistant crosslinked PVC. Other applications of TOTM include wetting agent, high-temperature insulation paint, coating materials for electrical parts, care wires and semiconductors, automotive seat cushions, synthetic leather, washing machine drain hoses, blinds, sealants, fillers, etc. [56].

5. Dioctyl Terephthalate (DOTP)

There are three synthetic methods for dioctyl terephthalate (DOTP) [57]: (1) esterification of terephthalic acid; (2) transesterification between dimethyl terephthalate (DMT) and polyterephthalate; (3) decomposition of waste polyesters. Figure 3.3 shows the synthetic protocol for DOTP via decomposition of a waste polyester.

6. Epoxidation and Esterification Catalysts

Industrial production of epoxidized soybean oil is shown in Fig. 3.4.

First, an organic carboxylic acid (typically formic acid or acetic acid) reacts with hydrogen peroxide with an acid catalyst to produce an organic peroxyacid. Then, double bond containing plant oil molecules react with the organic peroxyacid (i.e., peroxyformic acid or peroxyacetic acid) to produce epoxy compounds.

New epoxidation catalysts include methyltrioxorhenium(VII) (MTO) [58], molecular sieves [59], polyoxometalates [60], Hydrotalcite-like catalysts [61], biomimetic catalysts [62, 63], and lipase-like catalysts [64, 65].

Currently investigated esterification catalysts include solid superacids [66–70], polyoxometallic acids [71–74], ionic liquid catalysts [75–78], etc. Among them, acid-functionalized ionic liquids [50, 79–88], substrate-supported protic ionic liquids [85, 89, 90], and specialty ionic liquids [91–96] have been used to catalyze esterification reactions, but there is still a long way to go before commercialization.

With the emerging green chemistry, the key to the synthesis of environmentally friendly plasticizers is the preparation of epoxidation and esterification catalysts. The

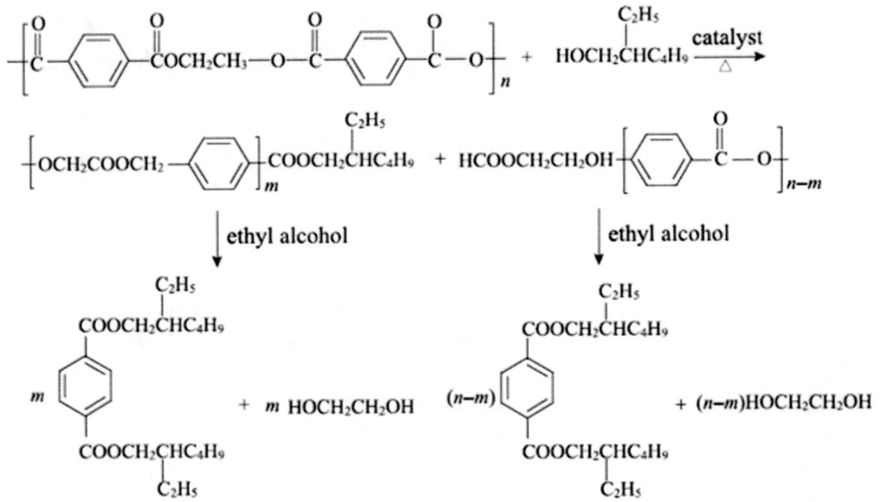

DOTP EG DOTP EG

Fig. 3.3 Production of DOTP via degradation of waste polyester

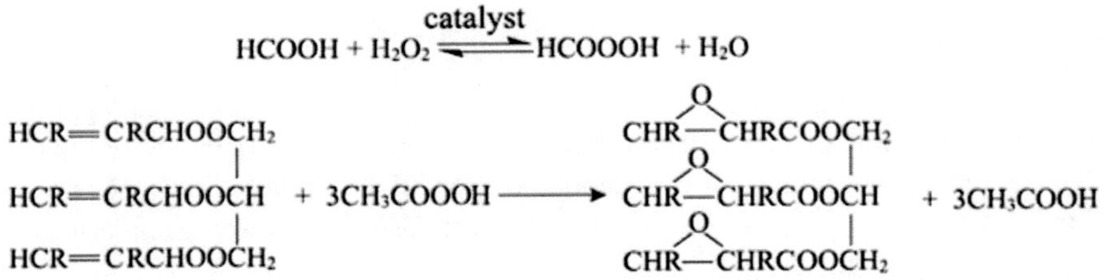

Fig. 3.4 Main principles of industrial production of epoxy soybean oil

elimination of protic acid catalysts and the development of pollution-free, energy-saving, environmentally friendly, high-efficiency epoxidation and esterification catalysts are the major research directions in the future.

Currently, the urgent task for the Chinese plasticizer industry and plastics additive companies is to adjust the product structure at a faster pace: decreasing the ratio of phthalate plasticizers, expediting the R&D of non-phthalate plasticizers, and promoting the research of processibility of environmentally friendly plasticizers in plastics and polymer materials. The purpose is to guide the direction of Chinese plastics additive industries toward environmental protection and energy saving. This will be the major task for Chinese plasticizer and soft PVC industries in the future.

3.7.3 *Innovations and Applications of Green Plasticizers [97]*

Phthalate plasticizers will still be the dominant type and production volume globally in the future, but those of long-chain (C_9–C_{10}) alcohol-based esters and non-phthalate plasticizers will increase largely. As important additives for functional polymers, the market and application of plasticizers are very bright. Therefore, China should expedite the R&D of environmentally friendly plasticizers and their performance in polymeric materials to provide reliable technological support to the national new material industries.

At present, the synthetic route via hydrogenation of phthalates is considerably advantageous, which not only can solve the "more production than sale" problem of domestic traditional plasticizers, but also can utilize the current facilities to scale up the hydrogenation production to enrich the available stocks and supply chains. These features can certainly provide important technological and product support for industry transformations.

There are three major technical bottlenecks for structural modification of phthalate molecules: catalytic synthesis, separation of products, and application of products.

1. Catalytic Synthesis

The major technical challenges of converting phthalates to cyclohexane-1,2-dicarboxylates via catalytic hydrogenation include the development of new catalysts and catalytic protocols.

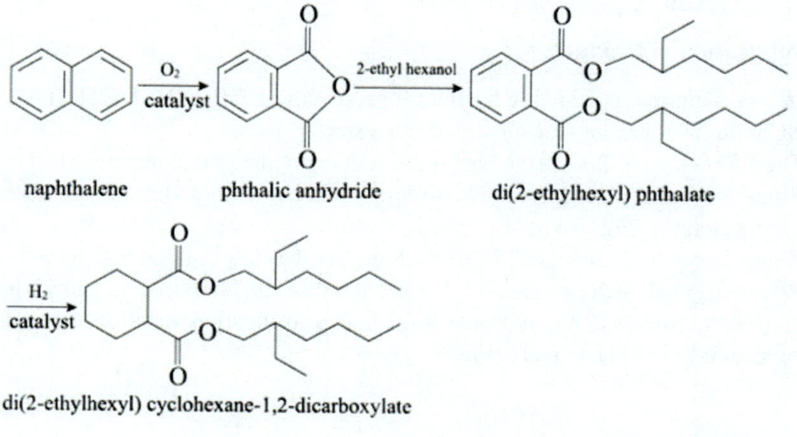

萘：naphthalene；苯酐 phthalic anhydride；邻苯二甲酸二（2-乙基己）酯：di(2-ethylhexyl) phthalate；环己烷二（2-乙基己）酯：di(2-ethylhexyl) cyclohexane-1,2-dicarboxylate

2. Separation of Products

The major product is DENCH.

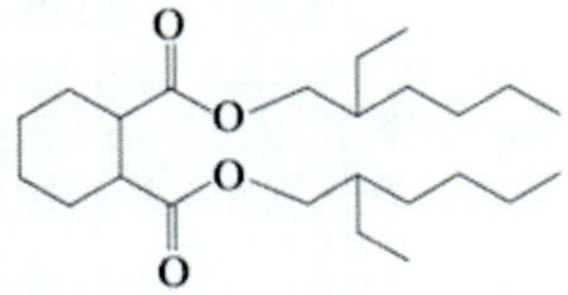

di(2-ethylhexyl) cyclohexane-1,2-dicarboxylate

The impurities are DEHP, 2-ethylhexanol, and hydrogenated carbonyl compounds.

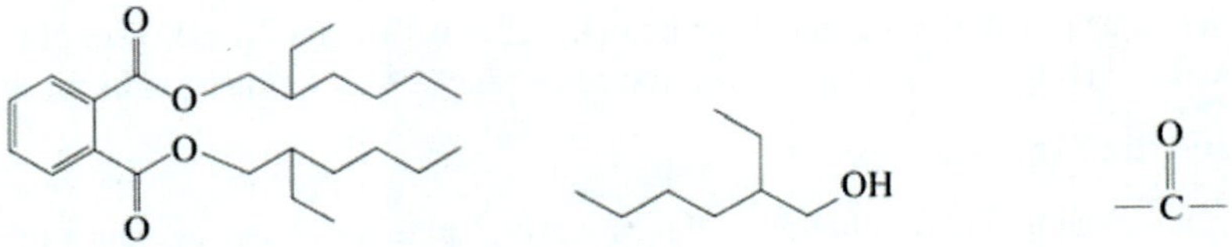

di(2-ethylhexyl) phthalate ; 2-ethylhexanol ; catalytic hydrogenation products

The final impurities should not exceed 500 ppm. But the DENCH and DEHP have very similar boiling points and are difficult to separate.

3. Application of Products

There are a number of existing barriers for replacing DOP, DBP, DOTP, TBC, and ATBC with cyclohexane-1,2-dicarboxylate esters.

The following methods have been used to modify the phthalate structures.

Route 1: benzene ring hydrogenation of phthalates to synthesize cyclohexane-based plasticizers (Fig. 3.5).

Route 2: synthesis from HHPA to cyclohexane-based plasticizers (Fig. 3.6).

Route 3: Diels–Alder reaction between butadiene and maleic anhydride to form HHPA, followed by hydrogenation and esterification reactions to produce cyclohexane-based plasticizers (Fig. 3.7).

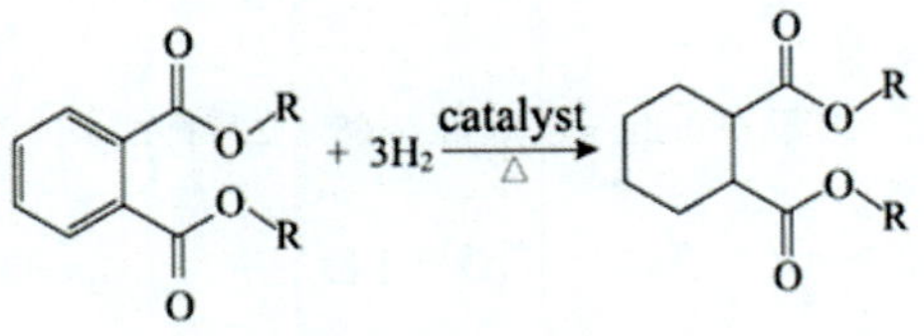

Fig. 3.5 Preparation of cyclohexane dicarboxylates via catalytic hydrogenation of phthalates

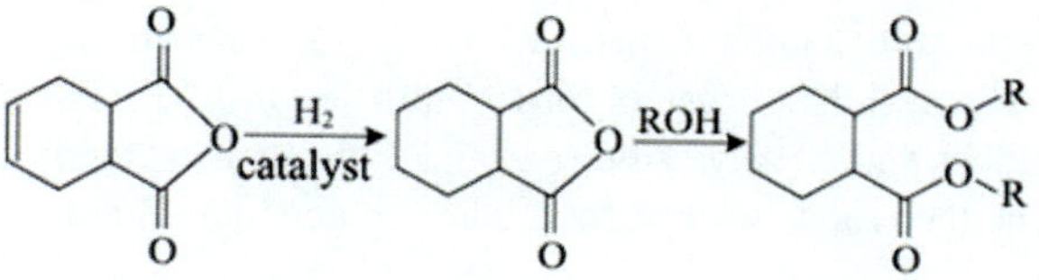

Fig. 3.6 Preparation of cyclohexane dicarboxylates via catalytic hydrogenation of tetrahydrophthalic anhydride and then esterification

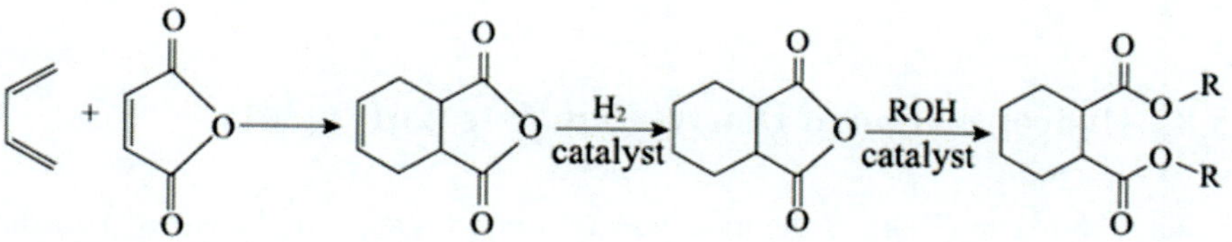

Fig. 3.7 Preparation of cyclohexane dicarboxylates via the olefin method. R can be same or different, usually C_5–C_{11} when used as plasticizers

3.7.4 Technological Innovation and Market Dynamics of Green Plasticizers (2017) [98]

China has an annual production volume of plastic articles of more than 60 million tons and is the largest plastics production and consumption country in the world. In 2015, the gross industrial output value of Chinese plastic articles reached as high as 3.87 trillions, about 4.25% of the GDP, making it a pillar industry of the light industry of China. In addition, the annual production volume increases at a rate of 10%, much higher than the world level of 4%.

Plasticizers are widely applied in food wraps and packaging, toys, PVC construction materials, automotive parts, rubber tires, electrical and medical devices, PVC cables, water pipes, plastic plates, soft PVC synthetic leather, etc. Plasticizers are the type of plastics processing additives that have the highest production capacity and consumption, ~60% of the overall additive production.

Since the "Taiwan plasticizer incident", phthalate plasticizers have raised increasing public concerns. In June 2011, the 551st Document signed by the Ministry of Health, i.e., Letter to Report the Maximum Phthalate Residue in Food and Additives, declared that the maximum residue of DBP was 0.3 mg/kg. In the "Chinese spirits plasticizer incident" in November 2012, the content of DBP in inspected spirit samples was 1.08 mg/kg, 260% times higher than the standard. This is similar to the "Melamine Incident" in 2008. After these occurrences, the Chinese government immediately strengthened the inspections and made more stricter inspection standards. But before the news reports, this problem did not induce sufficient attention. The Hygienic Standards for Uses of Additives in Food Containers and Packaging Materials (GB 9685-2008), effective since 2008, introduced the concept of "specific migration" to restrict the amount of migration of certain plasticizers into food (or model food). The 16th Announcement of 2011, published by Ministry of Health on June 1, 2011, put phthalate esters into the 6th set of the List of Potential Illegal Inedible Substances and Abusive Additives in Food and also clearly pointed out that phthalates are neither foodstuff nor food additive and should not be added in food and additives.

In the plasticizer production structure, phthalates have above 80% of the overall capacity, while environmentally friendly ones have less than 20%. Technological innovation and R&D of environmentally friendly plasticizers are under way.

3.8 Hydrogenation of Dioctyl Phthalate (2012) [34]

Dioctyl phthalate (DOP) is the most widely used plasticizer and has ~70% of the overall plasticizer consumption. The annual consumption in China only already reaches almost 1 million tons, mainly in PVC resins. The interactions between phthalate molecules and polymer chains are not chemical but physical forces. Thus, these molecules gradually leach out from the mother materials and pollute the air, soil, water, and food. Phthalate molecules can accumulate in the human body and decompose into related metabolites, resulting in damage to human health [99]. The toxicity lies mainly in the benzene ring. Therefore, the main research direction is currently the conversion of the benzene ring to cyclohexane while keeping the ester groups unchanged.

There are two major types of industrial catalysts for benzene ring hydrogenation. One is substrate-supported noble metal catalysts with Rh, Ru, Pd, or Pt as active component and with Al_2O_3, SiO_2, or activated carbon as the substrate. Their drawbacks include high cost and ease of poisoning [100, 101]. The other is common metal catalysts with Ni as a representative. The drawback is complicated technology [102].

The catalytic hydrogenation of DOP to prepare dioctyl cyclohexane-1,2-dicarboxylate is rarely reported. The authors prepared a Ni-based catalyst via complexation and used it to hydrogenate DOP in a self-made fixed-bed reactor at a catalyst loading of 50 mL. The effects of temperature, pressure, airspeed, and reaction time on the hydrogenation reaction to optimize the catalysis conditions, finally as 8.0 MPa, 180 °C, and 0.5 h^{-1} to obtain a ring hydrogenation selectivity as high as 99.2%. The lifetime tests indicated excellent catalyst stability.

3.9 Liquid-Phase Catalytic DOP Hydrogenation and Kinetics (2012) [50]

In recent years, higher health and safety requirements have been expected for plasticizers in food and drug packaging, daily necessities, and toy industries. Animal tests have demonstrated that long exposure to phthalate esters will induce peroxisome proliferation and liver tumors. Phthalate plasticizer molecules are easy to migrate and leach out of plastic articles. There have been increasing voices of restricting or prohibiting their uses in medical devices, food packaging, and toys. The USA, Germany, and Japan have made related laws and regulations [103, 104].

Di(2-ethylhexyl) cyclohexane-1,2-dicarboxylate (DEHCH) is a low-toxic, environmentally friendly plasticizer. The articles plasticized by DEHCH demonstrate comparable or better performance than those by DOP [2]. More importantly, DEHCH does not induce peroxisome proliferation and liver tumors, is biodegradable in natural environments, and thus has been approved for use in food packaging, medical devices, and toys. There are two major industrial production methods for DEHCH. First is the olefin method via Diels–Alder reaction between butadiene maleic anhydride to produce THPA, followed by hydrogenation and esterification reaction to yield DEHCH. The drawbacks of this method include too many reaction steps, competition for butadiene and maleic anhydride with other industries, and high production costs. The second is the hydrogenation method via catalytic hydrogenation of the benzene ring in DOP, or via hydrogenation of phthalic anhydride to form cyclohexane-1,2-dicarboxylic acid and then esterification to obtain DEHCH. This method has simple synthetic protocols and high yields. More importantly, this method can make good use of the current DOP production facilities. Compared to alkenes, aldehydes, ketones, and nitro compounds, benzene rings are more difficult to be hydrogenated, mainly because the conjugation structure leads to excellent chemical stability. In general, the hydrogenation of benzene is an electrophilic addition reaction. When an electron-donating functional group is present, the hydrogenation reaction proceeds more easily. Research [15] has indicated the following hydrogenation activity order of substituted benzenes: *t*-butyl benzene > isopropyl benzene > ethyl benzene > toluene > benzene; phenol > *p*-chloroaniline > chlorobenzene > nitrobenzene. There are two electron-withdrawing carbonyl groups, making the ring hydrogenation reaction able to proceed only under high temperature and high hydrogen partial pressure [105–108]. Non-noble metal catalysts such as Ni, Co, Mn, and Mo have been successfully used in the hydrogenation of benzene rings, but most of them are gas-phase catalytic hydrogenation reactions of volatile aromatic compounds under high temperature. There have been very few research reports on hydrogenation of high-boiling-point aromatic compounds, and even less on hydrogenation of phthalate esters [109].

This work examined 4 metal catalysts for hydrogenation of DOP to prepare DENCH and investigated the effects of reaction temperature, hydrogen pressure, reaction time, and amount of catalyst. The results demonstrated that Rh/C had the highest catalysis activity. The reaction temperature had little effect, but the hydrogen pressure and amount of catalyst had significant effects on the reaction rate. The optimal DOP hydrogenation conditions were determined to include temperature 170 °C, hydrogen pressure 2.0 MPa, and reaction time 4.0 h, under which DEHCH was the only product with a yield of >99.5%. The apparent kinetic analysis indicated that the hydrogen concentration adsorbed on the catalyst surface was the rate-determining factor.

This work systematically investigated the synthesis of DEHCH from DOP, compared the hydrogenation results of 4 catalysts, and finally determined the optimal synthetic protocols.

1. Preparation of the Catalysts

(1) Preparation of Rh/C Catalyst

Nitric acid-pretreated activated carbon (with a surface area of 400 m^2/g, 9.3 g) was added to 120 mL deionized water and the suspension was slowly heated to 80 °C, followed by the addition of $RhCl_3$ (0.96 g), HCl solution (10 mL), and 37 wt% formaldehyde solution (8 mL). 30 wt% NaOH solution was added to adjust the pH to be weakly alkaline. The mixture was stirred at 80 °C for 5 min. The solid was filtered and washed with deionized water, and this operation was repeated 10 times. The obtained solid catalyst was dried with NaOH for 24 h at room temperature and then stored in a closely sealed bottle for further use (9.3–10.0 g). The study carried out by Park et al. [110] also indicated that Rh/C was a highly efficient catalyst for benzene ring hydrogenation reactions.

(2) Preparation of Raney Ni catalyst

To a well-stirred mixture of 4 g Raney Ni–Al alloy (40 wt% Ni) in 10 mL deionized water at 50 °C was added 0.4 mL NaOH solution (20 wt%) under vigorous stirring. After about 1 h, the Raney alloy started to react with water and turned white. The reaction lasted for 1.5 h, before adding 6 mL NaOH solution (40 wt%). The stirring was continued until the top layer became white. The catalyst was washed 4 times with deionized water at 50 °C and then with the same amount of ethanol. Then it was stored for further use.

2. Experimental

The catalytic activity of 3 noble metals and Raney Ni catalysts was evaluated through hydrogenation of DOP (analytical, $\geq$98%) in an autoclave reactor with these catalysts. The reactor included a 100 mL autoclave, a magnetic stirrer, an electric thermal controller, and a pressure detector, as well as a high-pressure liquid sampling tube. DOP (20.0 g) and a certain amount of catalyst were added to the autoclave. After purging the autoclave with hydrogen 3 times, hydrogen was connected to it with the pressure fixed at 1.0 MPa. Hydrogen needed to be refilled frequently due to the consumption of hydrogen. Under vigorous stirring, the temperature was increased to 170 °C at 5 °C/min and was maintained for 240 min. Samples extracted from the autoclave every 60 min were analyzed.

The products were subjected to GC-MS analysis (VARIANCP-3800) under conditions: PEG-50M capillary column (50 m × 0.25 mm × 0.33 μm), sample injector and FID detector temperature of 250 °C, N_2 carrier gas flow rate 2.0 mL/min, and column temperature 250 °C, calculation using area normalization.

3. Conclusion

DEHCH is an exceptional low-toxicity, environmentally friendly plasticizer. China is the major producer of DOP. Production of DEHCH via hydrogenation of DOP is of important significance to the upgrading of plasticizers in China. High-performance catalyst is a key factor for the hydrogenation of DOP. Due to the high boiling point of DOP, gas-phase hydrogenation is difficult. While liquid-phase hydrogenation is slow, an autoclave-type reactor is needed. Rh/C is a high-performance catalyst for ring hydrogenation and can convert DOP to DEHCH at low hydrogen pressure, which is particularly important to the safe operations of hydrogenation reactions. Temperature is less influential than the hydrogen pressure and the amount of catalyst on the hydrogenation rate. The apparent reaction kinetic analysis indicates that the concentration of hydrogen absorbed on the catalyst surface is the rate-determining factor for the hydrogenation reaction. The optimal hydrogenation conditions include 170 °C, 2.0 MPa, and reaction time 4.0 h, under which DENCH is the only product with a yield of at least 99.5%.

Under a constant amount of catalyst and hydrogen pressure, a linear relationship between the DEHCH yield and reaction time is observed, which is characteristic of zero-order reactions, implying the reaction rate is independent of the DOP concentration. A small apparent activation energy E_a of 11.82 kJ/mol is obtained, in good agreement with the low impact of temperature on reaction rate.

The yield of DEHCH increases with temperature but not significantly. The yields of DEHCH are 78.3%, 90.3%, 98.7%, and 97.5% for the 240 min reactions at 130 °C, 150 °C, 170 °C, and 190 °C, respectively.

It is reported that hydrocarbon solvents and water can inhibit hydrogenation reactions [111]. In addition, the use of solvent increases the post-treatment cost. Therefore, solventless hydrogenation is desired.

The hydrogen pressure has a significant effect on the reaction rate. During the reaction, hydrogen gas needs to be dissolved in the liquid phase and then diffuse to the catalyst surface. The low hydrogen concentration in DOP results in a low hydrogen concentration at the catalyst surface, which is the primary reason for the zero kinetic order at a constant hydrogen pressure. The elevated hydrogen pressure increases the hydrogen concentrations both in DOP and at the catalyst surface, leading to faster reactions. Hu et al. [112] also observed similar results when investigating the catalytic benzene hydrogenation reaction with amorphous Ru/Zn catalyst.

3.10 Synthesis and Application of Environmentally Friendly DEHCH Plasticizer (2013) [38, 113, 114]

Plasticizers are the most important additive for PVC articles. The total production of plasticizers in China has reached 3 million tons per annum in 2011. Traditional phthalate plasticizers are the major products, ~80% of the overall use. Recently, phthalate plasticizers have been found toxic to infants, fetus, and reproductive systems. Many countries and organizations have made policies and regulations to limit the use of phthalate plasticizers. It has been a general trend to use environmentally friendly plasticizer alternatives.

The cyclohexane-1,2-dicarboxylate plasticizers, represented by diisononyl cyclohexane-1,2-dicarboxylate (Hexamoll® DINCH) produced by BASF, have comparable and even better properties and use value, plus other advantages of biodegradability, non-toxicity, and environmental friendliness. Thus, they are suitable alternatives to phthalate plasticizers.

With the financial support from the National Key Technology Research and Development Project 2012BAI22B07, the authors synthesized di(2-ethylhexyl) cyclohexane-1,2-dicarboxylate (DEHCH) using hexahydrophthalic anhydride (HHPA) and 2-ethylhexanol as the raw materials, with methanesulfonic acid as the catalyst. The synthetic route is shown as follows:

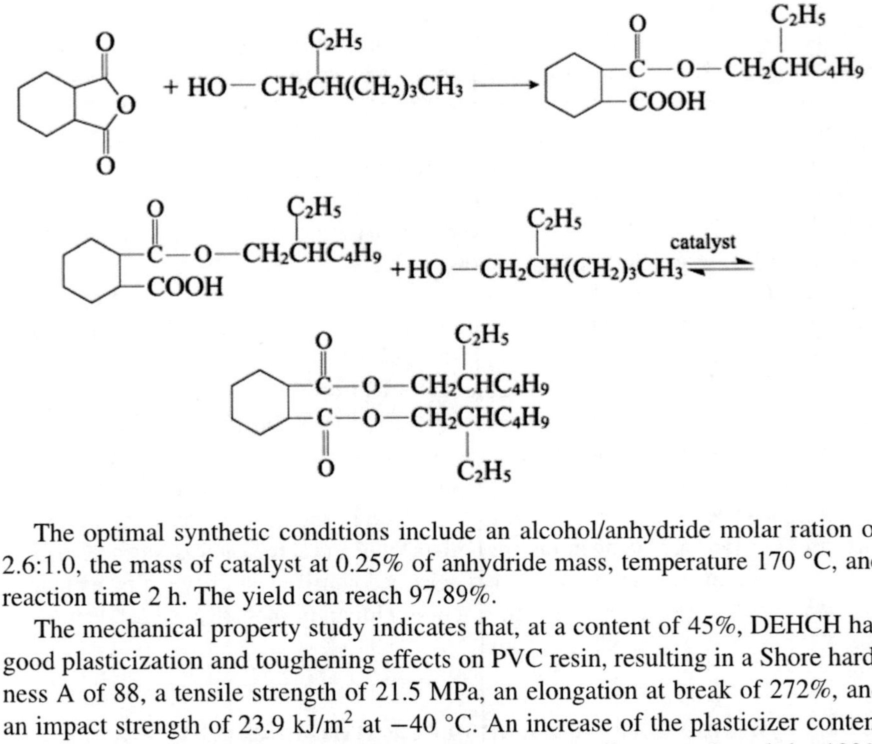

The optimal synthetic conditions include an alcohol/anhydride molar ration of 2.6:1.0, the mass of catalyst at 0.25% of anhydride mass, temperature 170 °C, and reaction time 2 h. The yield can reach 97.89%.

The mechanical property study indicates that, at a content of 45%, DEHCH has good plasticization and toughening effects on PVC resin, resulting in a Shore hardness A of 88, a tensile strength of 21.5 MPa, an elongation at break of 272%, and an impact strength of 23.9 kJ/m^2 at −40 °C. An increase of the plasticizer content from 30 to 70% leads to a gradual decrease in the tensile strength and the 100% tensile modulus and the embrittlement temperature, as well as a gradual increase in the elongation at break and the vaporization loss and the dissolution loss. As a whole, DEHCH has plasticization performance similar to DEHP and good evaporation and dissolution resistances, thus promising to replace the phthalate plasticizers.

Plasticizers are one of the most produced and consumed plastics additives. Despite the increasing use of DEHP and DOP in the food and medicine industry, their toxicity is being recognized. The US FDA has banned the industrial production of 6 types of phthalates, limited the use of DOP only in high-water-content food packaging, and required the use of non-toxic plasticizers in packaging materials for meat. Switzerland bans the use of DOP in children's toys. Germany restricts the use of DOP in plastic articles related to human health and food. Japan prohibits the addition of DOP to medical devices and permits DOP only in industrial plastics products [35–37, 115]. Dialkyl cyclohexane-1,2-dicarboxylate not only affords better properties in the plasticized PVC than phthalates but also shows no toxicity. Therefore, they are considered as the most promising phthalate alternatives for use in food packaging, medical devices, and children's toys [31, 39, 116, 117], and have been accepted in many countries as safe plasticizers.

Cyclohexane-1,2-dicarboxylates are synthesized mainly via hydrogenation or olefin conversion processes. Pingping Jiang et al. [22] successfully prepared DEHCH using HHPA and 2-ethylhexanol as the raw materials and activated carbon-supported methanesulfonic acid as the catalyst. Using the same raw materials, Jiping Shang et al. [39, 118, 119] synthesized DEHCH using titanium tetraisopropanolate and solid superacid $SO_4^{2-}/TiO_2–Al_2O_3$, respectively, as the catalyst.

The synthetic route for DEHCH is shown in Eq. (3.11).

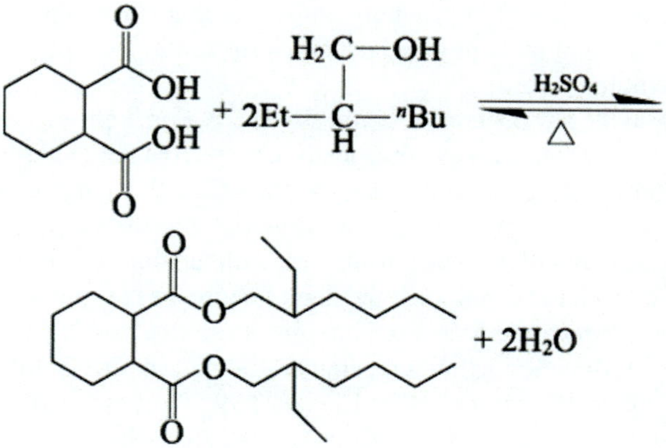

(3.11)

This reaction is reversible with excess 2-ethylhexanol. The concentrated H_2SO_4 catalyst absorbs water driving the equilibrium to the right side and increasing the yield of the cyclohexane-1,2-dicarboxylate esters.

The authors carried out the synthetic reactions and investigated the effects of reaction time, the alcohol/acid molar ratio, and temperature on the esterification ratio. The products were analyzed using ^{1}H-NMR, infrared (IR), thermogravimetry (TG), and gas chromatography. Under conditions of reaction time 6 h, reaction temperature 170 °C, and an alcohol/acid molar ratio of 2.5:1, a colorless oily liquid was obtained with an esterification ratio of 99.5% and a yield of 95%. The ^{1}H-NMR spectrum shows the multiplet peak at $\delta = 4.0$ ppm corresponding to the methylene group ($–CH_2–$) linked to the ester group. The IR spectrum shows the stretching peak of carbonyl (C=O) at 1735 cm^{-1}. These results indicate the formation of the desired product. The TGA curve of the product shows an extrapolated onset temperature of 215 °C. The retention times in gas chromatography are 14.310 and 14.505 min. Therefore, DEHCH can be easily separated from DOP.

3.11 Synthesis and Application of Environmentally Friendly DEHCH Plasticizer (2014–2015)

3.11.1 Preparation of DOP Hydrogenation Catalyst (2014) [120]

Plasticizers are the most used additive in plastics. DOP is currently the most widely used plasticizer, ~70% of the total consumption of plasticizers. In China, the annual consumption is close to 1 million tons. In recent years, the medicine, food packaging, daily supplies, and toys industries have raised more stringent requirements to plasticizers. Countries such as the USA, Germany, and Japan have made restrictions or prohibitions to plasticizer use. The product of DOP hydrogenation, i.e., DEHCH, is a low-toxic, environmentally friendly plasticizer that is biodegradable in nature. DEHCH has been approved in many countries for use in food packaging, medical devices, and children's toys.

Using DOP as the raw material to synthesize DEHCH in high-pressure hydrogenation reactors, the authors have systematically investigated the effects of the catalyst and support choices, the catalyst loading, and the calcination temperature on catalyst performance. The results favor macroporous γ-Al_2O_3 as the support, and Ru/Rh as the catalyst that is modified by La to decrease the amount of noble metals. Here are the optimal synthetic conditions: the mass fraction of the support 0.2–0.5% and the calcination temperature 550 °C. With a reaction time of 2 h at 220 °C, a reaction pressure 10 MPa, a reaction temperature 150 °C, an airspeed 0.75 h^{-1}, and a hydrogen/oil volumetric ratio of 1000:1, at least 99.5% of the DOP is converted to DEHCH.

3.11.2 DEHP Hydrogenation Product Analysis Using GC-MC (2015) [42]

Dialkyl cyclohexane-1,2-dicarboxylates prepared from ring hydrogenation of dialkyl phthalates are high-performance, environmentally friendly, low-toxic plasticizers. The toxicological data indicate that ring-hydrogenated phthalates are neither endocrine disruptors nor reproductive toxins, can afford better plasticization performance, and thus have been deemed as exceptional phthalate plasticizer alternatives.

The hydrogenation reaction products are analyzed using GC-MS, where the products are separated in HP-5MS capillary columns. The possible side reactions of hydrogenation are proposed through detection of reaction products and estimate of major molecular fragments.

Under heating, phthalates undergo intramolecular *cis*-elimination reactions to produce phthalate monoesters and alkene. The phthalate monoesters produce phthalic anhydride and respective alcohol via acyl oxygen cleavage.

The major parasitic side reactions in DOP hydrogenation are listed below.

1. Pyrolysis of DOP

DOP thermally decomposes into phthalic anhydride and 2-ethylhexanol (**A**). Then 2-ethylhexanol continues to dehydrate to form 2-ethylhexene (**B**). See Fig. 3.8.

2. Breaking of ester bond

One of the two ester bonds in DOP breaks to yield 2-ethylhexyl benzoate (**C**) and 3-ethylhexene (**D**); the former is hydrogenated to form 2-ethylhexyl cyclohexanate (**E**), while the latter is hydrogenated to 3-ethylhexane (**F**). See Fig. 3.9.

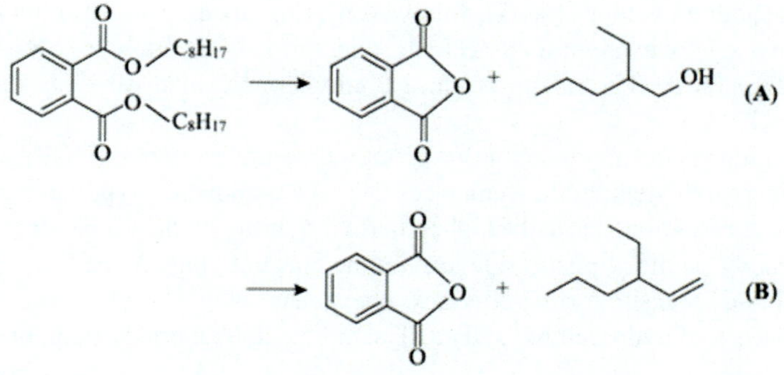

Fig. 3.8 Thermal degradation of DOP to produce phthalic anhydride, 2-ethylhexanol, and 2-ethylhexene

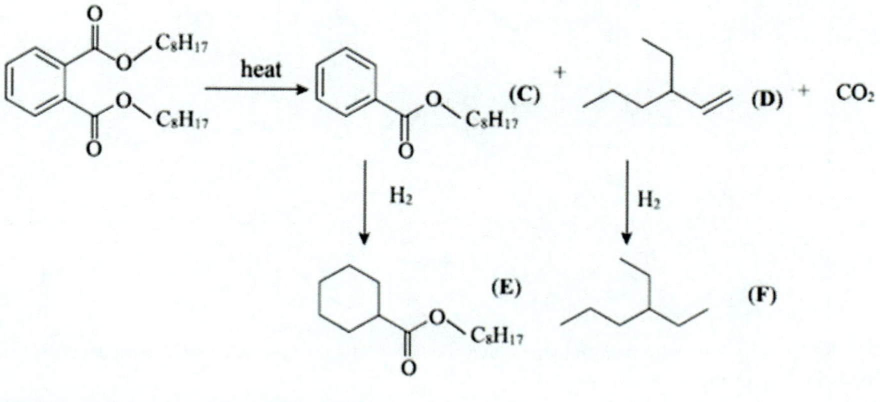

Fig. 3.9 DOP thermal degradation products

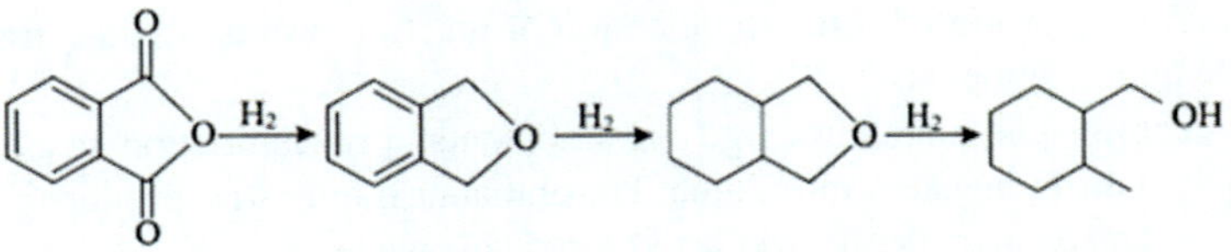

(G)

Fig. 3.10 Hydrogenation of phthalic anhydride to form methyl cyclohexyl methanol

3. Rearrangement of the carbon chain

The phthalic anhydride is hydrogenated to lose the two carbonyl oxygen atoms, further hydrogenated at the benzene ring to form cyclohexane, and finally hydrogenated to produce methyl cyclohexyl methanol (**G**). See Fig. 3.10.

4. Decarbonylation

The phthalic anhydride is hydrogenated to lose one carbonyl group, followed by ring hydrogenation to form 2-methyl cyclohexanoic acid (**H**). Further decarbonylation results in methyl cyclohexane (**I**), followed by ring opening to form heptane (**K**). The methyl cyclohexanoic acid (**H**) can undergo esterification reactions with 2-ethylhexanol (**A** in Fig. 3.8) to produce 2-ethylhexyl 2-methylcyclohexanate (**L**). See Fig. 3.11.

Temperature is the major factor that dictates the side reactions of DOP hydrogenation. An increase in temperature enhances DOP decomposition to phthalic anhydride and alkene. In addition, due to the higher thermal stability of the π bonds, the cleavage occurs mainly on the aliphatic side arm on the benzene ring. When DOP is heated, the ester bond breaks to form 2-ethylhexyl benzoate, 2-ethylhexyl hexane, and CO_2. In the presence of hydrogen gas and catalyst, all the double bonds are hydrogenated.

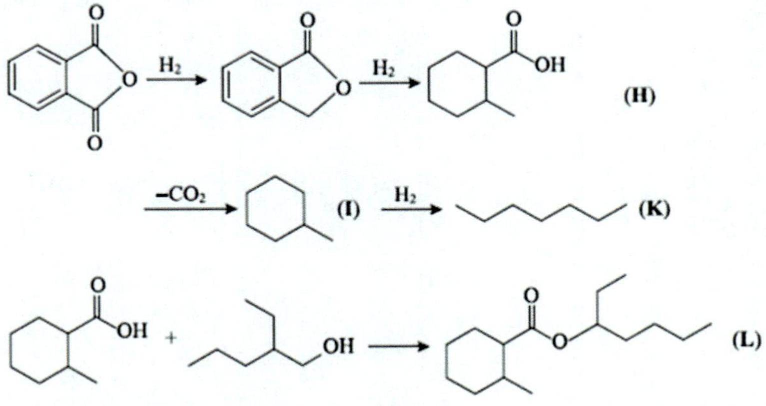

Fig. 3.11 Dehydrogenation of phthalic anhydride to form methyl cyclohexanoic acid, heptane, and methyl cyclohexanoate

Table 3.25 Components in DOP hydrogenation products

Peak #	Chemical	Molecular weight	Retention time/min
1	*n*-Heptane (K)	100	4.764
2	3-Ethylhexane (F)	114	5.339
3	2-Ethylhexanol (A)	130	8.025
4	2-Methyl cyclohexane methanol (G)	128	8.878
5	*cis*-2-Methyl cyclohexanoic acid (H)	142	9.668
6	*trans*-2-Methyl cyclohexanoic acid (H)	142	9.895
7	2-Ethylhexyl cyclohexanoate (E)	240	13.916
8	2′-Ethylhexyl 2-methylcyclohexanoate (L)	254	14.338
9	Di(2-ethylhexyl) cyclohexane-1,2-dicarboxylate (main product)	396	22.157
10	Di(2-ethylhexyl) phthalate (reactant)	390	23.921

Meanwhile, phthalic anhydride undergoes a hydrogenolysis reaction to produce 2-methyl cyclohexanoic acid that has a slightly stronger acidity ($pK_a = 5.43$) than H_2S ($pK_a = 6.89$) and is the major material that affects the product acidity.

The major product from DOP hydrogenation is di(2-ethylhexyl) cyclohexane-dicarboxylate. The side products mainly include those from DOP pyrolysis and ensuing hydrogenation, as shown in Table 3.25. Among the side products, 2-methyl cyclohexanoic acid has a relatively strong acidity and can corrode the facilities. Thus, attention should be paid to the selection of facility-related materials.

3.12 Synthesis of DEHCH via Direct Esterification (2015) [121]

PVC is the major raw material for disposable sterile medical devices. Plasticizers are important additives to PVC and the most widely used are phthalate types. In recent years, phthalate plasticizers are found to have liver toxicity, reproductive toxicity, and carcinogenic hazards. There have been multiple types of environmentally friendly plasticizers in the global market. In order to develop environmentally friendly plasticizers of independent intellectual property rights, the authors have synthesized DEHCH using HPPA and 2-ethylhexanol as the raw materials with the methanesulfonic acid catalyst, and have determined its molecular structure using GC-MS. We compare its physico-chemical properties, subchronic toxicity and plasticization performance with those of commercially available DEHCH, DINCH and DEHP. The results indicate that our synthetic DEHCH is comparable or even better than the commercial peer products, without significant difference in subchronic toxicity. The mechanical property analysis of plasticized PVC particles shows similar plasticization performance between DEHCH and DEHP. Therefore, DEHCH can be used as an alternative to phthalate plasticizers.

Because of the significant advantages of good physico-chemical properties, high biocompatibility, and reasonable cost, PVC is the most used material in disposable sterile medical devices. But PVC has a very close melting point and decomposition temperature, high glass transition temperature, and high hardness, making the direct processing challenging. Thus, PVC resins need plasticizers and other additives to decrease their melting point and glass transition temperature and increase their flexibility to be processed into thin films and tubes [122]. Currently, phthalates are the most commonly used plasticizers, with DEHP as the major one. However, numerous animal toxicity studies have revealed a wide range of side effects of DEHP including liver toxicity, reproductive toxicity, and carcinogenic hazards. Thus, many countries and organizations have made a series of policies to define the range of phthalate use [24, 123]. Environmentally friendly plasticizers such as epoxy esters and citrates have been accepted in the EU, the USA, Japan, and South Korea for use in medical devices to replace phthalates.

Cyclohexane-1,2-dicarboxylate plasticizers, represented by DINCH produced by BASF in 2002, are widely used in medical devices, food packaging, and toys because of their comparable plasticization performance, non-toxicity, and biodegradability [124, 125]. Due to the limitations in the production protocol, low purity, and high cost, the authors synthesized DEHCH, with a similar molecular structure and lower cost compared to DINCH, via a direct esterification route using HHPA and 2-ethylhexanol raw materials and methanesulfonic acid catalyst. We also compared its physico-chemical properties, subchronic toxicity, and plasticization performance with other environmentally friendly plasticizers.

1. Materials and Methods

(1) Lab synthesis procedure for DEHCH

The reaction between HHPA and 2-ethylhexanol proceeds stepwise: monoester is formed in the first step, followed by the formation of diester catalyzed by methanesulfonic acid [126].

(2) GC-MS analysis of DEHCH product

A medium polarity column is used. The injector and detector temperatures are both 250 °C. The column temperature conditions: maintained at 50 °C for 2 min, then increased to 250 °C at 20 °C/min, and maintained at 250 °C for 20 min.

(3) Physico-chemical property analysis of DEHCH

According to the Pharmacopoeia of the People's Republic of China or the Chinese Pharmacopoeia (2010), the refractive index, relative density, water content, ash content, dynamic viscosity, acid value, acidity, heating loss, thermal stability, and iodine value of DEHCH are analyzed.

(4) Subchronic toxicity test of DEHCH

45 SD rats (a type of white rats widely used in pharmacologic, toxicological, and medicinal effect tests) are randomly, averagely divided into 3 groups: normal saline, DEHCH (lab), and DEHCH (INNUA Inc.). The 15 rats in each group are further randomly divided into 3 subgroups. Each subgroup includes 5 male and female rats. High (0.2 mL/(kg d)), medium (0.1 mL/(kg d)), and low (0.05 mL/(kg d)) doses of samples are injected daily to the rats via tail vein for 4 weeks. After the rats are killed, blood samples are drawn from the abdominal aorta and are subjected to examinations of hematological parameters, blood clotting tetrachoric index, and blood biochemical index. Histopathological section analysis is performed on the livers and testes taken from the rats.

(5) Preparation of the plasticized PVC

A mixture of PVC, plasticizer, and Ca/Zn stabilizer (100:45:1.8 by weight) is stirred for 3 min at a rate of 200 rpm, followed by transferring to a double roll mill for compounding at 165 °C.

(6) Mechanical property analysis for PVC resins plasticized by DEHCH and DEHP

The tensile property is measured at a stretching rate of 50 mm/min, according to GB/T 1040-2006. Following GB/T 1843-2008, the simple bracket impact performance is measured with a sample size of 80 mm (L) × 10 mm (W) × 2 mm (H), a non-notched impact, a temperature −40 °C, and a pendulum speed 2.9 m/s. The hardness is measured according to GB/T 531.2-2009 with a sample thickness of 6 mm; an average value of 5 different sites no shorter than 1 cm from the sample edge is obtained.

2. Research Tasks

(1) Compositional analysis

The GC-MS results of synthetic DEHCH are shown in Table 3.26. The purity of DEHCH is close to 100%, with 99.68% *cis* structure and 0.32% *trans* structure.

(2) Physico-chemical property analysis

The refractive index, relative density, ash value, water content, and dynamic viscosity of lab synthetic DEHCH, INNUA DEHCH, and BASF DINCH are shown in Table 3.27. The acid value, acidity, heating loss, thermal stability, and iodine value are shown in Table 3.28.

Compared to the peer products, the independently developed DEHCH has comparable or even better physico-chemical properties than DINCH, and thus has great promise in substituting DEHP.

(3) Subchronic toxicity test

The liver pathological sections of the high-dose group show 1 light hepatic steatosis (+) and 4 normal in the normal saline subgroup. 2 light steatoses are observed in both

Table 3.26 GC-MS analysis results of products

Peak #	Retention time/min	Chemical name	Molecular formula	Molecular weight	Mass fraction/%	Major m/z
1	25.43	*trans*-DEHCH	$C_{24}H_{44}O_4$	396.61	0.32	27, 41, 57, 71, 84, 97, 113, 126, 149, 155, 173, 267, 285
2	25.5	*cis*-DEHCH	$C_{24}H_{44}O_4$	396.61	99.68	27, 41, 57, 71, 84, 97, 113, 126, 149, 155, 173, 267, 285

Table 3.27 Refractive index, relative density, ash value, water content, and dynamic viscosity test results

Tests	Test results			Test standards
	DEHCH (lab)	DEHCH (INNUA)	DINCH (BASF)	
Refractive index	1.4614	1.4601	1.460–1.466	China Pharmacopoeia (2010 II) Appendix
Relative density	0.9493	0.9561	0.9486	China Pharmacopoeia (2010 II) Appendix VI A (pyknometric method)
Ash value/%	0.0163	0.0147	No data	GB/T 2433-2001
Water content/%	0.0661	0.0676	0.1	China Pharmacopoeia (2010 II) Appendix VIII M (Karl Fischer titration)
Dynamic viscosity/(mm^2/s)	52.055 (20 °C)	47.505 (20 °C)	52	GB/T 1660-2008
	14.465 (50 °C)	13.005 (50 °C)	13	

the lab DEHCH and the INNUA DEHCH groups; the other 3 have organized hepatocyte arrangements with no observations of fibrous tissue hyperplasia, pseudolobule formation, macrophage hyperplasia, cholestasis in capillary and interlobular portal area bile ducts, or fibrous proliferation and inflammatory cell infiltration in interlobular portal areas. This indicates no obvious liver toxicity difference between DEHCH and commercial peer products.

The test results demonstrate low risks of liver toxicity and reproductive toxicity for DEHCH, without obvious difference compared to foreign peer products. The

Table 3.28 Acid value, acidity, heating loss, thermal stability, and iodine value test results

Tests	Test results			Test standards
	DEHCH (lab)	DEHCH (INNUA)	DINCH (BASF)	
Acid value/(mg KOH/g)	0.009270	0.009260	0.07	GB/T 1668-2008
Acidity/%	0.001421	0.001419	No data	GB/T 1668-2008
Heating loss (acid value change after heating)/%	0.1165	0.2249	No data	GB/T 1669-2001
	0.03937	0.04402	No data	GB/T 1670-2008
Thermal stability (acid value change)/(mg KOH/g)	0.03010	0.03476	No data	GB/T 1676-2008
Iodine value/%	0.05256	0.2974	No data	GB/T 1658-1982

human tolerances of the lab DEHCH and INNUA DEHCH are both 50 mg/(kg d), in sharp contrast to the 0.6 mg/(kg d) for DEHP. Therefore, DEHCH has significantly lower subchronic toxicity than DEHP and can replace DEHP in PVC plasticization.

(4) Mechanical property analysis

The mechanical properties of PVC plasticized with DEHCH and DEHP are shown in Table 3.29.

Seen from the table, the synthetic DEHCH has a comparable plasticization performance to DEHP and thus can be deemed as a DEHP substitute.

(5) Conclusion

The synthetic DEHCH, prepared using HHPA and 2-ethylhexanol raw materials and methanesulfonic acid catalyst, has comparable or even better physico-chemical properties and similar subchronic toxicity compared to foreign peer products. The mechanical property analysis indicates DEHCH has a similar plasticization performance to DEHP and can be used as a phthalate plasticizer substitute.

Table 3.29 Mechanical properties of plasticized PVC pellets

Plasticizer	Shore hardness A (HA)	Tensile strength/MPa	Elongation at break/%	100% tensile modulus/MPa	Impact strength (40 °C)/(kJ/m^2)
DEHCH	88	21.5	272	12.7	23.9
DEHP	87	20.2	281	11.4	21.8

3.13 Cable Industry's Response to EU's Newest Environmental Laws

With increasing attention paid to ecological and environmental protection and human health, higher and more requirements of environmental protection have been raised to electrical and electronic devices including cables. EU issued the ROHS, in effect on July 1, 2006, that is applied not only to electrical and electronic devices but also to their upstream and downstream industries including raw materials, parts, and packaging. In 2011, the EU upgraded the environmental protection act to ROHS 2.0. On June 4, 2015, the EU published the RoHS 2.0 Revision (EU) 2015/863 officially listing DEHP, BBP, DBP, and DIBP as restricted substances and totaling 10 mandatory controlled substances. Our country's related environmental protection acts have lagged behind the EU. The "Measures for the Administration of Pollution Control of Electronic Information Products", equivalent to EU's RoHS, was issued and implemented in 2007. The "Nationally Uniform Measures for Voluntary Certification of Pollution Control of Electronic Information Products" was implemented on November 1, 2022, and was revised in parallel to RoHS 2.0 Revision. Cable companies, especially those aiming to expand their international market, should closely watch the related laws and regulations. No one can win the competition without timely adjustment of product structure, promotion of technological innovation, or active development of green products that meet these policies.

3.14 Applications of Environmentally Friendly Plasticizers in Disposable Gloves (2016) [127]

PVC is currently one of the most widely used plastics. Because of the long durability and low cost, PVC is commonly used in industrial products, construction materials, daily supplies, artificial leather, tiles, cables, tubing, packaging thin films, foaming materials, sealants, etc.

Based on the variety of softness, function, processing, and end use of PVC articles, the plasticizer content generally varies: 0–5% in hard articles, 6–25% in semihard articles, 26–60% in soft articles, and 60–100% in paste articles.

Currently, 80–90% of the total produced plasticizers are used in PVC articles, among which 80–90% are used in soft PVC articles for applications in sealing tapes, soft tubing, cables, clothing, toys, medical devices, protection gloves, etc. Among them, the disposable glove industry is a major consumer.

In the production of disposable gloves, 50% of the total raw material weight belongs to the plasticizer. In PVC gloves produced in earlier times, phthalate plasticizers were used by 80% among the overall plasticizers, including DOP, DBP, DIBP, BBP, DINP, etc. DOP is called an all-round plasticizer because of its advantages of high plasticization performance, good PVC compatibility, low exudation, low volatility, softness, high brightness, etc.

However, recent research has revealed the mutagenicity, carcinogenicity, and teratogenicity of phthalate plasticizers. Long-time contact with them can cause damages to the livers and kidneys of humans and animals. The PVC glove industry is significantly impacted by the restrictive laws and regulations in many countries.

To protect the environment and solve the toxicity of phthalate plasticizers, seeking alternative plasticizers has been a hot research topic in the PVC glove industry. It will be an important direction to develop safe, reliable, high-performance, non-toxic, environmentally friendly plasticizers.

3.15 Development and Application of Environmentally Friendly Polyester Plasticizers (2016) [128]

With the stricter restrictions by EU policies, PVC article importers in the EU market adopt more stringent control on the type and concentration of plasticizers and even require zero tolerance to benzene ring-containing plasticizers. This situation has limited the use of terephthalate, phthalate, and phenyl polycarboxylate plasticizers. The R&D of green, environmentally friendly, biodegradable plasticizers has become a hot research direction.

Currently, commercial environmentally friendly plasticizers include aliphatic acid esters, citrates, epoxy, terephthalates, high-carbon phthalate esters, cyclohexane dicarboxylates (or tetrahydrophthalates), polyesters, phenyl polycarboxylates, etc.

Aliphatic acid esters can only be used as auxiliary plasticizers generally in cold-resistant conditions because of their small molecular weights, poor PVC compatibility, and risk of release.

Citrate plasticizers have better PVC compatibility but poor heat resistivity and easy aging, thus are generally used in PVC paste articles.

Epoxy plasticizers have moderate heat resistivity and upgraded comprehensive performance, leading to an increase in consumption in recent years, but still suffer from the technical hurdle of easy release in the case of large doses.

Cyclohexane dicarboxylates have comparable molecular weights to phthalates, but the poor PVC compatibility caused by the change of molecular polarity results in the issue of easy release.

Polyester plasticizers, also called permanent plasticizers, have significant resistances to high temperature, aging, extraction, and migration. Polyester plasticizers with different viscosities prepared from different raw materials typically have differences in performance.

Plasticizer performance tests include specifications including viscosity at 25 °C, density at 20 °C, chroma, acid value, and water content; processibility including plasticization efficiency, tensile strength, heat aging, low-temperature performance, fogging value, PVC compatibility, and UV aging. For the plasticization efficiency testing, DOP is usually used as the reference to obtain the ratio of their amounts needed to achieve the same Shore hardness A value. With the same formulation, the higher the Shore hardness A value, the worse the plasticization performance. The plasticization temperature and time can also be used. With the same formulation, the lower the plasticization temperature, the better the plasticization performance; the shorter the plasticization time, the better the plasticization performance.

The national or industrial standards for performance tests: acid value, GB/T 1668-2008; Shore hardness A, GB/T 2441-2004; tensile strength, GB 1040.3-2006; embrittlement temperature, ASTM D746-2007; fogging value, DIN 75201 B; PVC compatibility, ASTM D3291-1997 (2003); UV aging, ASTM G154-2006.

3.16 How to Accurately Assess and Use Benzoate-Based Plasticizers? (2016) [129]

Benzoate-based plasticizers globally have the earliest use, the most comprehensive performance, and the widest scope of application among all types of plasticizers. The amount of use in PVC industries is above 80% of the overall production. Benzoate-based plasticizers can also be applied in the following fields: polyvinylidene dichloride, polyvinyl butyral, polyvinyl acetate, polyacrylates, polyamide-12, and certain copolyamide–cellulose. Besides, they are also widely used in the paper and pulping industry, hydraulic oil, thermal exchange oil, lubricants, military industry, rocket industry, etc. Correct assessment and reasonable use of benzoate-based plasticizers are an important matter associated with the global development of plasticizers and their upstream and downstream products.

1. Relationship between the molecular structure and performance

Benzoate-based plasticizers include trimellitates such as TOTM; phthalates such as DOP, DBP, and DINP; terephthalates such as DOTP. The ester alkyl group can be from either straight-chain or branched-chain alcohols with long or short chains. Tables 3.30 and 3.31 show the structure–performance relationship of DOP, DBP, DINP, and DOA.

The ester group functions to produce high plasticization performance to PVC via "like dissolves like" intermolecular forces. A free benzene ring features a large, stable π structure with a uniform electron cloud. The ester group shifts the electron

Table 3.30 Effects of Temperature on Shore hardness A and D of plasticized PVC at 30%

Plasticizer	Shore hardness A (HA)/shore hardness D (HD)								
	10 °C	0 °C	10 °C	20 °C	30 °C	40 °C	50 °C	60 °C	70 °C
DBP	98/75	97/69	94/58	85/–	78/–	71/–	66/–	62/–	60/–
DOP	98/76	97/71	94/95	90/47	85/–	79/–	72/–	66/–	62/–
DINP	98/82	98/76	97/70	94/54	90/45	82/–	73/–	67/–	63/–
DOA	97/62	95/56	91/48	86/–	82/–	77/–	70/–	64/–	59/–

Table 3.31 Change in elongation at break of PVC with plasticizer content (20 °C)

Plasticizer	Elongation length (N/mm^2) and ratio (%) at break under each plasticizer content				
	25%	30%	35%	40%	45%
DBP	24/250	21/300	18/350	15/375	11/400
DOP	27/250	24/300	21/350	18/400	14/425
DINP	29/225	26/275	23/325	19/375	15/400
DOA	26/275	23/325	20/375	16/400	12/425

cloud producing a certain extent of polarization, and thus functions to increase the compatibility with the plasticized system and also keep the plasticizer molecule from being released. The alkyl group can enhance the plasticization performance and the flexibility of the plasticized article, taking advantage of the single bond's rotation. The molecular structures of benzoate-based plasticizers determine their comprehensive plasticization performance.

2. Effects of the alkyl structure on plasticization performance

(1) Plasticization: straight chains are better than branched chains; shorter chains are better than longer chains.
(2) Stability (no release or migration): longer chains with larger molecular weights are better than shorter chains with lower molecular weights; benzoate-based esters are better than aliphatic acid esters.
(3) Viscosity: with the same carbon number, branched chains are better than straight chains; with the same straight/branched chain, higher carbon numbers are better than lower carbon numbers.
(4) Degradability in nature: due to the stable aromatic ring, benzoate-based esters are more difficult to be degraded than aliphatic acid esters.
(5) Dissolution in water: benzoate-based esters are hardly soluble in water and thus less possible to be absorbed by plants; it is less likely that humans take in plasticizers via eating grains or vegetables and consequently are poisoned.

Benzoate-based plasticizers made from C_8–C_{12} alcohols have the following exceptional properties: (1) comprehensive plasticization performance with small side effects; (2) good compatibility with most synthetic resins and rubbers; (3) low release leading to high stability; (4) high resistance to heat and weather, low volatility, good low-temperature flexibility, low extraction, no tendency to turn soft or brittle; (5) good electrical performance; (6) excellent water resistance due to minimal solubility in water; (7) high chemical stability, low migration.

3. Toxicity analysis of benzoate-based plasticizers

(1) Harms to human. Numerous experimental data and patient case studies have proved that the benzene structure is harmful to humans, animals, plants, and environments. Benzoate-based plasticizers are more harmful than aliphatic acid

ester plasticizers. Among benzoate-based plasticizers, those with branched alkyl chains are more harmful than those with straight alkyl chains.

(2) Carcinogenicity. The tests are mostly performed on small animals, while not much has been done on humans. Therefore, there is still a lack of persuasive evidence and final conclusion on carcinogenicity.

(3) The harms of benzoate compounds to humans are possibly ascribed to the slow degradation of the π bonds in the benzene ring and accumulation of them in the human body causing chronic damages to organs.

4. Accurate assessments and reasonable use of benzoate-based plasticizers

We should be responsible for human health and environments and expedite the risk assessments of plasticizers. The government should provide strong support in this regard. The research institutes and industry should collaborate to start the toxicological testing and risk evaluation. The collected data should be used to guide the standard making, production, and end use of related industrial products. We should not blindly follow any blind marketing hypes without having scientific foundations to avoid negative impacts on the upstream and downstream products of plasticizers. Scientific analyses will help determine effective measures that can reduce the environmental damages to a minimum level and make related risk management strategies. These can avoid potential harm to humans and the environment by benzoate-based plasticizers.

(1) We should strictly follow the national and international policies and prohibit the use of benzoate-based plasticizers in close human-contacting areas such as drug and food packaging, and children's toys.

(2) Benzoate-based plasticizers should not be used in containers and appliances for cooking oil, fuels, solvents, and organic liquids that may dissolve the ester plasticizers.

(3) Low-quality additives should not be added to plastic articles driven by cost reduction, which may expedite the aging and breaking of the articles and bring plasticizers into the natural environment. Or such additives are poorly compatible with the plasticizer causing the release of the latter.

(4) We should use long-chain instead of short-chain alcohols to make benzoate-based plasticizers.

(5) Catalysts with high selectivity should be used to reduce side reactions and increase the final purity. This can decrease the detrimental effects of low-boiling-point aldehydes, ethers, alkenes, and alcohols on the plasticized system.

It has become a necessary trend to eliminate toxic plasticizers and develop non-toxic substitutes. Industry needs to devote more efforts to investigate new catalysis and separation techniques and commercialize the next-generation plasticizers with higher cost performance, more reliable safety, and environmental friendliness.

5. Suggestions to the future development

China is the Top 1 country in producing and using plasticizers with an annual production of more than 5 million tons and annual consumption of 3.6 million tons, about

1/5 of the global uses. Due to the significance of plasticizers, we must handle related issues in a serious and responsible manner.

(1) Based on the real situations in China, we should use scientific approaches and accurate experimental data to make our own developmental strategies.
(2) High attention should be paid to science and technology to push the advancement of the plasticizer industry.
(3) Industrial associations will be used as the platform to perform key research tasks. We need to understand the toxicity of benzoate-based plasticizers to humans and the environment, take measures to promote their advantages and avoid their disadvantages, offer accurate assessments, and develop new products to meet emerging social needs. These can effectively guide the R&D, production, and application of related industries.
(4) More efforts should be devoted to R&D to develop new materials and products. The technological level and product competitiveness need to be improved to expand the international market shares and enable adequate space for our plasticizer industry.

3.17 Production Study of Environmentally Friendly DEHCH Plasticizer (2016) [130]

Plasticizers are one of the most produced and consumed plastics additives and are the most used and critical additive for the PVC industry, which also finds important applications in fields such as rubber coating, adhesives, and sealants [131]. Phthalate plasticizers are widely used in our daily lives and industries, among which di(2-ethylhexyl) phthalate (DEHP) is the most used with a consumption equaling to ¾ of the total plasticizer production. Its content in plastics and rubber can reach as high as 40% [48].

Some plasticizer molecules are structurally similar to hormones and are thus called "environmental hormones". Long-time exposure can lead to reproductive system abnormalities and even the risks of teratogenesis and cancer [132, 133]. Trace residues of these compounds can enter the human body via the food chain and form false hormones, which can transmit false chemical signals, affect the content of real hormones, interfere with the original processes of the endocrine system, and result in general endocrine enhancement. Therefore, higher purity and health requirements have been requested by plastic articles including drug and food packaging, daily supplies, and toys to the major plasticizers such as DOP. The USA has stopped the industrial production of 6 phthalate plasticizers, has permitted the DOP use only in high-water-content food packaging, and has required the use of non-toxic plasticizers in meat packaging. The Swiss government has prohibited the use of DOP in children's toys. Germany prohibits the addition of DOP in all plastic articles related to human health and food. Japan requires that DOP cannot be used in medical devices and can only be used in industrial plastic articles.

Environmentally friendly plasticizers include synthetic plant sterol esters, citrates, cyclohexane-1,2-dicarboxylates, epoxy, etc. Diisopropyl cyclohexane-1,2-dicarboxylate has a low migration, is a transparent, colorless, slightly odorful, anhydrous liquid, is soluble in a variety of organic solvents, is readily miscible with common PVC plasticizers, and is almost water-insoluble [134]. Thus, it can be used in a number of soft medical materials such as stomach tubes and infusion bags, and provide an ideal safety solution for toy manufacturers, and meet the new EU laws and regulations.

Currently, the production of environmentally friendly, non-toxic plasticizers is facing two major technical barriers: (1) diverse side products and low yield; and (2) low production volume and high cost [135, 136]. Products such as citrates are a new type of non-toxic, odorless green phthalate plasticizer substitute and have been widely used in food and medical device packaging, cosmetics, daily supplies, toys, and military supplies, which meanwhile are important chemical intermediates. But because the expensive catalyst increases the plasticizer cost, the users still prefer the choices of inexpensive traditional plasticizers. As it's been an inevitable trend to eliminate toxic plasticizers and promote non-toxic substitutes, the producers must devote significant efforts in the investigations of new catalysis and separation techniques and the commercialization of next-generation plasticizers with higher performance–price ratios, better safety characteristics, and environmental friendliness.

The two-phase solvent extraction, first used in the oil industry [115], is a separation technique taking advantage of the different partition coefficients of the target substances in two immiscible or partially miscible solvents. Due to the similar solubilities of the target substances, a single solvent cannot separate them out. The two phases of immiscible or partially miscible solvents result in a larger difference between their partition coefficients and achieve effective separation.

The unique two-phase solvent extraction system separates substances based on their selective partitions in two phases. The separation of substances is based on the different concentrations of them in the top and bottle phases driven by surface properties, electrostatic forces, or bonding interactions (e.g., hydrogen bonding, ionic bonding, hydrophobic bonding). This method has been used in protein (enzyme) extraction, electrophoretic protein separation, antibiotic extraction, biological particle separation, temperature-induced phase separation, etc. [113].

Wei Guo et al. investigated the separation of methyl methacrylate (MMA) and methanol using hexane–water and cyclohexane–water two-phase systems. They studied the effects of the amount of hexane/cyclohexane–water on the partition coefficient and selectivity coefficient under different temperatures. The results indicated an exceptional MMA–methanol separation by the hexane/cyclohexane–water two-phase systems. An increase in the amount of hexane/cyclohexane–water led to the increased MMA solubility and partition coefficient. The temperature had an opposite effect on the hexane–water and cyclohexane–water systems. Multiple such two-phase extractions could effectively separate MMA and methanol [115].

Haoli Zhou et al. investigated the effects of temperature, time duration, ethanol, and butane concentrations on the extraction of oil and soy isoflavone from soy sauce residues using a butane–ethanol–water system. The optimal conditions were determined as follows: temperature 60 °C, ethanol concentration 80% (0.071 g/mL), butane concentration 0.1 g/mL, and time 2 h. The extraction ratios of oil and soy isoflavone were 99.2% and 94.5%, respectively [137].

The technical protocol of DEHCH production is divided into three steps: hydrogenation of DOP, two-phase solvent extraction, and regeneration of extractants.

1. Hydrogenation of DOP

Currently, the synthesis of DEHCH mostly uses the hydrogenation method. Jufeng Gu et al. investigated the performance of 4 metal catalysts for the synthesis of DEHCH via hydrogenation of DOP and studied the effects of reaction temperature, hydrogen pressure, reaction time, and amount of catalyst. The results revealed that Rh/C has the highest catalytic activity. Suitable conditions for DOP hydrogenation include temperature 170 °C, hydrogen pressure 2.0 MPa, and reaction time 4.0 h, under which DENCH is the only product with a yield of above 99.5%.

2. Separation via two-phase extraction

The different boiling points of DEHCH and DOP meet the requirement of distillation separation, but both compounds turn yellow at 180 °C and decompose. Thus, distillation or reduced-pressure distillation is not suitable for their separation. Experiments indicate that the two-phase extraction method can achieve separation of DEHCH and DOP at room temperature and avoid their decomposition. The hydrogenation products achieve a certain extent of separation via two-phase extraction leading to a collection of a light phase and a heavy phase for treatment in the next step.

3. Regeneration of extractants

The light and heavy phases contain considerable amounts of extraction solvents that must be removed to obtain high-quality products and recycle the extractant solvents.

The boiling points of DEHCH and DOP are 365 °C and 384 °C, respectively, which are significantly different than those of the extractant solvents. Therefore, reduced-pressure distillation can be used to recycle the two solvents. The distillation temperature should be controlled below 180 °C to avoid decomposition of DEHCH and DOP.

3.18 Co-existing Opportunities and Challenges for DOP Market (2017) [115]

Plasticizers are widely used in fields closely related to human life, including coatings, adhesives, dyes, inks, detergents, food packaging, cosmetics, electronics, plastics, rubber, car manufacturing, etc. The major hazards include hormonal disorders, immunity decrease, precocious puberty, gene damages, cardiovascular diseases, liver damages, and weakening of reproductive and urinary systems.

1. Multiple revisions of foreign laws and standards

During 1999–2010, countries and regions including the EU, the USA, Japan, Canada, Argentina, Denmark, Malaysia, etc. have made a series of laws and regulations on plasticizer restriction, including 1999/815/EC, 2002/72/EC (i.e., Regulations on Plastic Food-Contacting Materials), 2005/84/EC, REACH, 2007/19/EC, 2009/48/EC, "Consumer Product Safety Improvement Act (CPSIA)" (2008), AB1108 (2009), ASTM F963-2008 (2009), "Specification Standards for Food, Additives, Container and Packaging, Toys and Detergents" (2009), "The Order for Enforcement of the Food Safety Act" (2003), "Toy Safety Standards (ST2002, 8th Edition)" (2003), "Food Safety Act of Japan" (2003), etc.

2. More complete domestic environmental protection laws and standards

During 2008–2017, China has launched a series of laws and regulations including "Hygienic standards for uses of additives in food containers and packaging materials" (GB 9685-2016), "Determination of phthalate plasticizers in PVC plastics-based toys and children's supplies".

(GB/T 22048-2008), "Determination of phthalate in food" (GB/T 21911-2008), "Limits of hazardous substances in toy coatings" (GB 24613-2009), "Food Announcement NO 1001301729" (published by Taiwan Department of Health), "The 6th list of illegal non-edible substances and additives possibly added in food", "Determination of phthalates in cosmetics" (GB/T 28599-2012), "Toy safety" (GB 6675-2014)," Index of nationally encouraged substitutes to toxic, hazardous raw materials (products)" (2016 version), "The technical requirement for environmental labeling products-furniture" (2017), etc.

3. The structure adjustment direction of plasticizer industry

The future direction of the plasticizer industry in China must be based on principles of resource saving, green, low carbon, environmental friendliness, and safety to adjust the industrial structure and promote technological innovation. The development of high-performance, energy-saving, non-toxic, environmentally friendly, biodegradable plasticizers to substitute phthalate products has become an urgent task for the plasticizer and plastics industries.

Dioctyl terephthalate (DOTP): the domestic production capacity in 2016 was ~1.3 million tons with an actual production volume of ~0.9 million tons, ~27% share of the plasticizer market. Epoxy plasticizers: the domestic production capacity of epoxidized soybean oil in 2016 was ~0.9 million tons with an actual production volume of ~0.6 million tons, ~13% share of the plasticizer market. But, with the smaller price difference between epoxidized soybean oil and benzene-containing plasticizers, the dependence of downstream products on it has decreased in recent years. Citrates: the domestic production capacity in 2016 was ~0.25 million tons with an actual production volume of ~0.19 million tons, ~3% share of the plasticizer market. Due to their high prices, their markets are difficult to be widely expanded. Cyclohexane-1,2-dicarboxylates: there is no benzene ring in their molecular structures. Currently, they are used mainly in toys and medical devices, but will be more important in food-contacting PVC articles such as wraps and gaskets. These esters are miscible with a number of polymers such as PVC, polyurethane, acrylic resin, and phenolic resin. The mechanical properties of PVC plasticized by cyclohexane-1,2-dicarboxylate are comparable to those by phthalates. Because of these advantages, cycloalkane carboxylates are considered as a type of promising phthalate substitute plasticizers.

3.19 Conclusion

This chapter has reviewed the foreign and domestic research status since the 1990s. The major conclusions are as follows.

In foreign countries: The USA and Germany have early started investigations of the ring hydrogenation of benzoate-based compounds and the toxicity of phthalate plasticizers [1–8]. They converted benzene rings to cyclohexane via catalytic hydrogenation, which have filed patents in the USA, Germany, Europe, China, etc.

In 2002, BASF in Germany achieved ring hydrogenation of DINP. The product was recommended by the German Federal Institute for Risk Assessment (BfR) and approved by the European Food Safety Administration (EFSA), which has been exported to the EU, the USA, Japan, China, etc. In China, it is used in human-contacting PVC articles such as medical devices as a phthalate substitute.

In China: In 1999, Hongyan Shang et al. from China Petroleum University reacted 1,3-butadiene with maleic anhydride to synthesize tetrahydrophthalic anhydride that was then hydrolyzed to yield tetrahydrophthalic acid that then underwent an esterification reaction with isobutanol to produce diisobutyl cyclohexane-1,2-dicarboxylate (DIBCH). Under the conditions of 5% Pd/C catalyst, 40–80 °C, 0.56–1.01 MPa, a conversion ratio of 100% was achieved with a yield of above 97% [9].

In 2002, Ruina Yang et al. from Henan Academy of Science, Institute of Chemistry used self-invented Pd composite catalysts to convert tetrahydrophthalic anhydride to hexahydrophthalic anhydride in a 100% selectivity via hydrogenation under 170–180 °C, hydrogen pressure 1.4–1.6 MPa, and 4% catalyst [10].

In 2006, Haoran Guo et al. from the China Petroleum and Chemical Corporation (Sinopec), Beijing Research Institute of Chemical Industry, published a review paper on the properties and applications of a new plasticizer, di(2-propylheptyl) phthalate (DPHP). They pointed out that the method for synthesizing DPHP with inexpensive, abundant butane and 2-propylheptanol raw materials should be preferentially considered in producing DPHP [13].

In 2010, Wancong Shi from Hebei Light Industry Bureau published an article titled "New Plasticizers: Dialkyl Cyclohexane Dicarboxylates" that introduced the properties, preparation, and PVC plasticization performance of cyclohexane dicarboxylate plasticizers. They indicated that cyclohexane dicarboxylates have an exceptional property combination, are biodegradable in nature, have no eco-toxicity, and can be used in food packaging, medical devices, and children's toys [14].

In 2010 and 2011, Bozhang Qian from Shanghai Qingdu Information Technology Limited published a paper that summarized the current status of plasticizers, introduced environmentally friendly plasticizers, and suggested a healthy road to the future development of plasticizers in China. The suggestions included making up-to-date plasticizer standards, expediting risk assessments, promoting industrial structure adjustment, establishing application-based market segments, etc. [18, 19].

In 2010 and 2014, Duoren Wang from PetroChina Jilin Petrochemical Company published two articles titled "Progress of the Development and Applications of Ring Hydrogenation of Diisononyl Tetrahydrophthalate" and "Development and Applications of Diisononyl Cyclohexane-1,2-Dicarboxylate". He pointed out that, via a simple addition of hydrogenation facility, domestic phthalate producers can transit to produce a series of diverse ring-hydrogenated phthalates. This could increase the export sales of PVC, expand the market, generate enormous economic and social benefits, and advance the whole plasticizer industry out of dilemma [20, 21].

During 2011–2017, Pingping Jiang's research team from Jiangnan University carried out the investigations of green environmentally friendly catalysts.

In 2011, Pingping Jiang's team synthesized diisooctyl cyclohexane-1,2-dicarboxylate using HHPA and isooctanol as raw materials and activated carbon-supported methanesulfonic acid as the catalyst under the conditions of 160–180 °C, an isooctanol/HHPA molar ratio of 2.4:1.0, a catalyst mass of 2.1% isooctanol and HHPA total mass, and a reaction time 2.5 h. The esterification ratio was 99.82%. The spectral purity of the product was 99.15%. The product met the performance requirement for plasticizers [22].

In 2012, Pingping Jiang's team published an article titled "Progress of the Research and Applications of Catalytic Synthesis of Environmentally Friendly Plasticizers". This paper reviewed the catalytic synthesis methods for new plasticizers, new catalysis reaction protocols, and the types and preparations of catalysts. The focus was the introduction of the properties, applications, and research status of globally major environmentally friendly plasticizers including bio-based plant oils, polyesters, citrates, trimellitates, and terephthalates. After comparing with the traditional synthetic methods and catalysts, they pointed out the general trend of catalytic synthesis of plasticizers in China and the future applications of plasticizers in the national high-tech areas. A suggestion was made that the development of the plasticizer industry and new additives should be focused on production structure adjustments to expedite innovations of non-phthalate, plant oil-based plasticizers [15].

In 2008, Pingping Jiang's team synthesized a series of poly adipate ester plasticizers via reactions of adipic acid with 1,2-propanediol, 1,3-butanediol, 1,4-butanediol using 2-ethylhexanol as the capping agent [48].

In 2011, Pingping Jiang's team [49] synthesized diethylene glycol cyclohexane-1,2-dicarboxylate and diisooctyl cyclohexane-1,2-dicarboxylate using HHPA and diethylene glycol and isooctanol as the raw materials [22] on methanesulfonic acid, tetra(*n*-butyl) titanate, and activated carbon-supported methanesulfonic acid catalysts, respectively. They also obtained triethylene glycol adipate using adipic acid and triethylene glycol raw materials on tetra(*n*-butyl) titanate catalyst [50].

In 2016, in his presentation titled "Progress of Technical Innovations and New Applications of Green Plasticizers" at the International Forum of Green Plasticizer Development, Dr. Pingping Jiang [97] pointed out that the phthalate hydrogenation method was particularly advantageous because it could not only solve the oversupply issue of traditional plasticizers but also utilize the current manufacturing facilities to hydrogenate in large scales, as well as expand the available plasticizer types and product chain to support industrial upgrading and green development. Structural modification of phthalates is facing three technical bottlenecks: catalytic synthesis, production separation, and production use.

In 2017, in his presentation titled "Technological Innovations and Market Dynamics of Environmentally Friendly Plasticizers Suitable for Uses in PVC Paste Industry" at the 9th JinLianChuang Plasticizer Product Chain Market Summit and Trade Fair, Dr. Pingping Jiang [98] pointed out that phthalate plasticizers have become a public concern since the Taiwan Plasticizer Incident occurred. The 551st Announcement published by Department of Health in June 2011, i.e., "Letter Regarding the Maximum Amount of Phthalate Residues in Food and Food Additives", defined the maximum residual amount of dibutyl phthalate (DBP) was 0.3 mg/kg. On October 19, 2017, China implemented the new "Hygiene Standards of Additives for Uses in Food Containers and Packaging Materials" (GB 9685-2016).

In 2012, Kaijing Zhao et al. from the China Petroleum and Chemical Corporation (Sinopec), Beijing Research Institute of Chemical Industry, performed the research on hydrogenation of dioctyl phthalate [34]. They prepared the Ni-based catalysts using the coordination method and hydrogenated dioctyl phthalate in a house-made fixed-bed reactor. The selectivity for dioctyl cyclohexane-1,2-dicarboxylate reached as high as 99.2% under conditions of a catalyst loading of 50 mL, 8.0 MPa, 180 °C, and 0.5 h. The catalysts had exceptional stability and lifetime.

In 2012, Junfeng Gu et al. from Shandong University of Technology carried out a study on "Preparation of DEHCH via Liquid-Phase Catalytic Hydrogenation of DOP and Its Kinetics" [50]. They systematically investigated the catalytic performance of 4 metal catalysts and the effects of reaction temperature, hydrogen pressure, reaction time, and catalyst loading. According to the results, Rh/C had the highest catalytic activity. The reaction time had a much lower effect than the hydrogen pressure and catalyst loading on the hydrogenation reaction kinetics. At 170 °C, a hydrogen pressure 2.0 MPa, and a reaction time of 4 h, DEHCH was the only product with a yield above 99.5%. The apparent kinetic analysis indicated that the determining step was the hydrogen adsorption on the catalyst surface.

In 2013, Yuanhui Ou et al. from Beijing Chemical Industry University synthesized DEHCH using HHPA and 2-ethylhexanol raw materials, supported by the National Key Technology Research and Development Program (2012BA I22B07). Under conditions of 0.25% methanesulfonic acid catalyst, 2-ethylhexanol/HHPA molar ratio of 2.6:1.0, 170 °C, and 2 h, the esterification ratio was 97.89%. Based on the mechanical property measurements, DEHCH had a similar plasticization performance to DEHP and featured good evaporation and solvent resistances, thus considered as a phthalate plasticizer substitute [38, 113].

In 2013–2015, Xiaoguo Li et al. from the China National Offshore Oil Corporation (CNOOC), Tianjin Chemical Research and Design Institute Co., Ltd., published their research work titled "Synthesis of Environmentally Friendly Di(2-ethylhexyl) Cyclohexane-1,2-Dicarboxylate" [114], "Catalyst Preparation for Hydrogenation of Dioctyl Phthalate" [120], and "GC-MS Analysis of DEHP Hydrogenation Products" [42]. In 2013, this team synthesized DEHCH via the esterification reaction between cyclohexane-1,2-dicarboxylic acid and 2-ethylhexanol raw materials with a concentrated sulfuric acid catalyst. The esterification ratio and yield reached as high as 99.5% and 95%, respectively, under conditions of 170 °C, alcohol/carboxylic acid molar ratio 2.5:1.0, and reaction time 6 h. In 2014, using a high-pressure hydrogenation reaction evaluation device, this team systematically investigated the effects of the support, composition, loading, and calcination temperature of the catalysts on their performance. γ-Al_2O_3 was selected as the support. Rh and Ru noble metals were selected as the catalytic composition and were modified with La to decrease the amount of noble metals used. The optimal catalyst preparation conditions included loading mass ratio 0.2–0.5%, calcination temperature 550 °C, and reduction temperature at 220 °C for 2 h. The optimal reaction conditions included hydrogen pressure 10 MPa, 150 °C, airspeed 0.75 h^{-1}, and hydrogen/oil volume ratio 1000:1, leading to a DOP conversion ratio no less than 99.5%. In 2015, they analyzed the hydrogenation products using GC-MS. According to the products and side products, they proposed thermal decomposition and MS fragmentation mechanisms.

In 2015, Guoying Hu from Tianjin Medical University, Xuejia Ding from Beijing University of Chemical Industry, Qinan Wu from Nanjing University of Traditional Chinese Medicine, and Hongyu Luo from Jinan Quality Inspection Center for Medical Devices collaboratively carried out the study of "Synthesis of Environmentally Friendly Di(2-ethylhexyl) Cyclohexane-1,2-Dicarboxylate via Direct Esterification" [121]. Using direct esterification between HHPA and 2-ethylhexanol raw materials and methanesulfonic acid catalyst, they synthesized di(2-ethylhexyl) cyclohexane-1,2-dicarboxylate (DEHCH) that was structurally similar to DINCH but with a lower cost. The physico-chemical properties, subchronic toxicity, and plasticization performance of DEHCH were compared with those of other environmentally friendly plasticizers. The physico-chemical properties of DEHCH were comparable to and even better than its peer products internationally. The subchronic toxicity of DEHCH was no observable difference with its peer products. The mechanical property analysis showed that DEHCH had a similar PVC plasticization performance to DEHP and could be a phthalate plasticizer substitute.

In 2016, Ping Huai from Jiangsu Hengtong Wire & Cable Technology Co., Ltd. published an article titled "How Can the Cable Industry Respond to the EU's Newest Environmental Protection Act?" [128]. With increasing attention paid to ecological and environmental protection and human health, higher and more requirements of environmental protection have been raised to electrical and electronic devices including cables. EU issued the ROHS, in effect on July 1, 2006, that is applied not only to electrical and electronic devices but also to their upstream and downstream industries including raw materials, parts, and packaging. In 2011, the EU upgraded the environmental protection act to ROHS 2.0. On June 4, 2015, the EU published the

RoHS 2.0 Revision (EU) 2015/863 officially listing DEHP, BBP, DBP, and DIBP as restricted substances and totaling 10 mandatory controlled substances. Cable companies, especially those aiming to expand their international market, should closely watch the related laws and regulations. No one can win the competition without timely adjustment of product structure, promotion of technological innovation, or active development of green products that meet these policies.

In 2016, at the International Forum of Green Plasticizer Development, Yanhua Liu from Shandong Lanfan Chemical Industry Co., Ltd. gave a presentation titled "Applications of Environmentally Friendly Plasticizers in Disposable Gloves" [127]. Because of the hazards of phthalates to the environment and human health, many countries have made policies to restrict their use. The PVC glove industry is significantly impacted. Seeking alternative plasticizers has been a hot research topic in the PVC glove industry. It will be an important direction to develop safe, reliable, high-performance, non-toxic, environmentally friendly plasticizers.

In 2016, also at the International Forum of Green Plasticizer Development, Zhibo Shi, the Associate Director of Shandong Qilu Plasticizer Co., Ltd., presented a report titled "How to Accurately Assess and Use Benzoate-Based Plasticizers?" [129]. Based on the real situations in China, we should use scientific approaches and accurate experimental data to make our own developmental strategies. High attention should be paid to science and technology to push the advancement of the plasticizer industry. Industrial associations will be used as the platform to perform key research tasks. We need to understand the toxicity of benzoate-based plasticizers to humans and the environment, take measures to promote their advantages and avoid their disadvantages, offer accurate assessments, and develop new products to meet emerging social needs. These can effectively guide the R&D, production, and application of related industries. More efforts should be devoted to R&D to develop new materials and products. The technological level and product competitiveness need to be improved to expand the international market shares and enable adequate space for our plasticizer industry.

In 2016, Zhiyu Zhao from Qingdao University of Science and Technology carried out the "Investigations of Production of Environmentally Friendly DEHCH" [130]. The different boiling points of DEHCH and DOP meet the requirement of distillative separation, but both compounds turn yellow at 180 °C and decompose. Thus, distillation or reduced-pressure distillation is not suitable for their separation. Experiments indicate that the two-phase extraction method can achieve separation of DEHCH and DOP at room temperature and avoid their decomposition. The boiling points of DEHCH and DOP are 365 °C and 384 °C, respectively, which are significantly different than those of the extractant solvents. Therefore, reduced-pressure distillation can be used to recycle the two solvents. The distillation temperature should be controlled below 180 °C to avoid decomposition of DEHCH and DOP.

In 2017, at the 9th JinLianChuang Plasticizer Product Chain Market Summit and Trade Fair, Ping Han from JinLianChuang Internet Technology Co., Ltd. made a report titled "Co-existing Opportunities and Challenges for DOP Market in 2017" [115], which described in great detail the foreign and domestic environmental protection policy evolution and the future direction of plasticizer industry structure adjustments.

In conclusion, research on environmentally friendly plasticizers has been widely carried out internationally. It's been a general trend to replace the traditional phthalates with environmentally friendly plasticizers.

References

1. Bruna M, Botcher A, Brittschedel B. Hydrogenation of benzene polycarboxylic acid or its derivatives with macroporous catalyst: China. CN98813139, 18 Dec 1998.
2. Shi WC. Toxicity of plasticizers and related restrictions. Plast Addit. 2010;3:43–7.
3. Wolff MS, Teiteibaum SL, Mcgovern K. Phthalate exposure and pubertal development in a longitudinal study of US girls. Hum Peprod. 2014;29:1558–66.
4. Kim SH, Park MJ. Phthalate exposure and childhood obesity. Ann Pediatr Endocrinol Metab. 2014;19:69–75.
5. Virunya S, Jennifer L, Durham J. Derivation of an reference does (RfD) for the plasticizer di-(2-propyiheptyl) phthalate (Platinol* 10P). Regul Toxicol Pharm. 2014;79:26–34.
6. Bernard I, Dècaudin B, Lecoeur M. Analytical methods for the determination of DEHCH plasticizer alternatives present in medical devices: a review. Talanta. 2014;128:18–34.
7. Wei GL, Li DQ, Zhuo MN. Organophosphous flame retardants and plasticizers: sources, occurrence, toxicity and human exposure: a review. Environ Pollut. 2015;106:29–45.
8. Michelle M, Lori MA, Hoepner AH. Prenatal exposure to phthalates and childhood body size in an urban cohort. Environ Health Perspect. 2016;124:513–9.
9. Sang HY, Han ZY, Jiang ZY. Catalytic hydrogenation in the synthesis of diisobutyl hexahydrophthalate. Petrochem Eng. 1999;28(7):462–5.
10. Yang RN, Zhu Y, Xie Q. Preparation of hexahydrophthalic anhydride. Thermoset Resin. 2002;17(4):5–7.
11. Yang RN, Jin DM. Synthesis and properties of the chain-like polymeric copper (II) complexes bridged by bis(diphenylphosphino oxide) ethane. Chin J Chem. 2000;18:346–8.
12. Zhou YZ. Chemical commodity inspection manual. Beijing: Chemical Industry Press; 1996.
13. Guo HR, Zhu LQ. Review on application performance of new plasticizer DPHP. Plast Addit. 2006;6:36–9.
14. Shi WC. New plasticizer cyclohexane dicarboxylic acid dialkyl ester. Plast Addit. 2010;1:17–22.
15. Brian LW. An innovative plasticizer for sensitive applications. J Vinyl Addit Technol. 2003;9(4):172–6.
16. Gosse C, Larson TM, Leqrand PJP, et al. Plasticized polyvinyl chloride: USA. US74138, 19 Aug 2008.
17. Melanie B, Bottcher A, Breitscheidel B, et al. Method for hydrogenating benzene polycarboxylic acids or derivatives thereof by using a catalyst containing macropores: USA. US6284917, 04 Sept 2001.
18. Qian BZ. Development status of plasticizer at home and abroad (part I). Shanghai Chem Ind. 2010;36(1):36–8.
19. Qian BZ. Development status of plasticizer at home and abroad (part 2). Shanghai Chem Ind. 2011;36(2):37–9.
20. Wang DR. Development and application of diisononyl cyclohydride phthalate. Fine Chem Mater Intermed. 2010;9:28–32.
21. Wang DR. Development and application of diisononyl cyclohexane-1,2-dicarboxylate. Shanghai Plast. 2014;2:33–7.

22. Li CC, Jiang PP, Dong YM. Synthesis of environmental protection plasticizer cyclohexane-1, 2-diisooctyl dicarboxylate [J]. Fine Chem. 2011;28(11):1124–8.
23. Xiao KE, Mo CH, Cai QY. Concentrations of PAEs in vegetable fields of Pearl River Delta. Sichuan Environ. 2012;34(3):49–55.
24. Mustazur R, Christopher SB. The plasticizer market: an assessment of traditional plasticizers and research trends to meet new challenges. Prog Polym Sci. 2004;29:1223–48.
25. Jiang PP, Cong MB, Zhang DP, et al. Accelerate the research and application of citrate ester, a new environmental friendly plasticizer in China. Plast Addit. 2003; 4(5):1–8.
26. Rodrigo N, Monica PP, Myriam GT, et al. Phthalate plasticizers covalently bound to PVC: plasticization with suppressed migratio. Macromolecules. 2010;43:2377–81.
27. Annika L, Minna H. Migration resistant polymeric plasticizer for poly(vinyl chloride). J Appl Polym Sci. 2007;104:2458–67.
28. Bo Y, Min NH. Oligomeric isosorbide esters as alternative renewable resource plasticizers for PVC. J Appl Polym Sci. 2011;119:2400–7.
29. Zhang L. Market status and prospect of citric acid ester plasticizer. Plast Addit. 2008; (1):10–1.
30. Mlanie B, Lucien T, Boris B, et al. Use of cyclohexanepolycarboxylic acids as plasticizers for preparing toxicologically advantageous plastics. China, TW 266784B, 21 Nov 2006.
31. Wadey BL. An innovative plasticizer for sensitive applications. J Vinyl Addit Technol. 2003;94:172–6.
32. Regine N, André C, Christiane S, et al. Diisononyl-1,2-cyclohexanedicar-boxylic acid (DINCH) and di(2-ethylhexyl) terephthalate (DEHT) in indoor dust samples: concentration and analytical problems. Int J Hyg Envir Heal. 2011;214:26–35.
33. Crespo JE, Balart R, Sanchez L, et al. Substitution of di(2-ethylhexyl) phthalate by di (isononyl) cyclohexane-1,2-dicarboxylate as a plasticizer for industrial vinyl plastisol formulations. J Appl Polym Sci. 2007;104:1215–20.
34. He YC. Discussion on plasticizer in food packaging materials. Grand View Weekly. 2012;51(5):141–2.
35. Singh S, Li SSL. Phthalate: toxicogenomics and inferred human diseases. Genomics. 2011;97(3):148–57.
36. Edmundo B, Jesús DM. Deregulation of the Sod1 and Nd1 genes in mouse fetal oocytes exposed to mono-(2-ethylhexyl) phthalate (MEHP). Reprod Toxicol. 2010;30(3):387–92.
37. Peter K. How dangerous are phthalate plasticizers? Integrated approach to toxicity based on metabolism, electron transfer, reactive oxygen species and cell signaling. Med Hypotheses. 2010;74(4):626–8.
38. Li CC, Jiang PP, Dong YM. Synthesis of environmental protection plasticizer cyclohexane 1,2-diisooctyl dicarboxylate. Fine Chem. 2011;28(11):1124–8.
39. Hu SC, Chen YW. Partial hydrogenation of benzene: a review. J Chin Inst Chem Eng. 1998;29(6):387–96.
40. Muetterties EL, Bleeke JR. Catalytic hydrogenation of aromatic hydrocarbons. Acc Chem Res. 1979;12(9):324–31.
41. An Y, Ding YM, Tan J, et al. Influences of polyester plasticizers on the properties of oil resistant flexible poly(vinyl chloride) and powder nitrile butadiene rubber blends. Adv Sci Lett. 2011;4(3):875–9.
42. Mccusker-Orth JE. Process for a cyclohexanedimethanol using raney metal catalysts: USA. US 6919489. 19 July 2005.
43. Chaeles ESJ, Gustafson BL. Hydrogenation of phthalic acids to cyclohexane-dicarboxylic acid: USA. US 6291706. 18 Sept 2001.
44. Oriol-Hemmerlin C, Pham QT. Poly 1,3-butylene adipate reoplex as high molecular weight plasticizer for PVC-based cling films-microstructure and number-average molecular weight studied by ^{1}H and ^{13}C NMR. Polymer. 2000;41(12):4401–7.
45. Farahat MS. On the heat stabilization efficiency of polyester plasticizers for poly(vinyl chloride) (PVC). Macromol Mater Eng. 2001;286(2):88–93.
46. Lindström A, Hakkarainen M. Migration resistant polymeric plasticizer for poly(vinyl chloride). J Appl Polym Sci. 2007;104(4):2458–67.

47. Lindström A, Hakkarainen M. Environmentally friendly plasticizers for poly(vinyl chloride)-improved mechanical properties and compatibility by using branched poly(butylene adipate) as a polymeric plasticizer. J Appl Polym Sci. 2006;100(3):2180–8.
48. Shi WC, Shi ZB, Jiang PP. Plasticizer and its application. Beijing: Chemical Industrial Press; 2002.
49. Shi WC, Si JJ, Liu WG. Practical handbook for plasticizers. Beijing: Chemical Industrial Press; 2009.
50. Jiang PP, Zhou YF. Environmental protection plasticizer. Beijing: National Defence Industry Press; 2009.
51. Liu SC, Luo G, Han ML, et al. Characterization of the catalyst prepared by impregnation for selective hydrogenation of benzene to cyclohexene. Chin J Catal. 2001;22:559–62.
52. Liu SC, Zhu BZ, Luo G, et al. Characterization of amorphous Ru-M-B/ZrO_2 catalysts for partial hydrogenation of benzene to cyclohexene. Chin J Molec Catal. 2002;16(3):217–22.
53. Takasu A, Makino T, Yamada S. Polyester synthesis at moderate temperatures via the direct polycondensation of dicarboxylic acids and diols catalyzed by rare-earth perfluoroalkanesulfonates and bis(perfluoroalkanesulfonyl)imides. Macromolecules. 2010;43(1):144–9.
54. Takasu A, Iio Y, Oishi Y, et al. Environmentally benign polyester synthesis by room temperature direct polycondensation of dicarboxylic acid and diol. Macromolecules. 2005;38(4):1048–50.
55. Liu SC, Wu YM, Wang Z, et al. Study of a Ru-La/ZrO_2 catalyst prepared by precipitation method for selective hydrogenation of benzene to cyclohexene. J Natur Gas Chem. 2005; 14:226–32
56. Si RJ, Liu SC, Liu ZY, et al. Study on Ru-M1-M2 catalyst prepared by precipitation method for selective hydrogenation of benzene to cyclohexene. Chin J Molec Catal. 2005;19:141–5.
57. Huang ZX, Liu ZY, Wu YM, et al. Selective hydrogenation of benzene to cyclohexene by a novel Ru-Zn catalyst. Chin J Molec Catal. 2006;20(3):226–9.
58. Liu SC, Liu ZY, Zhao SH, et al. Study on the amorphous alloy Ru-Fe-B/ZrO_2 nanosized catalyst for benzene selective hydrogenation to cyclohexene. J Natur Gas Chem. 2006;150:1–9.
59. Ye X, Jiang PP, Zhang PB. Novel Ti and Mn mesoporous molecular sieves: synthesis, characterization and catalytic activity in the epoxidation of vegetable oil. Catal Lett. 2010;137:88–93.
60. Jiang PP, Chen M, Dong YN. Novel two-phase catalysis with organometallic compounds for epoxidation of vegetable oils by hydrogen peroxide. J Am Oil Soc. 2010;87:83–91.
61. Angleescu E, Pavel OD, Bîrjega R, et al. The impact of the "memory effect" on the catalytic activity of Mg/Al; Mg, Zn/Al; Mg/Al, Ga hydrotalcite like compounds used as catalysts for cycloxene epoxidation. Appl Catal A Gen. 2008;341:50–7.
62. Liu CJ, Yu WY, Li SG, et al. Ruthenium meso-tetrakis(2,6-dichlorophenyl) porphyrin complex immobilized in mesoporous MCM-41 as a heterogeneous catalyst for selective alkene epoxidations. J Org Chem. 1998;63(21):7364–9.
63. Arends IW. Metal-catalyzed asymmetric epoxidations of terminal olefins using hydrogen peroxide as the oxidant. Angew Chem Int Ed. 2006;45(38):6250–2.
64. Rüsch GKM, Warwei S. Chemoenzymatic epoxidation of unsaturated fatty acid esters and plant oils. J Am Oil Chem Soc. 1996;73(11):1453–7.
65. Wendylene SDS, Alexande AML, Paulo AZS, et al. Enzyme-mediated epoxidation of methyl oleate supported by imidazolium-based ionic liquids. J Mol Catal B: Enzym. 2011;68(1):98–103.
66. Liu ZY, Sun HJ, Wang DB, et al. The modifiable character of a novel Ru-Fe-B/ZrO_2 catalyst for benzene selective hydrogenation to cyclohexene. Chin J Chem. 2010;28:1927–34.
67. Sun HJ, Jiang HB, Li SH, et al. Selective hydrogenation of benzene to cyclohexene over nanocomposite Ru-Mn/ZrO_2 catalysts. Chin J Catal. 2013;34:684–94.
68. Jing TS, Zhao Q, Li M, et al. Preparation of mesoporous titania solid superacid and its catalytic property. J Hazard Mater. 2008;159:204–9.

69. Yu GX, Zhou XL, Li CL, et al. Esterification over rare earth oxide and alumina promoted SO_4^{2-}/ZrO_2. Catal Today. 2009;148:169–73.
70. Liao Y, Huang X, Liao XP, et al. Preparation of fibrous sulfated zirconia (SO_4^{2-}/ZrO_2) solid acid catalyst using collagen fiber as the template and its application in esterification. J Mol Catal A Chem. 2011;347:46–51.
71. Tropecêlo AI, Casimiro MH, Fonseca IM, et al. Esterification of free fatty acids to biodiesel over heteropolyacids immobilized on mesoporous silica. Appl Catal A Gen. 2010;390(1–2):183–9.
72. Riccardo T, Martino DS, Luca C, et al. Acid exchange resins deactivation in the esterification of free fatty acids. Chem Eng J. 2010;161:212–22.
73. Masato K, Atsuko N, Jyu-Suke H. Preesterification of FFA in plant oil transeste-rified into biodiesel with the help of solid acid catalysis of sulfonated cation-exchange resin. Appl Catal A Gen. 2011;405:36–44.
74. Bamoharram FF, Heravi MM, Ebrahimi J, et al. Catalytic performance of nano-SiO_2-supported preyssler heteropolyacid in esterification of salicylic acid with aliphatic and benzylic alcohols. Chin J Catal. 2011;32(5):782–8.
75. Chaturvedi D. Recent developments on task specific ionic liquids. Curr Org Chem. 2011;15:1236–48.
76. Olivier-Bourbigou H, Magna L, Morvna D. Ionic liquids and catalysis: recent progress from knowledge to applications. Appl Catal A Gen. 2010;373:1–56.
77. Dupont J, Scholten JD. On the structural and surface properties of transition-metal nanoparticles in ionic liquids. Chem Soc Rev. 2010;39:1780–804.
78. Fang D, Zhou XL, Ye ZW, et al. Brøsted acidic ionic liquids and their use as dual solvent-catalysts for fischer esterifications. Ind Eng Chem Res. 2006;45:7982–4.
79. Cole AC, Jensen JL, Ntai I, et al. Novel Brønsted acidic ionic liquids and their use as dual solvent-catalysts. J Am Chem Soc. 2002;124(21):5962–3.
80. Xing H, Wang T, Zhou Z, et al. Novel Brønsted-acidic ionic liquids for esterifications. Ind Eng Chem Res. 2005;44:4147–50.
81. Xie C, Li H, Li L, et al. Synthesis of plasticizer ester using acid-functionalized ionic liquid as catalyst. J Hazard Mater. 2008;151:847–50.
82. Zhang S, Lefebvre H, Tessiera M, et al. Influence of Brøsted acid ionic liquid structure on hydroxyacid polyesterification. Green Chem. 2011;13:2786–93.
83. Li X, Eli W. A green approach for the synthesis of long chain aliphatic acid esters at room temperature. J Mol Catal A Chem. 2008;279:159–64.
84. Xu J, Jiang J, Zuo Z, et al. Synthesis of tributyl citrate using acid ionic liquid as catalyst. Process Saf Environ Prot. 2010;88(1):28–30.
85. Miao JM, Wan H, Guan GF. Synthesis of immobilized Brønsted acidic ionic liquid on silica gel as heterogeneous catalyst for esterification. Catal Commun. 2011;12:353–6.
86. Zhang H, Xu F, Zhou X, et al. A Brønsted acidic ionic liquid as an efficient and reusable catalyst system for esterification. Green Chem. 2007;9:1208–11.
87. Zhou H, Yang J, Ye L, et al. Effects of acidity and immiscibility of lactam-based Brøsted-acidic ionic liquids on their catalytic performance for esterification. Green Chem. 2010;12:661–5.
88. Gu Y, Li G. Ionic liquids-based catalysis with solids: state of the art. Adv Synth Catal. 2009;351:817–47.
89. Qiao K, Hagiwara H, Yokoyama C. Acidic ionic liquid modified silica gel as novel solid catalysts for esterification and nitration reactions. J Mol Catal A Chem. 2006;246(1–2):65–9.
90. Xu Z, Wan H, Miao J, et al. Reusable and efficient polystyrenesupported acidic ionic liquid catalyst for esterifications. J Mol Catal A Chem. 2010;332:152–7.
91. Bourlinos AB, Raman K, Herrera R, et al. A liquid derivative of 12-tungsto-phosphoric acid with unusually high conductivity. J Am Chem Soc. 2004;126:15358–9.
92. Rickert PG, Antonio MR, Firestone MA, et al. Tetraalkylphosphonium polyoxometalates: electroactive, "task-specific" ionic liquids. Dalton Trans. 2007; 529–31.
93. Leng Y, Wang J, Zhu D, et al. Heteropolyanion-based ionic liquids: reaction-induced self-separation catalysts for esterification. Angew Chem Int Ed. 2009;48(1):168–71.

94. Leng Y, Wang J, Zhu D, et al. Sulfonated organic heteropolyacid salts: recyclable green solid catalysts for esterifications. J Mol Catal A Chem. 2009;313(1–2):1–6.
95. Zhang W, Leng Y, Zhu D, et al. Phosphotungstic acid salt of triphenyl(3-sulfopropyl) phosphonium: an efficient and reusable solid catalyst for esterification. Catal Commun. 2009;11:151–4.
96. Li H, Qiao Y, Hua L, et al. Imidazolium polyoxometalate: an ionic liquid catalyst for esterification and oxidative esterification. Chem Cat Chem. 2010;2(9):1165–70.
97. Lu H, Bi YL, Liu YL. Development status and prospect of epoxy plasticizer. Grain Oil Process. 2010;6:26–30.
98. Tang YF. Development and application of green environmental protection polyester plasticizer. Nanjing: Nanjing Publishing Media Group; 2016.
99. Frados J. Plastics engineering handbook of the society of the plastics industry. New York: Van Nostrand Reinhold; 1976.
100. Thomas RF. Process for reacting an organic compound in the process of a supported ruthnium catalyst: USA. US6248824, 19 June 2001.
101. Hansjürgen S. International plastics handbook: for the technologist, engineer and user. Munchen: Hans Publishers; 1987.
102. Földes E. Study of the effects influencing additive migration in polymers. Angew Makromol Chem. 1998;261(1):65–76.
103. EFSA. Opinion of the scientific panel on food additives, flavourings, processing aids and materials in contact with food (AFC) on a request related to a 12th list of substances for food contact materials. EFSA J. 2006; 395–401:1–21.
104. Union Official Journal of the European Commission Directive. 2007/19/EC amending directive 2002/72/EC relating to plastic materials and articles intended to come into contact with food and council directive 85/572/EEC laying down the list of simulants to be used for testing migration of constituents of plastic materials and articles intended to come into contact with foodstuffs; 2007.
105. Bohnen H, Klein T, Bergrath K. Process for preparing cyclohexane-dicarboxylic esters: USA. US6740773, 09 Jan 2003.
106. Charles ES, Gustafson BL. Hydrogenation of phthalic acids to cyclohexanedicarboxylic acid: USA. US6291706, 18 Sept 2001.
107. Gustafson BL, Tennant BA, Kuo YJ, et al. Low pressure process for the hydrogenation of dimehtyl benzenedicatboxylates to the corresponding dimethyl cyclohexane- dicarboxylates: USA. US5286898, 15 Feb 1994.
108. Mccusker-Orth JE, Kingsport TN. Process for a cyclohexanedimethanol using raney methal catalysts: USA. US6919489, 19 July 2005.
109. Liu YX, Xing TF, Wei ZJ, et al. Liquid phase selective hydrogenation of phthalic anhydride to phthalide over titania supported gold catalysts. Catal Commun. 2009;10(15):2023–6.
110. Park KH, Jang K, Kim HJ. Near-monodisperse tetrahedral rhodium nanoparticles on charcoal: the shape-dependent catalytic hydrogenation of arenes. Angew Chem Int Ed. 2007;119(7):1170–3.
111. Shi GJ, Zhao HY, Song LG, et al. Effect of solvents on the hydrogenation and isomerization of 1-hexene over sulfided Co-Mo/γ-Al_2O_3 catalysts for hydrodesulfurization. Energy Fuel. 2008;22(4):2450–4.
112. Hu SC, Chen YW. Effect of preparation on ruzn ultrafine catalysts in partial hydrogenation of benzene. Ind Eng Chem Res. 2001;40(14):3127–32.
113. Ding YJ, Diao YJ, Ma LX. The invention discloses a method for preparing 1,2-cyclohexane diformic acid ester: China. CN 200810224912. 1, 27 Oct 2008.
114. Jiang PP, Shen ML, Dong YM. Homogeneous catalytic synthesis of polyhexa hydrophthalic anhydride diethylene glycol ester by titanium complex. Chemical Prog. 2011;30(3):634–7.
115. Zhao J, Xue M, Huang Y, et al. Hydrogenation of dioctyl phthalate over supported Ni catalysts. Catal Commun. 2011;16(1):30–4.
116. Wilhelm G. Method for producing ester plasticizers: USA. US6310235, 30 Oct 2001.

117. Brunner M, Amd B, Boris B, et al. Method for hydrogenating benzene polycarboxylic acids or derivatives thereof by using a catalyst containing macropores: USA. US6284917 B1, 04 Sept 2001.
118. Harman WD. The activation of aromatic molecules with pentaammine-osmium (II). Chem Rev. 1997;97(6):1953–78.
119. She RS, Steven SL. Phthalate: toxicogenomics and inferredhuman diseases. Genomics. 2011;97:148–57.
120. Moyes RB, Wells PB. The chemisorption of benzene. Adv Catal. 1973;23:121–56.
121. Sumner J, Gustafson BL. Hydrogenation of phthalic acids to cyclohexane-dicarboxylic acid: USA. US 6291706, 18 Sept 2001.
122. Wilkes CE, Summers JW, Daniels CA. PVC handbook. Munich: Hanser Publications; 2005.
123. Joel A, Ted S, Tee G, et al. Health risks posed by use of di-2-ethylhexyl phthalate (DEHP) in PVC medical devices: a critical review. Am J Ind Med. 2001;39(1):100–11.
124. Schutze A, Palmke C. Quantification of biomarkers of environmental exposure to di(isononyl) cyclohexane-1,2-dicarboxylate (DINCH) in urine via HPLC-MS/MS. J Chromatogr B. 2012;895:123–30.
125. Ou YH, Ding XJ, Zhang L, et al. Synthesis and application of an alternative plasticizer di(2-ethylhexyl)-1,2-cyclohexane dicarboxylate. J Appl Polym Sci. 2014;131(2):39763–7.
126. Baxter. Histopathological evaluation of testes from neonatal male rats and rabbits treated with saline or approximately 62 mg/kg di-(2-ethylhexyl)phthalate (DEHP) in 4% bovine serum albumin (BSA) during postnatal days 3–21 (rats) or 14–42 (rabbits); 2000.
127. Lu F, Liu J, Xu J. Fast aqueous/organic hydrogenation of arenes, olefins and carbonyl compounds by poly (*N*-vinylpyrrolidone)-Ru as amphiphilic microreactor system. Adv Syn Catal. 2006;348(7–8):857–61.
128. Lu F, Liu J, Xu J. Synthesis of PVP-Ru amphiphilic microreactors with Ru nanocatalysts and their application in the fast hydrogenation of unsaturated compounds in aqueous media. J Mol Catal A Chem. 2007;271(1):6–13.
129. Li TX, Zhang XF. Study on amorphous ultrafine nickel alloy catalyst ii. Kinetics of hydrogenation of benzene over Ni-B catalyst. Chin J Catal. 1995; 16(5):399–402.
130. Hileman B. FDA suggests replacing DEHP in plastics. Chem Eng News. 2002;80(37):6–9.
131. Shi WC, Zhao CY. Toxicity of plasticizers and related regulations. Plast Addit. 2007;2:46–51.
132. Wang DR. Formulation process and application of plasticizer. Beijing: Printing Industry Press; 2013.
133. Bohnen H, Klein T, Bergrath K. Process for preparing cyclohexanedicarboxylic esters: USA. US 6740773, 25 May 2004.
134. Chaudhary BI, Sxzekalla B, Schiller K, et al. Acetylated polyglycerine fatty acid ester and a PVC insulator plasticised therewith: USA. US 2011041372, 07 April 2011.
135. George AA, Harrell JLE, Toy FR, et al. Plural stage hydrogenation of dial kylterephthalate using palladium and then copper chromite: USA. US 3334149, 01 Aug 1967.
136. John S, Michel WM, Tuck MA. Wood, process for the preparation of alcohols and diols: USA. US5387753, 07 Feb 1995.
137. Rothwell IP. A new generation of homogeneous arene hydrogenation catalysts. Chem Commun. 1997;15:1331–8.

Chapter 4
Foreign and Domestic Inventions and Patents

The phthalate ring hydrogenation methods mostly exist in the form of patents. This chapter reviews foreign and domestic inventions and patents, grouped by research teams, following a chronological order since 1990s. The technical background, claim, and positive effect of these inventions and patents are emphasized.

4.1 Foreign Inventions and Patents

4.1.1 Hydrogenation Methods

In 1998, Bruno et al. from BASF, Germany, filed a patent application in China titled "Method for hydrogenating benzene polycarboxylic acids or derivatives thereof by using a catalyst containing macropores" [1]. This patent disclosed a method to prepare cyclohexane carboxylates via hydrogenation of benzene polycarboxylic acids and their derivatives in the presence of macropore-containing catalysts and also introduced their applications as plasticizers in plastics.

1. Technical Background

Patents US5286898 and US5319129 disclosed hydrogenation of dimethyl terephthalate to prepare corresponding dimethyl hexahydroterephthalate under the conditions of >140 °C, 5.0–17.0 MPa, and supported Pd catalysts pretreated with Ni/Pt/Ru. Patent DE-A2823165 disclosed a method for preparing cyclohexane carboxylates from aromatic carboxylates under the conditions of 70 –250 °C, 3.5–10.5 MPa, and supported Ni, Ru, Rh, and/or Pd catalysts. Patent US3027398 disclosed a method for hydrogenating dimethyl terephthalate to prepare dimethyl cyclohexane-1,4-dicarboxylate under the conditions of 110–140 °C, 3.5–0.5 MPa hydrogen pressure and supported Ru catalysts. But all these patents contained no uses of macroporous supports.

Z. Liu, *Green Catalytic Hydrogenation of Phthalate Plasticizers*,
https://doi.org/10.1007/978-981-16-9789-0_4

Phthalates such as DBP, DOP or DINP are used as plasticizers in plastics such as PVC but are hazardous to human health. Their uses in children's toys are increasingly being criticized and are even prohibited in certain countries.

Patent DE-A2823165 disclosed the use of dimethyl cyclohexane-1,2-dicarboxylate (DMCH) or diethyl cyclohexane-1,2-dicarboxylate (DECH) as plasticizers in plastics. Patent EP-A07-011,074 disclosed the use of DINCH as a plasticizer in plastics. DE-A1263296 disclosed the use of DEHCH as a plasticizer in plastics.

2. Main Content

A benzene polycarboxylic acid or a derivative thereof, especially benzene dicarboxylate, is hydrogenated with a specific catalyst to achieve a ring-hydrogenated product, especially cyclohexane dicarboxylate, at a high selectivity and yield without obvious secondary reactions. The product can be used as a plasticizer in plastics.

A benzene polycarboxylic acid or its derivative, or a mixture of two or more thereof, is hydrogenated in the presence of a catalyst which comprises as active metal at least one metal of transition group VIII of the Periodic Table alone or together with at least one metal of transition group I or VII of the periodic table applied to a support which contains macropores. If dimethyl terephthalate is hydrogenated, the hydrogenation using a catalyst which comprises active metal ruthenium either alone or together with at least one metal of transition group I, VII or VIII of the Periodic Table applied to a support, where the support has a mean pore diameter of at least 50 nm and a BET surface area of at most 30 m^2/g and the amount of the active metal is from 0.01 to 30% by weight, based on the total weight of the catalyst. From 10 to 50% of the pore volume of the support is formed by macropores having a pore diameter in the range of 50 to 10,000 nm and from 50 to 90% of the pore volume of the support is formed by mesopores having a pore diameter in the range of 2 to 50 nm.

In principle, the active metal can be any of the metals in transition group VIII of the Periodic Table. Preference was given to Pt, Rh, Pd, Co, Ni, Ru, or two or more thereof. A particular preference was Ru. Any of the metals in main or transition Group I and VII can be used, preferably Cu and/or Rh.

The term "macropore" and "mesopore" refer to those having a pore diameter larger than 50 nm or in the range from 2 to 50 nm, respectively.

In all cases, the amount of the active metal is from 0.01 to 30% by weight, based on the total weight of the catalyst, preferably from 0.01 to 5%, and particularly preferably from 0.1 to 5%.

The term "benzenepolycarboxylic acid or the derivative thereof" includes single benzene polycarboxylic acid or its derivative, especially monoalkyl, dialkyl, trialkyl or tetraalkyl esters, and its anhydride, where the alkyl groups can be linear or branched and each have from 1 to 30 carbon atoms, preferably from 2 to 20 carbon atoms, particularly preferably from 3 to 18 carbon atoms. The esters include: terephthalate, phthalate, isophthalate, trimellitate, trimesate, hemimellitate, pyromellitate, and anhydrides of phthalic acid, trimellitic acid, hemimellitic acid, pyromellitic acid, and mixtures of two or more thereof.

The product is the corresponding cyclohexane polycarboxylic acid or the derivative thereof including:

(1) diisopentyl cyclohexane-1,2-dicarboxylate, obtained via hydrogenation of diisopentyl phthalate, CAS 84777-06-0.
(2) diisoheptyl cyclohexane-1,2-dicarboxylate, obtained via hydrogenation of diisoheptyl phthalate, CAS 71888-89-6.
(3) diisononyl cyclohexane-1,2-dicarboxylate, obtained via hydrogenation of diisononyl phthalate, CAS 68515-48-0.
(4) diisononyl cyclohexane-1,2-dicarboxylate, obtained via hydrogenation of diisononyl phthalate, CAS 28553-12-0, based on n-butene.
(5) dinonyl cyclohexane-1,2-dicarboxylate, obtained via hydrogenation of dinonyl phthalate, CAS 8515-46-8.
(6) diisodecyl cyclohexane-1,2-dicarboxylate, obtained via hydrogenation of diisodecyl phthalate, CAS 68515-49-1.
(7) di(C_7–C_{11}) cyclohexane-1,2-dicarboxylate, obtained via hydrogenation of di(C_7–C_{11}) phthalate, CAS 68515-42-4.
(8) di(C_7–C_{11}) cyclohexane-1,2-dicarboxylate, obtained via hydrogenation of di(C_7–C_{11}) phthalate, CAS 111381–89-6, 111,381-90-9, 111,381-91-0, 68,515-44-6, 68,515-45-7, and 3648-20-7.
(9) di(C_9–C_{11}) cyclohexane-1,2-dicarboxylate, obtained via hydrogenation of di(C_9–C_{11}) phthalate, CAS 98515-43-5.
(10) di(2-propylheptyl) cyclohexane-1,2-dicarboxylate, obtained via hydrogenation of di(2-propylheptyl) phthalate.
(11) di(C_7–C_9) cyclohexane-1,2-dicarboxylate, obtained via hydrogenation of di(C_7–C_9) phthalate, including branched and linear C_7–C_9 alkyl groups. Single phthalate raw materials include: di(C_7–C_9) phthalate, CAS 111381-89-6; di(C_7–C_9) phthalate, CAS 68515-44-6; di(C_7–C_9) phthalate, CAS 68515-45-70.

This invention also introduced the application of cyclohexane polycarboxylates, especially those obtained from this invented method, as plasticizers in plastics, preferably dialkyl and trialkyl esters with C_3–C_8 carbon atoms, more preferably with C_3–C_{18} carbon atoms, even more preferably C_5-, C_7-, C_9-, C_{10}-, C_7–C_{11}-, C_9–C_{11}-, C_7–C_9- cyclohexane-1,2-dicarboxylates mentioned above. For example, the following hydrogenation products of benzene polycarboxylates incuding DINCH (CAS 68515-48-0), DIDCH(CAS 68515-49-1), Palatinol 9-P, TOTM-I(CAS 3319-31-1), etc., are used as plasticizers. These compounds or their mixtures are used as plasticizers in mass produced plastics, such as PVC and polyolefin (polyethylene, polypropylene, polyisoprene, and copolymers of ethene or propene with 1-butene, 1-hexene, 1-octene).

An example of this invention: 1.4L supported Ru catalyst was loaded in a vertical high-pressure steel tube reactor. The tube had an inner diameter of 30 mm and a length of 2.2 m. Under an average temperature of 125 °C and a hydrogen pressure of 20 MPa, a mixture of Palatinol N(CAS 28553-12-0) slurry and pure hydrogen gas was circulated through the reactor from bottom to top at a speed of 0.45 kg/h. After leaving the reactor, a part of the reaction product was recirculated through the reactor with fresh Palatinol N, while the rest was moved to a container. The hydrogenation reaction proceeded with a 20% excess hydrogen and exhaust control. GC analysis indicated that 99.5% of Palatinol N has reacted to produce the corresponding cyclohexane dicarboxylate with a selectivity of 99.2%. To get rid of the 0.5% unreacted Palatinol N, the reaction product was circulated through the reactor from bottom to top at a speed of 1 kg/h with the second hydrogen gas feeding. No residual Palatinol N was detected in the product. After the second hydrogenation, the selectivity for the cyclohexane dicarboxylate was 99% with about 1% lower-boiling-point side products that were removed via steam distillation at 170 °C and 5 kPa. The purity of the obtained cyclohexane dicarboxylate was 99.7%.

3. Positive Effect

Compared to the dominant phthalate plasticizers, the cyclohexane polycarboxylic acid derivatives derived from this invention have lower densities and viscosities. When used in the same amount as phthalates, the plastics plasticized by cyclohexane polycarboxylates have improved low-temperature softness while comparable Shore Hardness and mechanical strength. The cyclohexane polycarboxylates can also enhance the processibility of the blend leading to higher production efficiency.

From 2001 to 2011, a number of hydrogenation invention patents were filed in the USA.

Patents US6284917 [2] and US6888021 [3] disclosed a method for hydrogenating phthalates to produce cyclohexane dicarboxylate plasticizers. The catalyst included at least one metal of transition group VIII of the Periodic Table either alone or together with at least one metal of transition group IB or VIIB of the Periodic Table. The reactions proceeded at 50–250 °C and 2–30 MPa hydrogen pressure. For example, diisodecyl phthalate (DIDP) was converted to diisodecyl cyclohexane dicarboxylate (DIDCH) with a conversion ratio of 100% and a yield of 99.5% under conditions of 80 °C, 20 MPa hydrogen pressure, supported Ru catalyst, and reaction time 10 h. This method used a two-step esterification followed by the hydrogenation route.

Patents WO2004/046078 [4] and US7893295 [5] disclosed a method for ring-hydrogenating orthophthalic polycarboxylic acids and derivatives. The catalyst included at least one metal of transition group VIII of the Periodic Table either alone or together with at least one metal of transition group IB or VIIB of the Periodic Table. The support could be one or more ordered mesoporous materials. For example, DINP was converted with a conversion ratio of above 99% at 120 °C, 5.8 MPa hydrogen pressure, supported Ru catalyst, and reaction time 7.5 h.

Patent US7319161 [5] disclosed a method for preparing cyclohexane dicarboxylic acids and their derivatives. First, a Diels–Alder reaction of butadiene and maleic anhydride produced tetrahydrophthalic anhydride (THPA). Then, THPA was converted to hexahydrophthalic anhydride (HHPA) via hydrogenation in the presence of Pd/C catalyst. Finally, HHPA reacted with an alcohol to yield the esterification product.

Patent US7361714 [6] disclosed a method for converting orthophthalic polycarboxylate polyesters to cyclohexane polycarboxylate polyesters via ring hydrogenation. The catalyst included at least one metal of transition group VIII of the Periodic

Table either alone or together with at least one metal of transition group IB or VIIB of the Periodic Table. The reactions proceeded at 70–150 °C and 8–10 MPa hydrogen pressure.

Patent US7355084 [7] disclosed a method for converting DINP to DINCH. The catalyst was 3% Ru/SiO_2, and the reaction conditions were 120 °C and 20 MPa hydrogen pressure. The conversion ratio reached 100% and the selectivity for DINCH was above 99%.

Patent US7632961 [8] disclosed a method for preparing diisononyl cyclohexane dicarboxylate (DINCH) via three steps: monoesterification-hydrogenation-diesterification. An equal amount of isononanol and phthalic anhydride reacted to form monoisononyl phthalate, followed by hydrogenation at 120 °C and 20 MPa hydrogen pressure with Ru catalyst to obtain monoisononyl cyclohexane carboxylate; finally, the product DINCH was obtained via additional esterification with tetrabutyl titanate.

Patent US7786201 [9] disclosed a DINCH preparation method. The catalyst contained at least one active metal from transition group VIII together with at least one metal from transition groups IIB, IIIB, IVB, VB or VIB. The reaction proceeded at 60–200 °C, 3–25 MPa hydrogen pressure with a DINP conversion ratio of 99.9%.

The hydrogenation method for preparing cyclohexane dicarboxylate plasticizers has advantages of high product yields and good product quality, but suffers from high capital and operational costs. Currently, only BASF in Germany has commercialized this method.

4.1.2 Esterification Methods

Patent US6310235 [10] used cyclohexane dicarboxylic acid or anhydride and alcohol as raw materials with Ti-, Zr-, or Sn-based metal catalysts. There were two stages of the esterification reaction: monoesterification at a lower temperature (120–160 °C), then diesterification at 250 °C with a catalyst. After the reaction was completed, an aqueous alkali metal or alkali earth metal hydroxide solution was used to neutralize the reaction mixture, followed by removal of the excess alcohol, drying, and filtration to obtain the final product.

4.2 Domestic Inventions and Patents

4.2.1 A Catalyst for Hydrogenating Dimethyl Terephthalates to Dimethyl Cyclohexane-1,4-Dicarboxylates and the Preparation Thereof (2001) [11]

1. Technical Background

Dimethyl cyclohexane-1,4-dicarboxylate is an environmentally friendly plasticizer and also the intermediate for preparing cyclohexane-1,4-dimethanol. The saturated polyester resin prepared using cyclohexane-1,4-dimethanol raw material is very advantageous for paint and coating applications. First, cyclohexane-1,4-dimethanol has a symmetric structure, making the polymer chains well organized, which facilitates the formation of inter-chain H-bonds and limits the mobility of polymers, leading to high hardness. Second, the chair-like and boat-like movement of the para-substituted cyclohexane ring provides flexibility to coatings. The high symmetry of cyclohexane-1,4-dimethanol increases the crystallinity of polyester resin and thus improves the stability and sintering resistance of powder coatings.

2. Main Content

The catalysts for hydrogenating dimethyl terephthalate (DMT) to produce dimethyl cyclohexane-1,4-dicarboxylate (DMCD) are composed of the active composition, additive, and support. The active composition is metallic Pd. The additive is a combination of two types of elements. The first additive element is from Group IIA metals such as Ca, Ba, Mg, etc., and the second additive element is from Group VIII metals such as Rh, Ru, Ir, Pt, etc. The support uses Al_2O_3 with a surface area of 50–200 m^2/g. Of the total mass of the catalysts, Pd is 0.5–5.0%, the first additive element is 0.01–0.5%, and the second additive element is 0.01–0.5%.

Catalyst preparation method: the support was impregnated with aqueous solutions of Pd and additive elements such as nitrate or chloride, which, after drying, was reduced with H_2 for 0.5–14 h at 100 –500 °C.

The hydrogenation reaction used a small fixed bed, catalyst loading 5 mL, reaction pressure 6.0 MPa, reaction temperature 184 °C, 18 g DMT/500 mL ethyl acetate, feedstock speed 50 mL/h, molar ratio H_2/DMT 100. The results are shown in Table 4.1.

3. Positive Effect

Compared to current technologies, the catalysts obtained in this invention can obviously decrease the reaction pressure for hydrogenating DMT to DMCD.

Table 4.1 Catalyst evaluation results

No.	Catalyst composition	DMT conversion ratio/%	DMCD selectivity/%
1	1%Pd,0.03%Ru/Al_2O_3	98.16	93.41
2	1%Pd,0.5%Ca/Al_2O_3	88.52	98.63
3	1%Pd,0.03%Ru,0.5%Ca /Al_2O_3	96.51	95.48

4.2.2 A Plasticizer Containing Cyclohexane Multi-carboxylate and its Preparation Method (2006) [12]

1. Technical Background

Plasticizers are widely used, but must satisfy the requirements for odorless, colorless, UV resistant, low-temperature resistant, heat resistant, hydrolysis resistant, non-flammable, non-volatile, and non-hazardous characteristics. In addition, the plasticizer preparation methods must be simple, meet the need for environmental protection, and produce no side products or polluted wastewater.

In current technologies, the catalysts for esterification reactions include Sn, Ti, Zr or their salts, oxides, soluble organometallic compounds. The esterification reactions occur at >180 °C.

2. Main Content

Cyclohexane multi-carboxylates are synthesized via single-step high-temperature esterification reactions between cyclohexane multi-carboxylic acids or derivatives and linear or branched alcohols having 4–13 carbon atoms. The cyclohexane multi-carboxylic acid or derivative uses 1 mol and the alcohol uses 2.2–3.8 mol under conditions of reaction temperature 200–250 °C, reaction time 5–8 h, reaction pressure 0.5–76 kPa, and with the catalyst. The formed water and unreacted alcohol are removed via azeotropic distillation at 90–180 °C. When the acid value of the reaction mixture turns <1 mg KOH/g, in the reactor are the target di- or multi-carboxylates, as well as partially esterified di- or multi-carboxylic acids, excess alcohol, and catalyst.

Completion of the esterification reaction is followed by a neutralization reaction, i.e., neutralizing the residual acids with alkali metal hydroxide solution to form corresponding salts. The concentration of the neutralizing agent is 5–25% (mass fraction), more preferably 9 –16%. The added amount is 4–5 times the acid value of the reaction mixture. NaOH is the preferred neutralizing agent. The formed salt will precipitate as solid crystals that can easily be removed through filtration. When the acid value is <0.08 mg KOH/g, the reaction mixture is distilled till the alcohol content reaches below 300 ppm. To facilitate the removal of hydrolyzed catalyst compound during filtration, high surface area adsorbents such as activated carbon can be added.

After removal of excess alcohol, the product is dried and filtered to remove unreacted carboxylic acid, hydrolyzed catalyst, and adsorbent. The filtration can occur at room temperature or elevated temperature. Suitable filter materials include cellulose, diatomite, wood dust powder, etc. Then, the product is purified at room temperature or at 100 °C to remove the neutralized salts, catalyst metal salts, and impurities. The product has >99.8% purity and 10APHA (standard color code) chromaticity and can be used as plasticizer.

3. Positive Effects

This invention can effectively shorten the reaction time, save energy, decrease the cost, and obtain high-quality esters. The products such as plasticizers can be used in food packaging, children's toys and supplies, blowing articles, strollers, bottle caps, home gloves, medical devices, etc.

4.2.3 A Catalyst for Preparing Cyclohexane-1,2-Dicarbolxylates (2008) [13]

1. Technical Background

Aliphatic polycarboxylates such as cyclohexane-1,2-polycarboxylates can be used as additives for lubricants and metal processing and as plasticizers for polyolefin and PVC. They have rather low migration, excellent toxicological advantages, and more importantly, are suitable for PVC sensitive applications, which make them a type of high-performance phthalate substitute plasticizers for toys, medical devices, and food contacting PVC articles. More and more consumers have shown strong interest in them.

2. Main Content

The catalysts are composed of the active composition, additive, and support. The active composition can be noble metals Ru, Pt, Pd, and/or Rh, more preferably Ru and Pd, with mass fractions of 0.05–1.5%. The additive can be one or more of Fe, Co, Ni, Cu metals or oxides, with mass fractions of 0.5–5.0%. The support can be Al_2O_3, ZrO_2 or TiO_2. Al_2O_3 has a surface area of 5–100 m^2/g and an average pore size of 5–100 nm. ZrO_2 has a surface area of 5–50 m^2/g and an average pore size of 5–90 nm. TiO_2 has a surface area of 10–100 m^2/g and an average pore size of 5–80 nm.

Before use, the catalyst can be activated via H_2 treatment under conditions: hourly space velocity 2400 h^{-1}, ambient pressure, 300 °C, reduction time 5 h.

The catalyst can be used for hydrogenating phthalates (DINP, DOP, DBP, etc.) to prepare cyclohexane-1,2-dicarboxylate, with tubular trickle bed reactor or slurry bed reactor, more preferably trickle bed reactor. The trickle bed reactor is operated under conditions: catalyst loading 10 mL, no added solvent, 200 °C, H_2 pressure 7.0 MPa, hourly phthalate liquid volume space velocity 0.5 h^{-1}, H_2/phthalate molar ratio 150. Phthalates (DINP, DEHP, DBP) can be converted to cyclohexane-1,2-dicarboxylates (DINCH, DEHCH, DBCH) with high conversion ratio and high selectivity.

Table 4.2 Phthalate hydrogenation reaction test results

Example # phthalate conversion ratio/% cyclohexane	1,2	dicarboxylate selectivity/%
1	98.9	97.9
2	98.0	98.5
3	96.5	97.9
4	99.7	99.8
5	98.8	96.4
6	94.8	92.8
7	99.2	95.9
8	98.9	99.8
9	99.9	98.6
10	98.8	97.5
11	98.6	95.3

Note Example 1–10: DINP reactant and DINCH product; Example 11: DBP reactant and DBCH product

Table 4.2 lists the conversion and selectivity of Examples 1–11 in this invention.

3. Positive Effect

This invention can result in obviously decreased reaction pressure and energy consumption, and increased selectivity.

4.2.4 A Method for Preparing Cyclohexane-1,2-Dicarboxylates (2008) [14]

1. Main Content

This invention provides a method for preparing cyclohexane-1,2-dicarboxylates via hydrogenation of phthalates raw materials.

The reaction system is composed of phthalate, H_2, and catalyst under conditions: 100–250 °C (more preferably 160–200 °C), H_2 pressure 3.0–12.0 MPa (more preferably 5.0–9.0 MPa), H_2/phthalate molar ratio 50–450 (more preferably 120–200), hourly phthalate mass space velocity 0.1–2.5 h^{-1} (more preferably 0.5–1.5 h^{-1}).

The catalysts use Al_2O_3, ZrO_2 or TiO_2 or SiO_2-Al_2O_3 as the support, and one or several metals of Ru, Pt, Pd, Rh, Fe, Co, Ni, Cu as the active composition. The active composition mass loading is 1–3% of the total catalyst mass.

The phthalate is DINP, DEHP or DOP with corresponding products as DINCH, DEHCH or DBCH.

The catalyst is composed of the support and active composition, and/or additive. The support can be Al_2O_3, ZrO_2 or TiO_2 or SiO_2–Al_2O_3. The active composition is one or several metals of Ru, Pt, Pd, Rh, Fe, Co, Ni, Cu. The catalyst can be prepared via regular methods such as precipitation, impregnation, hydrothermal, or a combination thereof.

Liquid phthalates (DINCH, DEHCH or DBCH) can be directly pumped into the preheater and then the fixed-bed reactor. No solvent is needed for dissolution or dilution.

The reaction system in this invention can use a tubular trickle bed reactor or slurry reactor, more preferably the former.

Before use, the catalyst can be activated via H_2 treatment under optimal conditions: hourly space velocity 1200–2400 h^{-1}, ambient pressure 1.0 MPa, 200–400 °C, reduction time 5–12 h.

The catalyst used was Pd–Ru–ZrO_2/Al_2O_3 with mass fractions: Pt 1%, ZrO_2 2%, Ru 0.3%, Al_2O_3 the rest. The Al_2O_3 support was purchased from Tianjin Chemical Technology Institute. The active composition Pd and additives Ru, ZrO_2 were introduced onto the Al_2O_3 support (size 20–40 mesh) using the regular impregnation method. Reaction conditions: tubular trickle bed reactor, catalyst 5.0 g, 200 °C, H_2 pressure 5.0 MPa, H_2/phthalate molar ratio 150, hourly DINP mass space velocity 0.5 h^{-1}, reaction time 20 h, sampling for analysis. Liquid chromatography (LC) was used to analyze DINP and DINCH with operation conditions: pure methanol eluent, flow rate 0.5 mL/min, injected sample volume 10μL, with differential refractive index detector, C_{18} column, column temperature 35 °C. Table 4.3 shows the

Table 4.3 Phthalate hydrogenation reaction test results

Example #	Temperature/°C	H_2 Pressure/MPa	H_2/ester molar ratio	Mass space velocity/h^{-1}	Reactant conversion/%	Product selectivity/%
1	200	5.0	150	0.5	99.9	99.9
2	200	5.0	150	0.3	~100	98.5
3	200	5.0	150	0.8	95.5	99.9
4	180	5.0	150	0.5	90.7	99.5
5	210	5.0	150	0.5	~100	96.4
6	200	3.0	150	0.5	50.8	95.8
7	200	7.0	150	0.5	~100	99.9
8	200	12.0	150	0.5	~100	~100
9	200	5.0	100	0.5	93.9	98.6
10	200	5.0	200	0.5	~100	~100
11[a]	200	5.0	150	0.5	97.6	96.3
12[b]	200	5.0	150	0.5	99.9	99.6
13[c]	200	5.0	150	0.5	99.7	99.7

Note [a] 200 mL stirred slurry bed reactor, 5.0 g catalyst (100–200 mesh) and 5 mL DINP introduced in the 200 mL slurry bed reactor, then introduction of circulating H_2 to 5.0 MPa, other conditions same as Example 1. The rest were trickle bed reactor. [b] Example 1–11: DINP reactant and DINCH product; Example 12: DEHP reactant and DEHCH product; [c] Example 13: DBP reactant and DBCH product phthalate hydrogenation reaction results using the method in this invention.

2. Positive Effects

The method in this invention can obviously decrease the reaction pressure and energy consumption and increase the selectivity of phthalate hydrogenation reactions.

4.2.5 A Method for Preparing Environmentally Friendly Di(2-ethylhexyl) Cyclohexane-1,2-Dicarboxylate (DEHCH) (2009) [15]

1. Technical Background

With the increasing awareness of environmental protection globally, plastic articles such as medicine packaging, medical devices, food packaging, daily supplies, toys, etc., have raised higher hygienic requirements to the primary plasticizers. The whole world has expedited the research and applications of environmentally friendly plasticizers and plastic articles. However, the current domestically produced primary plasticizers cannot satisfy the environmental requirements, especially in terms of non-toxicity. DEHCH plasticizer meets all the strict requirements of the REACH law.

2. Main Content

The preparation protocol is stepwise feeding-esterification-alcohol removal-neutralization and washing-rectification-adsorptive decolorization-pressure filtration-product.

Preparation of DEHCH used the direct esterification between cyclohexane-1,2-dicarboxylic anhydride and 2-ethylhexanol. To the reactor were added cyclohexane-1,2-dicarboxylic anhydride, 2-ethylhexanol, and catalyst with stirring. The temperature was increased from 185 °C to 220 °C and then kept for 3–4 h. The excess alcohol was removed to obtain the crude product, which was purified to afford DEHCH of ≥ 99.5% purity via the following steps: neutralization with NaOH solution at 85–95 °C, water washing at 90–95 °C, steam rectification 3–4 h, addition of activated carbon (at 0.1–0.2% of the total product) to decolorize for 20–30 min, and pressure filtration.

3. Positive Effect

During granulation or plasticization, DEHCH has high material compatibility, can be applied to specific injection molds, is even suitable for use with very delicate designs, and features long lifetime and superior performance.

4.2.6 A Method for Preparing Environmentally Friendly Di(2-ethylhexyl) Cyclohexane-1,2-Dicarboxylate (DEHCH) Plasticizer (2011) [16]

1. Technical Background

In recent years, EU enacted a number of laws and regulations to strictly limit the use of phthalate plasticizers in food packaging, medical supplies, children's toys, and other close human contacting articles. This largely impacts the plasticizer industry and downstream industries in China and brings limitations and challenges to the development of related industries. Therefore, the development of green, environmentally friendly plasticizers has become an urgent need. Di(2-ethylhexyl) hexahydrophthalate plasticizer is a non-phthalate plasticizer specially developed for sensitive applications such as medical equipment, food packaging, children's toys, etc. It has similar structure and properties to di(2-ethylhexyl) phthalate, with even better performance including small migration, colorlessness, transparency, completely environmental friendliness, miscibility with common organic solvents and plasticizers, excellent elasticity, and low-temperature performance. More importantly, it has superior toxicological advantages and can be used in medical equipment, food packaging, children's toys, and other close human contact articles.

2. Main Content

The environmentally friendly di(2-ethylhexyl) hexahydrophthalate plasticizer can be prepared as follows:

A mixture of 1 mol hexahydrophthalic anhydride (HHPA), 2.0–2.2 mol 2-ethylhexanol, and 0.4–0.6% solid superacid SO_4^{2-}/TiO_2-Al_2O_3 catalyst was added to a reactor, heated to 210 °C with stirring, esterified for 1 h, and dealcoholized to obtain the crude product. Then, after the steps of reduced pressure steam rectification for 3–4 h, addition of 0.1 ~ 02.% activated carbon to decolorize for 20–30 min, followed by pressure filtration, the purified di(2-ethylhexyl) hexahydrophthalate was obtained.

3. Positive Effect

During granulation or plasticization, di(2-ethylhexyl) hexahydrophthalate has high material compatibility, low migration in plasticized articles, and long lifetime and superior performance.

4.2.7 *A Method for Preparing Cyclohexane-1,2-Dicarboxylates Using Organic Acid Catalysts (2010) [17]*

1. Technical Background

Cyclohexane-1,2-dicarboxylates are a type of novel non-phthalate plasticizers that meet EU's "Limitations for using phthalates in toys and children products", are recognized by German, European and Japanese food contact products, and have advantages of low element migration, good compatibility, high plasticization efficiency, non-toxicity, low volatility, good water resistance, etc. Therefore, cyclohexane-1,2-dicarboxylates are green, environmentally friendly products, that can be used to substitute phthalates in children's toys, medical equipment, paste resin articles, plastic particles, wires and cables, etc., and will become increasingly important in applications in food contact materials such as food packaging and wraps. However, the production volume of cyclohexane-1,2-dicarboxylates is almost zero in China.

2. Main Content

Esterification reactions were performed using hexahydrophthalic anhydride (HHPA), C_8 or C_9 alcohol, and organic acid catalyst, followed by dealcoholization, decolorization, neutralization, water washing, and drying to obtain cyclohexane-1,2-dicarboxylates with an esterification ratio >98%.

The HHPA/alcohol molar ratio was 1:3. The organic acid catalyst was methanesulfonic acid or sulfamic acid at a loading of 0.05–0.5% of the total mass of fed raw materials. The reaction occurred at 160–230 °C for 3–7 h. The acid value was lower than 2 mg KOH/g. The decolorant was activated carbon or hydrogen peroxide.

Cyclohexane-1,2-dicarboxylate products include dinonyl cyclohexane-1,2-dicarboxylate, diisononyl cyclohexane-1,2-dicarboxylate, dioctyl cyclohexane-1,2-dicarboxylate or diisooctyl cyclohexane-1,2-dicarboxylate. The reaction steps are as follows:

(1) Esterification reaction: HHPA, alcohol, and catalyst were mixed and heated under reflux to remove the formed water from the water trap. When the acid value, measured according to GB/T 1668–1995, was below 2 mg KOH/g, the reaction was completed.
(2) Product purification: the reaction mixture was dealcoholized using steam distillation to recycle the excess alcohol. After cooling at room temperature, the 5wt% decolorant was added at 50 °C for 1 h. Then 5% NaOH solution was added to neutralize the crude ester to weakly alkaline (pH = 8–9) to remove the unreacted HHPA and formed cyclohexane monoacid monoester. The separated top organic layer was washed with water 2–3 times to attain neutral pH and then dried to obtain cyclohexane-1,2-dicarboxylates in >99.5% purity.

3. Positive Effect

The catalyst is inexpensive, readily accessible, efficient (in small loading), non-toxic, environmentally friendly, and cost-effective. The product quality is superior. One plan can produce a series of products, making it economically advantageous.

4.2.8 A Method for Preparing Cyclohexane-1,2-Dicarboxylates Using Ionic Liquid Catalysts (2010) [18]

1. Technical Background

Currently, esterification reactions mostly use concentrated sulfuric acid as the catalyst but suffer from drawbacks including equipment corrosion, acid waste, side reactions, complicated post-treatments, serious water pollution, etc. On the contrary, ionic liquids have almost no vapor pressure, excellent chemical stability, and adjustable positive/negative ions and acidity. Using ionic liquids to catalyze esterification reaction between cyclohexane-1,2-dicarboxylic acid and n-butanol provides a new, industrially promising, green synthetic method for dibutyl cyclohexane-1,2-dicarboxylate, because of the advantages of high catalyst activity, light product color, high product quality, less side reactions, facile post-treatment, etc.

2. Main Content

Cyclohexane-1,2-dicarboxylic acid and n-butanol are raw materials. After the esterification reaction is completed, the reaction mixture is neutralized, washed, dried, dealcoholized, vacuum distilled to obtain dibutyl cyclohexane-1,2-dicarboxylate. The esterification ratio is >98%. The preparation involves no aromatic compounds and the product contains no aromatic rings. The synthetic steps are as follows:

(1) Esterification reaction

To a three-necked flask equipped with magnetic stirrer, thermometer, condenser, and water trap were added 0.2 mol HHPA, 0.5–0.7 mol n-butanol, and ionic liquid catalyst at a loading of 0.5–2% of HHPA mass. The mixture was refluxed for 5~8 h in an N_2 atmosphere with the temperature maintained at 110–150 °C to remove the formed water from the water trap. The acid value was measured every hour following GB/T 1668–2008. When the acid value turned <3 mg KOH/g, the reaction was terminated, at which the reaction temperature did not exceed 150 °C.

The ionic liquid catalyst from commercial sources or self-synthesis remains liquid at room temperature. The ionic liquid can select 1-methyl-2-pyrrolidonium hydrogen sulfate, 1-(3'-sulfonic acid)propyl-2-pyrrolidonium hydrogen sulfate, 1-methyl-2-pyrrolidonium methanesulfonate, 1-(3'-sulfonic acid)propyl-2-pyrrolidonium methanesulfonate, (3-sulfonic acid)propyl triethylammonium hydrogen sulfate, and 1-(3'-sulfonic acid)propyl-3-methylimidazolium hydrogen sulfate.

(2) Product purification

To the esterification reaction mixture was added 5 wt% NaOH solution until weakly. alkaline pH (8–9) at 70 °C to remove the unreacted HHPA and form cyclohexane monoacid monoester. The crude product was washed with water 2–3 times to neutral, followed by reduced pressure distillation to remove n-butanol and water and obtain a light-yellow, transparent liquid.

(3) Reduced pressure distillation

Reduced pressure distillation was used at 230–250 °C and 0.09 MPa to obtain the product distillate, followed by removal of colored substances, to the pure dibutyl cyclohexane-1,2-dicarboxylate with a chromaticity (Pt–Co method) of 20–100 Hazen. This decoloration method substituted the traditional activated carbon or hydrogen peroxide method, achieving the product purity of >99.6% and yield of >98%.

3. Positive Effects

The new ionic liquid catalyst is non-corrosive, non-toxic, non-toxic, environmentally friendly, and efficient (in small loading) leading to high esterification ratio, mild reaction conditions, and short reaction time. The direct use of reactant n-butanol as the azeotropic agent avoids toxic toluene.

4.2.9 A Type of Catalysts for Hydrogenating Benzenecarboxylates to Cyclohexanecarboxylates and the Preparation Method Thereof (2012) [19]

1. Technical Background

EU 2005/84/EC lists 6 types of restricted phthalates: DEHP, DBP, BBP, DINP, DIDP, and DNOP. The former three are prohibited in children's toys and products and are limited to <0.1% in plastics. The latter three are prohibited in bitable children's toys and products and the content of each cannot exceed 0.1%.

Then, USA, Canada, Japan, and South Korea released similar laws and standards. In 2011, Taiwan, China announced that the food additive "clouding agent" used by a certain spice company contained DEHP for making drinks which caused serious hazards to our society. After this incident, the Main Land, China reinforced the supervision of benzoate-based plasticizers.

2. Main Content

The catalysts in this invention are composed of the active composition, additive, and support.

The mass fraction of the active composition Ni is 5–50%, more preferably 10–30%. The mass fraction of additive is 0.1–25%, more preferably 1%~10%. The rest is the support. The additive is at least one of K, Mg, Ca, Sr, Ba, Co, Fe, La, Ac, Mo, Zr, Ti, Mn, Cu.

The support is Al_2O_3, SiO_2, Al_2O_3–SiO_2, or activated carbon. Al_2O_3 can be α, θ or γ types with a surface area of 20–300 m^2/g, pore volume of 0.2–1.2 cm^3/g, porosity of 30–80%, and pore size of 10–100 nm; more preferably γ-Al_2O_3 with a surface area of 100–200 m^2/g, pore volume of 0.5–1.0 cm^3/g, porosity of 30–80%, and pore size of 10–100 nm. SiO_2 has a surface area of 50–500 m^2/g, pore volume of 0.3–1.3 cm^3/g, porosity of 30–80%, and pore size of 10–100 nm; more preferably a surface area of 100–300 m^2/g, pore volume of 0.5–1.0 cm^3/g, porosity of 30–80%, and pore size of 10–100 nm. Al_2O_3-SiO_2 has a Si mass fraction of 1–5%, surface area of 50–500 m^2/g, pore volume of 0.3–1.3 cm^3/g, porosity of 30–80%, and pore size of 10–100 nm; more preferably a surface area of 100–300 m^2/g, pore volume of 0.5–1.0 cm^3/g, porosity of 30–80%, and pore size of 10–100 nm. Activated carbon has a surface area of 40–400 m^2/g, pore volume of 0.4–1.5 cm^3/g, porosity of 30–80%, and pore size of 10–100 nm; more preferably surface area of 100–200 m^2/g, pore volume of 0.5–1.0 cm^3/g, porosity of 30–80%, and pore size of 10–100 nm.

The catalysts were prepared using the impregnation method as follows:

(1) The soluble salts of the active composition and additive were dissolved in deionized water to make a mixed solution with a total metal concentration of 0.9–5.5 g/mL. The solution was maintained at a temperature within 20–80 °C.
(2) Polyethylene glycol was added to the solution at a mass fraction of 0.001–2% and was well stirred.
(3) The support was soaked in the stirred solution for 1–6 h.
(4) The soaked sample was dried at 80–150 °C for 5–24 h. When the support was Al_2O_3, SiO_2, or Al_2O_3-SiO_2, the sample was baked at 180–250 °C for 1–10 h and then at 251–500 °C for 2–8 h. When the support was activated carbon, the sample was baked at 180–250 °C for 1–10 h and then at 251–500 °C for 2–8 h in an inert gas (e.g., N_2) atmosphere.

Before use, the catalyst was activated via H_2 reduction with programmed heating under conditions: pressure 0.5–5 MPa, hourly volume space velocity 1000–10000 h^{-1}, 150–250 °C for 2–10 h, then 300–500 °C for 2–8 h, rate of heating 10–100 °C/h.

Hydrogenation reaction conditions: 100–300 °C, 2–10 MPa, 20% benzenecarboxylate in ethyl acetate, benzenecarboxylates mass space velocity 0.1–5 h^{-1}, H_2/benzenecarboxylates molar ratio (10–300):1.

Benzenecarboxylates can be *para*-, *meta*-, or *ortho*-benzenedicarboxylates. *para*-Benzenedicarboxylates can be dimethyl and diethyl terephthalates; *ortho*-benzenedicarboxylates can be diisooctyl phthalate, diisopropyl phthalate, diethyl phthalate, diisobutyl phthalate, diisohexyl phthalate; *meta*-benzenedicarboxylates can be diisononyl isophthalate, diheptyl isophthalate, diisopentyl isophthalate, etc.

After activation, the catalysts prepared in Examples 1–10 of this invention were evaluated in fixed-bed reactors. The raw material was 20% benzenecarboxylate in ethyl acetate. After the reactions were completed, the GC analysis results are shown in Table 4.4.

Table 4.4 The reaction conditions, activity and selectivity of catalysts prepared via impregnation

#	Catalyst composition	Raw materials	Pressure/MPa	Temperature/°C	Space velocity/h^{-1}	H_2/ester molar ratio	*C* /%	*S*/%
1	5%Ni-2.5%Co/$Al_2O_3(\alpha)$	diisooctyl phthalate	10	160	0.1	10	90.5	98.1
2	50%Ni-0.1%Mn/$Al_2O_3(\gamma)$	diisononyl isophthalate	5	200	2	300	100	100
3	20%Ni-3%Ba-2% Mo-5%Fe/$Al_2O_3(\theta)$	dimethyl terephthalate	8	150	3	100	99.8	100
4	10%Ni-10%Cu-1%Ti/SiO_2	diisopropyl phthalate	2	300	5	200	97.3	96.5
5	20%Ni-5%La-5% Zr/SiO_2	diheptyl isophthlate	3	160	1	150	100	98.4
6	40%Ni-1%Ca-10%Mg/SiO_2	diethyl phthalate	4	180	4	60	100	100
7	15%Ni-1%Cu-2% Mn-5%Th/C	diisobutyl phthalate	6	220	0.5	30	99.5	99.3
8	25%Ni-2%Cu-2% Mn-3%K/C	diisopentyl isophthalate	7	240	1	250	100	98.2
9	8%Ni-20%Cu-2% Ba/SiO_2-Al_2O_3	diisohexyl phthalate	5	160	4	60	95.6	95.1
10	18%Ni-2%Co/SiO_2-Al_2O_3	diethyl terephthalate	6	180	2	90	100	100

Note C is the benzenecarboxylate conversion ratio; *S* is the cyclohexane carboxylate selectivity

3. Positive Effect

This invention provides an impregnation method for preparing the catalyst used for hydrogenating benzenecarboxylates to synthesize cyclohexanecarboxylates. The catalysts are non-noble metal catalysts with low cost and high activity and selectivity.

4.2.10 A Type of Benzenecarboxylate Hydrogenation Catalyst and the Preparation and Application Thereof (2012) [21]

1. Technical Background

Cyclohexanecarboxylates are aliphatic polycarboxylates that can be easily processed using current facilities. They not only have excellent ductility, plasticity, light and heat stability, but also have very low element migration and toxicological advantages. They are promising substitutes for benzenecarboxylate plasticizers that are more suitable for toys, medical equipment, and food contact plastics.

2. Main Content

The catalysts are composed of the active composition, additive, and support. The mass fraction of the active composition Ni is 5%–50. The additive is at least one of Mg, Ca, Sr, Ba, Co, Fe, La, Lu, Ac, Mo, Zr, Ti, Mn, Cu, with the mass fraction of 0.1–25%. The rest is the support. The catalysts have a surface area of 20–300 m^2/g, pore volume of 0.2–1.2 cm^3/g, porosity of 30–80%, and pore size of 10–100 nm.

The catalysts were prepared using the impregnation method as follows:

(1) The soluble salts or precursors of the active composition and additive were dissolved in deionized water to make a mixed solution with a total metal concentration of 0.5–2.0 g/mL. Polyethylene glycol or non-ionic surfactant was added to the solution at a mass fraction of 0.001–5% and was well stirred to obtain Solution A. The soluble salts or precursors of the support were aluminum nitrate, aluminum sulfate, aluminum chloride, or silica gel.
(2) The precipitating agent was dissolved to make 0.5–2.0 mol/L Solution B. The precipitating agent can be single or mixture of sodium carbonate, sodium hydroxide, ammonium carbonate, ammonium bicarbonate, urea, and ammonia.
(3) Solution A was added dropwise to Solution B, or Solution B was added dropwise to Solution A, or Solutions A and B were co-precipitated. The temperature was maintained within 40–90 °C and the pH value was adjusted to within 7.0–9.0. Then, the mixture was aged for 1–5 h.
(4) The obtained precipitate was filtered and washed 2–8 times with deionized water.
(5) The washed precipitate was dried at 80–150 °C for 5–24 h, baked at 180–250 °C for 1–10 h, and then at 250–500 °C for 2–8 h.

Before use, the catalyst was activated in an H_2 atmosphere under the following conditions: 150–250 °C and 0.5–5 MPa H_2 pressure for 2–10 h, then 300–500 °C for 2–8 h, rate of heating 10–100 °C/h, volume space velocity 1000–10000 h^{-1}.

Hydrogenation reaction conditions: 100–300 °C and 2–10 MPa for 2–5 h, H_2/benzenecarboxylates molar ratio (10–300):1, catalyst loading 1–2wt% of benzenecarboxylates.

After activation, the catalysts prepared in Examples 1–18 were evaluated in high-pressure autoclave reactors. After the reactions were completed, the GC analysis results are shown in Table 4.5.

3. Positive Effect

This invention provides a precipitation method for preparing the catalyst used for hydrogenating benzenecarboxylates to synthesize cyclohexanecarboxylates. The catalysts are non-noble metal catalysts with low cost and high activity and selectivity.

4.2.11 Catalysts for Hydrogenating Benzenecarboxylate to Cyclohexanecarboxylates and the Preparation and Application Thereof (2012) [20]

1. Technical Background

In the 1980s, the numerous animal tests performed by the toxicity test team at the United States National Institute of Environmental Health Sciences (NIEHS) demonstrated that DOP could cause lesion and cancer to animal liver tissues, and that phthalate plasticizers including DOP, DINP, DBP, DIDP, DIOP, DNOP, and DCP are harmful to human to different extents. Sweden banned the sale of soft plastic toys since August 1, 1999. Denmark and Austria banned the sale of plastic toys since April and January 2001 to avoid the entry of phthalate plasticizer molecules via soft plastics into the human body.

To address the contradiction between the importance of phthalate plasticizers and the potential hazards to humans, the conversion of phthalates to environmentally friendly plasticizers via ring hydrogenation is a promising solution.

2. Main Content

The active compositions in the catalysts are the oxides of Ni, the additive is P, and the support is Al_2O_3 with a surface area of 100–350 m^2/g. The optimal mass fraction is 25–30% for the active composition and is 1–5% for the additive. The optimal surface area of the support is 200–300 m^2/g. The preparation method for the catalysts are as follows:

(1) Support: $Al(OH)_3$ dry powder, acetic acid, and sesbania powder were mixed at a mass ratio of 1: (0.02–0.028): (0.025–0.043). With deionized water added, the mixture was kneaded to form spheres of $\Phi = 2.5$ mm granularity, which

Table 4.5 The catalysts, reaction conditions, activity, and selectivity of Examples 1–8 BET (m2/g) (m3/g) /% /MPa /oC /h H2/ C/% S/%

#	Catalyst composition	BET surface area / (m^2/g)	*V*n / (cm^3/g)	Pore ratio (size 10–100 nm) /%	Raw materials	Pressure/ MPa	Temperature/ °C	Time/ h	H_2/ ester molar ratio	Catalyst loading (wt%)	*C*/ %	*S*/ %
1	5%Ni-2.5%Co/Al_2O_3	20	0.2	30	DIOP	8	200	2	10	10	90.5	98.1
2	50%Ni-0.1% Mn/Al_2O_3	80	1.0	45	*m*-DINP	6	100	6	300	8	100	100
3	20%Ni-3%Ba-2%Mo-5%Fe/ SiO_2	100	1.1	55	*p*-DMP	10	150	8	100	4	99.8	100
4	10%Ni-10% Cu-15%Ti/ SiO_2- Al_2O_3	200	0.8	70	DIBP	4	300	10	200	15	97.3	96.5
5	20%Ni-5% La-5%Zr/SiO_2	250	0.4	80	*m*-DHP	3	180	12	150	9	100	98.4
6	40%Ni-1%Ca-10%Mg/SiO_2	300	0.6	60	DEP	2	160	15	60	6	100	100
7	30%Ni-5%Mg/Al_2O_3	150	1.0	65	*p*-DEP	7	170	9	30	20	100	100
8	25%Ni-1%Ba/SiO_2	180	0.9	75	*m*-DBP	6	190	7	80	1	100	99.5

Note *C* is the phthalate conversion ratio; *S* is the cyclohexane dicarboxylate selectivity was air-dried at room temperature for 5 h, then thermally dried at 110–140 °C for 4 h, and finally sintered at 680–900 °C for 2–3 h to obtain the spherical Al_2O_3 support of $\Phi = 2.5$ mm.

(2) Catalyst: The obtained $\Phi = 2.5$ mm spherical Al_2O_3 support was impregnated in a solution of citric acid and phosphoric acid at a mass ratio of 1:(0.00503–0.00714) in water, dried at 125–130 °C for 2–3 h, baked at 500–550 °C for 3–6 h to obtain the P/Al_2O_3 supporter. Then, the P/Al_2O_3 supporter was impregnated in a $Ni(NO_3)_2$ solution (130.3 g $Ni(NO_3)_2$ per 100 g water) in a 1:1 mass ratio, air-dried, then thermally dried at 120–130 °C for 3 h, baked at 500 °C for 2.5–6 h to obtain the catalyst.

A continuous fixed-bed reactor was used to perform the hydrogenation reactions under conditions: 150–200 °C, 3.0–15.0 MPa H_2 pressure, volume space velocity 0.1–0.5 h^{-1}, H_2/benzenecarboxylates volume ratio (300–1000):1.

The catalysts are suitable for hydrogenation reactions of DOP, DINP, DBP, DIDP, DIOP, DNOP, and DCP. The properties of the catalyst are shown in Table 4.6.

Small fixed-bed hydrogenation conditions: catDOP-1 and catDOP-2 were used as the catalyst that was 1:1 mixed with the diluent and put in the isothermal bed layer. The mixture of heated raw material and H_2 entered the reactor to catalytically hydrogenate the benzene rings under well-controlled conditions. After the reaction was completed, the gas and liquid materials were separated via the splitter. The gas was recycled and the liquid was the product. The raw material and product were analyzed with a GC2008A gas chromatographer. Experiments # 1–5 used DOP and #6 used DINP. The catalyst, conversion ratio, and side product contents are shown in Table 4.7.

3. Positive Effects

The catalysts in this invention use inexpensive non-noble metals as the active composition leading to low cost and increased economic benefits. When used for hydrogenation reactions, the benzene rings are converted to non-toxic cyclohexane with high conversion ratios and low side product contents.

Table 4.6 Properties of catalysts

Items	catDOP-1	catDOP-2	catDOP-3	catDOP-4	catDOP-5
NiO content/%	25–30	30–33	15–20	≥40	≥40
Additive	–	–	–	–	–
Morphology	sphere	sphere	sphere	sphere	sphere
GranularityΦ/mm	2.5	2.5	2.5	2.5	2.5
Pore volume/(mL/g)	0.35–0.40	0.35–0.38	≥ 0.4	0.28–0.30	0.25–0.30
Surface area/(m^2/g)	≥170	≥165	≥180	≥125	≥140
Packing density /(g/mL)	0.80–0.85	0.85–0.95	0.78–0.82	0.98–1.15	0.78–0.85
Confined compressive strength /(N/mm)	≥10	≥10	≥10	≥10	≥10

Table 4.7 Experiments 1–6: catalysts, reactant conversion ratio, and side product contents

#	Catalyst	Reactant	Reactant conversion/%	Side product content/%
1	catDOP-2	DOP	99.63	0.37
2	catDOP-2	DOP	99.74	0.26
3	catDOP-2	DOP	98.93	1.07
4	catDOP-1	DOP	73.94	22.76
5	catDOP-1	DOP	63.63	36.37
6	catDOP-2	DINP	100	0

4.2.12 A Method for Preparing Diisooctyl Cyclohexane-1,2-Dicarboxylate Plasticizer (2012) [21]

1. Technical Background

Since the 1990s, with the deeper research on the xenoestrogen pollution of phthalate plasticizers, the call for their limited uses has been becoming stronger globally. Currently, EU, USA, Switzerland, etc., have enacted related laws to limit the use of phthalate plasticizers in food packaging, cosmetics, and children's toys. It has become a focus of attention to develop new non-toxic, environmentally friendly plasticizers. Cyclohexane-1,2-dicarboxylate plasticizers not only have comparable plasticization performance, but also are non-toxic and environmentally friendly. Therefore, they are a new type of environmentally friendly phthalate substitute plasticizers for use in food packaging, medical equipment, and children's toys.

2. Main Content

The readily accessible diisooctyl phthalate was directly hydrogenated under high pressure to prepare the diisooctyl cyclohexane-1,2-dicarboxylate. The isooctanol was used as the solvent with an alcohol/ester mass ratio of 2.5:1. Ru/C was the catalyst with a loading of 5% of diisooctyl phthalate. The reaction was operated under conditions of 170–180 °C, 6–7 MPa H_2 pressure, 6–8 h to obtain the product diisooctyl cyclohexane-1,2-dicarboxylate.

Example 1: To a 1L autoclave were added diisooctyl phthalate 60 g, isooctanol 150 g, and Ru/C catalyst 3 g. After 3 H_2 exchanges, the mixture was maintained at 170–180 °C and 6–7 MPa H_2 pressure for 6–8 h. A sample was drawn and analyzed with GC to detect the purity of 98.8% (not including isooctanol), and then the reaction was stopped. 57.8 g diisooctyl cyclohexane-1,2-dicarboxylate was obtained with a yield of 95% and GC purity of 99.33% after isooctanol recycling and vacuum distillation.

Example 2: To a 1L autoclave were added diisooctyl phthalate 100 g, isooctanol 250 g, and Ru/C catalyst 5 g. After 3 H_2 exchanges, the mixture was maintained at 170–180 °C and 6–7 MPa H_2 pressure for 6–8 h. A sample was drawn and analyzed with GC to detect the purity of 98.9% (not including isooctanol), and then the reaction was stopped. 96.7 g diisooctyl cyclohexane-1,2-dicarboxylate was obtained with a yield of 95.2% and GC purity of 99.31% after isooctanol recycling and vacuum distillation.

Example 3: To a 1L autoclave were added diisooctyl phthalate 200 g, isooctanol 500 g, and Ru/C catalyst 10 g. After 3 H_2 exchanges, the mixture was maintained at 170–180 °C and 6–7 MPa H_2 pressure for 6–8 h. A sample was drawn and analyzed with GC to detect the purity of 99.0% (not including isooctanol), and then the reaction was stopped. 194.3 g diisooctyl cyclohexane-1,2-dicarboxylate was obtained with a yield of 96.2% and GC purity of 99.50% after isooctanol recycling and vacuum distillation.

Example 2: To a 1L autoclave were added diisooctyl phthalate 250 g, isooctanol 625 g, and Ru/C catalyst 12.5 g. After 3 H_2 exchanges, the mixture was maintained at 170–180 °C and 6–7 MPa H_2 pressure for 6–8 h. A sample was drawn and analyzed with GC to detect the purity of 99.0% (not including isooctanol), and then the reaction was stopped. 245.0 g diisooctyl cyclohexane-1,2-dicarboxylate was obtained with a yield of 96.5% and GC purity of 99.54% after isooctanol recycling and vacuum distillation.

3. Positive Effect

This invention uses readily accessible diisooctyl phthalate directly hydrogenated under high pressure to prepare the plasticizer diisooctyl cyclohexane-1,2-dicarboxylate with >98% selectivity and purity of >99%.

4.2.13 *A Method for Preparing the Catalysts for Hydrogenation-Based Cyclohexane-1,2-Dicarboxylates (2012) [22]*

1. Technical Background

In recent years, the toxicity, carcinogenicity, and biodegradability of phthalate plasticizers have been investigated. Due to the benzene ring, phthalates have drawbacks of carcinogenicity, teratogenicity, accumulation in human body, poor biodegradability, etc. Therefore, the application fields of traditional plasticizers have been limited. It's become a research focus to R&D non-toxic, green plasticizer products.

2. Main Content

The supported noble metal Ru and/or Pd catalysts contain the active component Ru at 0.05–3.0wt% and Pd at 0.03–2%. The support is a mesoporous oxide with suitable pore structure and surface area (BET surface area 150–350m^2/g, pore volume 0.3–2.0 mL/g, pore size 7–15 nm); it can be Al_2O_3, amorphous SiO_2-Al_2O_3, SiO_2 or a combination thereof. The preparation procedure is as follows:

(1) The mesoporous oxide support with BET surface area 150–350m^2/g, pore volume 0.3–2.0 mL/g, pore size 7–15 nm was thermally dried.

Table 4.8 The activity test results and average grain sizes of catalysts

#	Reactant conversion/%	Product selectivity/%	Average grain size/nm
1	99.92	99.97	6
2	99.94	99.91	7
3	99.95	99.93	8
4	100	100	8
5	99.91	99.95	5
6	99.96	100	7
7	100	100	5
8	100	100	7
9	99.12	99.92	20

Note the average grain sizes of catalysts were analyzed using Hitachi 5–4800 SEM

(2) Stoichiometric Ru salt HCl solution and Pd salt HCl solution were mixed together and diluted with deionized water.
(3) The noble metal solution was impregnated to the mesoporous support for 3–8 h using the isovolumetric impregnation method.
(4) The catalyst was dried at 110 °C for 10 h and then baked at 250–400 °C for 3 h.
(5) The catalyst was washed 2–7 times with 0.2–3 wt% NaOH solution.
(6) The catalyst was air-dried and then baked at 120 °C for 10 h.

The catalyst was loaded into a 20 mL fixed-bed reactor for hydrogenation at 250 °C for 4 h. Test conditions: 130 °C, 3 MPa H_2 pressure, volume space velocity 0.5 h^{-1}, H_2/oil volume ratio 1000:1.

Test method: Agilent HP-7890 chromatographer, HP-5 column, N_2 carrier gas, flow rate 1.5 mL/min, column temperature 260 °C, FID detector.

The catalyst performance test results (from stable small-scale tests within 1000 h) and average granularity are shown in Table 4.8.

3. Positive Effects

The catalysts obtained in this invention ensure high noble metal dispersity and provide suitable pore structure for adsorption and desorption of reactant and product molecules, leading to remarkably improved conversion ratio and selectivity (both as high as 99%) with decreased reaction temperature and energy consumption.

4.2.14 A Method for Preparing Diisooctyl Cyclohexane-1,2-Dicarboxylates via Hydrogenation (2012) [23]

1. Technical Background

Cyclohexane-1,2-dicarboxylates have excellent heat-, cold-, UV-, and flame-resistant properties, and are compatible with PVC, polyurethane, acrylic resin, acetal resin, etc. Animal tests indicate no risk of peroxisome hyperplasia or liver tumors. Many countries have recognized their use in human contact food packaging, medical equipment, and children's toys. Cyclohexane-1,2-dicarboxylates are a new type of safe, environmental plasticizer products that are biodegradable in nature and cause no environmental pollution.

2. Main Content

The hydrogenation reaction system is composed of diisooctyl phthalate, H_2, and catalyst. Reaction conditions: 100–250 °C, 5–18 MPa H_2 pressure, liquid space velocity 0.1–2.5 h^{-1}, H_2/oil volume ratio (2000–200):1.

The catalyst uses Al_2O_3 support, one or several of Ru, Pd, Ni as the active composition at 1–3wt% mass fraction. The mass composition of Pd-Ru/Al_2O_3 is Pd 0.5%, Ru 1%, and Al_2O_3 the rest. The Pd and Ru compositions are loaded onto the Al_2O_3 support via the regular impregnation method. Before use, the catalyst is activated via in situ H_2 reduction under conditions: volume space velocity 1000–2000 h^{-1}, ambient pressure ~1.0 MPa, 250–450 °C, reaction time 5–12 h.

The final reaction conditions: trickle bed reactor, 175 °C, 15 MPa H_2 pressure, H_2/ester molar ratio 600, volume space velocity 0.4 h^{-1}, 24 h.

GC analysis of reactants and products: N_2 carrier gas, flow rate 1.5 mL/min, injected sample 4μL, FID detector, HP-5 column, column temperature 260 °C.

Table 4.9 shows the test results of hydrogenating diisooctyl phthalate to prepare diisooctyl cyclohexane-1,2-dicarboxylate in this invention.

3. Positive Effects

The invention can remarkably decrease the reaction temperature and energy consumption and improve the conversion ratio and selectivity.

Table 4.9 Diisooctyl phthalate hydrogenation reaction test results

#	Temperature/°C	H_2 Pressure/MPa	H_2/ester molar ratio	Volume space velocity/h^{-1}	Reactant conversion/%	Product selectivity/%
1	175	15	600	0.4	99.95	99.98
2	175	15	600	0.3	99.96	99.98
3	175	15	600	0.5	99.94	99.98
4	160	15	600	0.4	99.94	99.98
5	190	15	600	0.4	99.93	99.98
6	175	13	600	0.4	99.92	99.98
7	175	16	600	0.4	99.92	99.98
8	175	15	800	0.4	99.95	99.98
9	175	15	400	0.4	99.91	99.97

4.2.15 A Method for Preparing Cyclohexane-1,2-Dicarboxylates via Hydrogenation (2012) [24]

1. Technical Background

Since 2007, EU, USA, Japan, etc., have enacted policies to prohibit the use of phthalate plasticizers in packaging, medical, and children's toys. Therefore, the R&D of non-toxic green plasticizer products has currently become a hot research topic.

2. Main Content

The preparation of diisooctyl cyclohexane-1,2-dicarboxylate is based on the hydrogenation of diisooctyl phthalate.

The hydrogenation reaction system is composed of diisooctyl phthalate, H_2, and catalyst. Reaction conditions: 100–250 °C, 5–18 MPa H_2 pressure, liquid space velocity 0.1–2.5 h^{-1}, H_2/oil volume ratio (2000–200):1. The reactant conversion ratio and product selectivity are high.

The catalyst uses Al_2O_3 support, one or both of Ru or Pd as the active composition at 1–3wt% mass fraction. The active composition is loaded onto the support via precipitation, impregnation, hydrothermal, or a combination thereof. Before use, the catalyst is activated via in situ H_2 reduction under conditions: volume space velocity 1000–2000 h^{-1}, ambient pressure ~ 1.0 MPa, 250–450 °C, reaction time 5–12 h.

The catalyst was loaded into a 20 mL fixed-bed reactor for hydrogenation at 250 °C for 4 h. Test conditions: 130 °C, 3 MPa H_2 pressure, volume space velocity 0.5 h^{-1}, H_2/oil volume ratio 1000:1.

This invention directly pumps liquid isooctyl phthalate into the preheater and then into a fixed-bed reactor. No solvent is needed for dissolution or dilution.

This invention can use fixed-bed, fluidized bed or trick bed reactors, more preferably fixed-bed reactor.

3. Positive Effect

The invention can remarkably decrease the reaction temperature and energy consumption and improve the conversion ratio and selectivity.

4.2.16 A Method for Preparing Environmentally Friendly Cyclohexane-1,2-Dicarboxylates (2013) [25]

1. Technical Background

The plasticizers used in China are mainly phthalates that is ~90% of the total consumption. With the wider use of phthalates, people pay increasing attention to their toxicity. Because of the potential carcinogenicity, many countries have implemented strict restrictions on their uses. However, ring-hydrogenated phthalates have excellent processibility with PVC, comparable and even better performance, superior heat-, cold-, UV- and flame-resistant properties, and exhibit no toxicity or genotoxicity. Currently, they are considered as the most possible phthalate substitute plasticizers with promising market potential for use in food packaging, medical products, and children's toys.

2. Main Content

The hydrogenation reaction system is composed of phthalate, H_2, and catalyst. Reaction conditions: H_2/phthalate molar ratio (30–5):1, 100–250 °C, 5–18 MPa H_2 pressure, liquid volume space velocity 0.1–2.5 h^{-1}.

The catalyst uses Al_2O_3–SiO_2 support, one or several of Ru, Pd, Ni as the active composition at 0.05–3wt% fraction. The mass composition of Pd-Ru/Al_2O_3–SiO_2 is Pd 0.5%, Ru 1%, and Al_2O_3–SiO_2 the rest. The active composition is loaded onto the support via the impregnation method. Before use, the catalyst is activated via in situ H_2 reduction under conditions: volume space velocity 1000–2000 h^{-1}, ambient pressure ~1.0 MPa, 200–300 °C, reaction time 1–5 h.

Hydrogenation method: H_2/ester molar ratio (20–10):1, 120–160 °C, 12–16 MPa, liquid volume space velocity 0.2–0.8 h^{-1}, reaction time 24 h.

Table 4.10 shows the test conditions and results of hydrogenating phthalates to cyclohexane-1,2-dicarboxylates in this invention.

3. Positive Effects

The invention remarkably decreases the energy consumption and especially reduces the H_2 consumption by 90%, expands the feasible phthalate types, and improves the conversion ratio and selectivity.

Table 4.10 Phthalate hydrogenation reaction test conditions and results

#	Temperature/°C	H_2 Pressure/MPa	H_2/ester molar ratio	Volume space velocity/h^{-1}	Reactant conversion/%	Product selectivity/%
1	140	15	15	0.4	99.95	99.98
2	140	15	15	0.3	99.94	99.98
3	140	15	15	0.5	99.92	99.98
4	130	15	15	0.4	99.93	99.98
5	150	15	15	0.4	99.98	99.98
6	140	13	15	0.4	99.93	99.98
7	140	16	15	0.4	99.98	99.98
8	140	15	18	0.4	99.95	99.98
9	140	15	12	0.4	99.94	99.97
10	140	15	15	0.4	99.96	99.97
11	140	15	15	0.4	99.97	99.98
12	140	15	15	0.4	99.92	99.96
13	140	15	15	0.4	99.98	99.97
14	140	15	15	0.4	99.94	99.96

Note 1–9: DIDP reactant and DIDCH product; 10: DEHP reactant and DEHCH product; 11: DINP reactant and DINCH product; 12: DBP reactant and DBCH product; 13: DNOP reactant and DNOCH product; 14: BBP reactant and BBCH product

4.2.17 A Method for Preparing Phthalate Selective Hydrogenation Catalysts (2013) [26]

1. Technical Background

The current technologies have drawbacks such as high reaction temperature, low conversion ratio or selectivity. In addition, due to the low surface area (30–70m^2/g) of macroporous supports, the noble metal active compositions are distributed on the support surface decreasing the active surface area and reaction rate.

2. Main Content

This invention provides a method for preparing the catalysts for selectively hydrogenating phthalates, especially for hydrogenating phthalates to synthesize cyclohexane-1,2-dicarboxylates. The primary active composition can be Pd (precursors H_2PdCl_4 or $H_2Pd(NO_3)_4$), Ru (precursors $RuCl_3$, $RuBr_3$, RuI_3, $Ru(OAc)_3$, or $Ru(NO_3)_3$), or both of them; the mass fraction is 0–0.8% for Pd, 0–1.0% for Ru, and 0.1–1.5% for the total of Pd and Ru. The support is Al_2O_3 or Al_2O_3–SiO_2 or both, with the mass fraction of SiO_2 less than 10%. The preparation of catalysts uses the metal gel-impregnation method as follows:

(1) To a vigorously stirred aqueous surfactant solution of polyvinylpyrrolidone (PVP) was added an active composition salt solution, followed by dropwise addition of a reductant (methanol or formaldehyde) to form a stable metal gel.
(2) The support was put in a rolling granulator and the metal gel solution was sprayed to impregnate the support under rolling conditions.
(3) After the impregnation was completed, the temperature was raised in an oxygen-free environment via programmed heating with 30–80 °C for 1–6 h and then 80–100 °C for 2–10 h.
(4) The sample was baked at 200–300 °C for 1–5 h.

This method is applicable to the selective hydrogenation of phthalates to hydrogenate the benzene ring and leave the ester bonds untouched. The reactant can be dimethyl phthalate, diethyl phthalate, dibutyl phthalate, dioctyl phthalate (diisooctyl phthalate), diisononyl phthalate, benzyl butyl phthalate, diisodecyl phthalate, ditridecyl phthalate. Among them, dioctyl phthalate (diisooctyl phthalate), diisononyl phthalate, benzyl butyl phthalate, and diisodecyl phthalate are the major phthalate plasticizers.

The catalyst was loaded in a 20 mL fixed-bed reactor and slowly heated to 250 °C for hydrogenation for 4 h. The reactant DOP and H_2 were fed into the reactor to undergo non-homogenous reactions. The product (diisooctyl phthalate) flowed out through the high pressure and low-pressure separation tank at the reactor bottom. Reaction conditions: 130 °C, 3 MPa, volume space velocity 0.5 h^{-1}, H_2/ester volume ratio 1000:1.

Table 4.11 shows the catalyst test results (from stable test data after 144 h).

3. Positive Effects

The invention remarkably increases the utilization ratio of noble metals, plus excellent catalyst performance, simple preparation protocols, and high scalability.

Table 4.11 Catalyst evaluation results

Example	Metal dispersity/%	C_{DOP}/%	S_{DEHCH}/%	Y_{DEHCH} /%
1	37	99.13	99.65	98.47
2	30	98.76	99.71	98.47
3	33	99.28	99.71	98.99
4	39	99.32	99.52	98.84
5	32	99.26	99.69	98.96
6	38	99.88	99.64	99.52
7	31	99.83	99.64	99.48
Control	13	96.43	98.76	95.23

Note C_{DOP} is the DEHP conversion ration; S_{DEHCH} and Y_{DEHCH} are the selectivity and yield of DEHCH

4.2.18 A Method for Preparing Phthalate Hydrogenation Catalysts (2015) [27]

1. Technical Background

Patent CN201510083129.8 uses azeotropic distillation to obtain supported Ni catalysts with carbonated support for ring hydrogenation of phthalates. The active metal composition is highly dispersed and can decrease reaction temperature and pressure to a certain extent. However, the preparation protocol is too complicated for scaled production.

2. Main Content

This invention provides a method for preparing the catalysts for hydrogenating phthalates. The catalysts have optimized surface area and pore structure advantageous for the adsorption and desorption of reactant and product molecules. The halide content in the catalysts is lower than 0.05wt%. The catalyst catalyzes the hydrogenation reaction in a fixed-bed reactor with the conversion ratio and selectivity >99%.

The support is amorphous Al_2O_3 or SiO_2 or a combination thereof, with the surface area 80–400 m^2/g, pore volume 0.2–3.0 mL/g, pore size 5–20 nm. The active composition is noble metal Ru at a mass fraction of 0.2–1.3%.

The catalysts were prepared as follows:

(1) Dilute ammonia was slowly added to an aqueous Ru salt solution. The pH was adjusted to 7.0–12.0.
(2) The support was impregnated with the solution prepared in (1) in saturated adsorption and then was dried at 60–150 °C for 1–8 h.
(3) The sample was baked at 150–300 °C for 1–6 h.

The features of the catalysts prepared using this invention: solid Ru salts can be ruthenium acetate or ruthenium nitrosylnitrate or a mixture of both; the pH value is 8.5–11.5; before use, activation via H_2 reduction is necessary; the application is for hydrogenating phthalates to produce cyclohexane-1,2-dicarboxylates.

Table 4.12 shows the catalyst test results.

Table 4.12 Catalyst evaluation results

#	Reactant conversion/%	Product selectivity/%	Product yield/%	Conversion at 800 h/%	Selectivity at 800 h/%
1	99.93	99.70	99.63	99.90	99.70
2	99.95	99.74	99.69	99.92	99.74
3	99.98	99.52	99.50	99.99	99.51
4	99.91	99.63	99.54	99.90	99.64
5	99.92	99.71	99.63	99.93	99.69

4.2.19 An Integrated Method for Preparing Cyclohexane-1,2-Dicarboxylates (2015)

1. Technical Background

Phthalate plasticizers, mainly DOP, DEHP, and DBP, are widely used in China. It's been known that aromatic-bearing plasticizers have potential carcinogenic and environmental risks. Thus, related laws are increasingly stricter. But because of the large production volume and numerous manufacturers, upgrading these plasticizers is necessary to address the environmental protection need. The best strategy is to hydrogenate these plasticizers to saturate the aromatic ring to cycloalkane.

Preparation of phthalates via esterification needs a series of refining treatments. Similarly, hydrogenation of phthalates also needs refining treatments due to possible thermal decomposition side reactions. To address the technical issues of complicated protocols and high energy consumption, this invention provides an integrated method that eliminates the need for post-esterification impurity removal, but proceeds directly to the hydrogenation stage after water washing to remove the catalyst.

2. Main Content

The integrated method to prepare cyclohexane dicarboxylates (including dibutyl cyclohexane dicarboxylate, dioctyl cyclohexane dicarboxylate, diisooctyl cyclohexane dicarboxylate, dinonyl cyclohexane dicarboxylate, didecyl cyclohexane dicarboxylate, etc.) is described as follows:

(1) Esterification: The raw materials were introduced into the esterification unit for esterification with non-acid catalysts, preferably tetraisopropyl titanate or tetrabutyl titanate, to obtain the esterification product with an acid value lower than 0.5 mg KOH/g. The esterification product was then moved to the washing water tank to remove the catalyst. The organic phase was dehydrated and dried.

(2) Hydrogenation: The dried product in Step (1) was introduced into the hydrogenation reactor. The primary catalyst was covered with a protectant that used Ni as the active composition at 0.1–5 wt% to avoid the titanium compound residue in the esterification product covering the noble metal catalyst surface and decreasing the activity. The hydrogenation catalyst contains Pt, Pd, Ru, Rh, etc. Reaction conditions: 80–250 °C, 3.0–17.0 MPa, H_2/phthalate molar ratio 10–300, liquid volume space velocity 0.1–0.5 h^{-1}.

(3) Refining treatments: The product passed successively the high-pressure splitter, low-pressure splitter, and crude dealcoholization tower to remove the alcohol to <1wt%. Then, the product passed the base wash tank, water wash tank, and vacuum steam stripper to eliminate alcohol and low-boiling-point side product residues. Finally, the product entered the drying tower and decoloration tower to obtain the final product.

When the cyclohexane dicarboxylates were the target products, the esterification product from raw materials phthalic anhydride and isooctanol had an acid value lower than 0.5 mg KOH/g. After water washing, it entered the hydrogenation reactor with a composition: diisooctyl phthalate 10t/h, isooctanol 0.5t/h, and trace phthalic acid ignored. The hydrogenation reactor was loaded with the catalyst and a protectant; the protectant was Ni/Al_2O_3 with 2% Ni content; the catalyst was Ru/Pd. Technical parameters: 150 °C, 15 MPa, volume space velocity 0.1–0.5 h^{-1}, H_2/ phthalate molar ratio 600. The crude hydrogenation product entered the high-pressure splitter to separate the gas and liquid phases; the liquid composition: diisooctyl cyclohexane dicarboxylate 9.95 t/h, isooctanol 0.51 t/h, isooctyl (methyl-cyclohexane) carboxylate 0.03 t/h. The liquid flowed into the low-pressure splitter to separate out a small amount of gas side products, then the reduced pressure dealcoholization tower to dealcoholize until the isooctanol content was less than <1%, and then the base wash tower with 3% Na_2CO_3 solution, 90 °C, liquid/oil mass ratio 1:10. The washing water tower conditions: 90 °C, liquid/oil mass ratio 2:10. Then the product entered the steam distillation to be dealcoholized under conditions: feed material temperature 140 °C, bottom temperature 155 °C, top temperature 135 °C, steam/oil mass ratio 1:20. The upper product in the steam distillation tower was separated from the organic phase, merged with the crude dealcoholization tower top product, and flowed into the esterification reactor. The distillate at the bottom of the steam distillation tower was moved to the drying tower under conditions: raw material temperature 150 °C. The dehydrated product was decolored in the decoloration tower and then filtered to obtain the final product. Its physico-chemical properties are shown in Table 4.13.

3. Positive Effects

Compared with the current methods, the invention simplifies the post-esterification dealcoholization, base wash, steam distillation dealcoholization, drying, decoloration steps, with greatly decreased capital cost and energy consumption. In this integrated method, the hydrogenation unit is combined with the traditional plasticizer production, making it possible to switch between aromatic-bearing esters and cyclohexane-based esters. If one wants to prepare aromatic-bearing esters, using the single esterification unit can easily achieve it.

Table 4.13 Physico-chemical properties of product

Item	Property metrics
Appearance	Transparent, no visible impurity
DEHCH content/%	≥99.0
Chromaticity (Pt–Co)/Hazen	≤40
Density(20°C)/(kg /m^3)	954–956
Acidity (based on phthalic acid)/%	≤0.015
Water content/%	≤0.1
Flash point/°C	≥192

4.2.20 A Type of Efficient Plasticizer Hydrogenation Catalysts and their Preparation Method (2013)

1. Technical Background

DEHCH is a low-toxic, environmentally friendly plasticizer with similar or even better plasticization performance compared to DEHP. More importantly, DEHCH does not induce peroxisome hyperplasia or liver tumor and is biodegradable in nature. Thus, it has been approved for use in food packaging, medical equipment, and children's toys.

There are few reports on DOP hydrogenation, but the hydrogenation of structurally similar DINP can be used as a reference. The technique can be described as a continuous method: synthesis of DINP from phthalic acid and isononanol, then hydrogenation with an efficient catalyst. Currently, direct catalytic phthalate hydrogenation still suffers from issues such as difficult catalyst selection and high cost. For DOP hydrogenation to prepare DEHCH, only research reports are seen, but no scaled production is reported.

2. Main Content

This invention provides a type of low cost, active, selective, efficient hydrogenation catalysts. As the special features, the catalysts are composed of the primary active component, additive, and support. The primary active component is Fe and/or Ni at a mass fraction of 0.1–25%. The additive is Na, K, Ca, Cu, and their oxides at a mass fraction of 0.1–20%. The support is N-doped carbon nanotubes with a surface area 15–500m^2/g, pore size 5–200 nm, N/C atomic ratio 0.01–45.Preparation of catalysts: the nitrate salt of the active composition was mixed with N-doped carbon nanotubes, was dried naturally and then at 120 °C for 4 h, was naked at 350 °C for 5 h, and finally was activated in H_2 at 350 °C for 2 h.

Hydrogenation conditions: 200 °C, 7.0 MPa H_2 pressure, H_2/ester molar ratio 150, DEHP mass space velocity 0.5 h^{-1}, 20 h.

Analytical methods: HPLC, pure methanol as the eluent, flow rate 0.5 mL/min, injected sample volume 10 mL, 1362A differential refractive index detector, C_{18} column, column temperature 35 °C. The following results were obtained with different catalysts: (i) DEHP conversion 97.2%, DEHCH selectivity 98.0%; (ii) DBP conversion 96.8%, DBCH selectivity 95.6%; (iii) DINP conversion 89.6%, DINCH selectivity 98.5%; (iv) DINP conversion 92.5%, DINCH selectivity 94.1%; (v) DINP conversion 96.0%, DINCH selectivity 95.2%; (vi) DINP conversion 97.6%, DINCH selectivity 98.5%.

3. Positive Effect

The catalysts have advantages of low-cost active composition, widely accessible additive, better catalyst effects with the N-doped carbon nanotube support, simple preparation steps, low preparation temperature, high materials utilization, low wastewater discharge, energy saving, and environmental friendliness.

4.2.21 A Type of Non-linear Plasticizers and the Preparation and Applications Thereof (2013)

1. Technical Background

The safety of phthalate plasticizers has long been questioned. In recent years, these materials have been increasingly produced and have become a type of globally common environmental pollutants. On October 28, 2008, ECHA added 3 phthalates (DBP, DEHP, BBP) to the candidate SVHC list. The REACH prohibits the use of SVHC in offered products or materials with a total content <0.1%. Currently, the restrictions have applied to almost all phthalates, including but not limited to DBP, DEHP, BBP, DINP, DIDP, DNOP, DINP, DNP, DCHP, DHP, DIOP, DMP, DEP, DPRP, DPP, etc. China has put DMP, DBP, and DOP into the blacklist of environmental priority pollutants. Therefore, there is a need to provide the phthalate substitute plasticizers.

2. Main Content

Non-linear plasticizers are prepared via the esterification reactions between non-linear $C_{8\sim10}$ alcohols and cyclohexane-1,4-dicarboxylic acid.

Linear alcohols (straight-chain alcohols) are alcohols with linear or almost linear chains, such as n-heptanol, n-octanol, n-nonanol, etc. Non-linear alcohols (branched-chain alcohols) are alcohols with certain numbers of side chains or cyclic structures, such as isoheptanol, isooctanol, isononanol, isodecanol, etc.

Preferable preparation conditions: non-linear alcohol and cyclohexane-1,4-dicarboxylic acid at molar ratio (2–3):1, 130–220 °C, 4–5 h, finish point acid value $\leq$0.3 mg KOH/g. The non-linear alcohols can be isooctanol, isononanol, 2-propylheptanol. When the non-linear alcohol is isooctanol, the product is diisooctyl cyclohexane-1,4-dicarboxylate with molecular formula $C_{24}H_{42}O_4$ and molecular weight 394.6. When the non-linear alcohol is isononanol, the product is diisononyl cyclohexane-1,4-dicarboxylate with molecular formula $C_{26}H_{48}O_4$ and molecular weight 422.6. When the non-linear alcohol is 2-propylheptanol, the product is di(2-propylheptyl) cyclohexane-1,4-dicarboxylate with molecular formula $C_{28}H_{52}O_4$ and molecular weight 450.7.

The preparation method for non-linear plasticizers is as follows:

(1) The non-linear alcohol and cyclohexane-1,4-dicarboxylic acid were stirred in the reactor and heated.
(2) When the temperature reached 130 °C, the catalyst was added, followed by heating to 220 °C to undergo the esterification reaction under stirring.
(3) The acid value was measured. The reaction was stopped when the acid value was 0.3 mg KOH/g.
(4) The reaction mixture was neutralized with Na_2CO_3 and then was dealcoholized in vacuum.
(5) Deionized water was added to wash off the catalyst, followed by vacuum evaporation of water and filtration.

Table 4.14 Physical properties of non-linear plasticizers obtained from this invention

#	Molecular weight	Water content/%	Chromaticity/APHA	Purity/%	Heated loss/%
1	450.7	0.02	20	99.5	≤0.1
2	422.6	0.02	20	99.5	≤0.1
3	394.6	0.02	20	99.5	≤0.1

The catalyst can be single or several tetraisopropyl titanate (TPT), sulfuric acid, organotin (tin oxalate), or triphenylphosphine.

Non-linear plasticizers are widely used in PVC articles, dyes, inks, rubber or coatings.

The non-linear plasticizers obtained in this invention are shown in Table 4.14.

3. Positive Effecta

The non-linear plasticizers described in this invention have excellent evaporation-, heat-, and aging-resistances, contain no phthalate compounds or benzene ring, are non-toxic and environmentally friendly, and thus have wide applications. The preparation has advantages of a simple procedure, low cost, safety, and environmental friendliness.

4.2.22 A Type of Apparatus for Preparing Dioctyl Cyclohexane-1,4-Dicarboxylate (2013)

1. Technical Background

Plasticizers are everywhere in our daily, such as plastic containers, utensils, daily supplies, toys, and even food wrap films. Plasticizers are widely used in food packaging, medical products, paintings, etc. A tiny amount of phthalates such as DEHP also exist in the environment and drinking water. The plasticizer is the most used one among all plastics additives and the annual consumption is more than 8 million tons, 75% of which are phthalates with the best cost performance. However, phthalates tend to be released to the environment, which has been detected in air, water, soil and dust, and enter the human body via the digestive system, respiratory system, and skin to form environmental endocrine disruptors to adversely affect the reproductive and developmental systems. Such harms are hidden and long-lasting. The recent research focus is on phthalate exposure of pregnant rats and the intergenerational influences via placenta and lactation and also on the productive system of male offspring. Although there is no sufficient human toxicity data, keeping phthalates away from our life has become a global consensus.

2. Main Content

To overcome the drawbacks of current technologies, this invention provides a set of industrial-scale mass production facilities for preparing cyclohexane-1,4-dicarboxylates, which include the esterification kettle, refining kettle, filtration tank, and hydrogenation reaction unit.

The hydrogenation reaction unit includes a relay tank with a solution heater, H_2 compressor, and hydrogenation reactor connected to the relay tank. The H_2 compressor has an outlet connected to the catalyst bed at the bottom of the hydrogenation reactor and the other outlet connected to the recycled H_2 preheater and heater. The recycled H_2 heater is connected to the hydrogenation reactor, which is also connected to a high-pressure flash distillation tank. The flash distillation tank has an outlet connected to a 4.5 MPa separation tank and the other outlet connected to a water cooler for the reactor. The water cooler is connected to a recycled H_2 separation tank with an outlet connected to the H_2 compressor, an outlet to the boiler system and an outlet to the 4.5 MPa separation tank. The separation tank has an outlet connected to the boiler system and the second outlet to the low-pressure flash distillation tank. The low-pressure flash distillation tank has an outlet connected to the boiler system and the other outlet to the decolorization tank that is connected to the product cooler. The esterification kettle is equipped with a stirrer shaft with blades with uniformly distributed holes. The invented facilities are particularly applicable to industrial-scale mass production with low cost. During stirring, the blade holes make one part of liquid stirred by the stirring blades and one other part of liquid under different speeds to form turbulence resulting in stronger stirring and more reactions.

Industrial-scale mass production of dioctyl cyclohexane-1,4-dicarboxylate includes the following steps:

(1) Esterification: N_2 was introduced to the esterification kettle to replace the air inside. After the stirring started, octanol, catalyst, and terephthalic acid were fed successively at 1:3 acid/alcohol molar ratio. Then, the feeding port and N_2 were closed and the mixture was heated to 220–230 °C to undergo the esterification reaction. When the acid value reached below 1 mg KOH/g, the reaction was stopped and cooled down to 100 °C. The reaction product was vacuum transferred to the refining kettle.
(2) Neutralization and water washing: The stirrer in the refining kettle was turned on. The alkaline solution was added to the refining kettle to neutralize the esterification product. After fully mixed, the stirring was paused and the settled water layer was removed. When the acid value dropped below 0.05 mg KOH/g, the water formed in the esterification reaction and steam condensation was introduced to the refining kettle under stirring for 10 min. Then the stirring was paused and the settled water layer was removed.
(3) Refining: The vacuum pump in the refining kettle was turned on. When the vacuum reached above −0.09 MPa, the material was heated to 170–175 °C. Steam was introduced to start the steam distillation for 240 min. Samples were drawn to measure their flash points. When the flash point reached an appropriate value, the N_2 at the kettle bottom was turned on to dry the product for 30 min.

The water content and acid value of the product were measured; after they reached appropriate levels, the product was pumped to the filtration tank to be filtered.

(4) Hydrogenation: The filtered product was introduced into the relay tank, pressurized to 12.9 MPa, heated to 200 °C by the solution heater with 3.0 MPa steam, and entered the hydrogenation reactor. One part of the recycled deoiled H_2 from the H_2 compressor was heated to 180 °C in the preheater, then heated to 200 °C by the solution heater with 3.0 MPa steam, and then entered the hydrogenation reactor; the other part as cold H_2 passed through the catalyst bed at the reactor bottom. The gas and liquid in the hydrogenation product were separated. The liquid was cooled to 100 °C, decolorized in the decolorization tank, and cooled down to 40 °C by the product cooler to obtain the final product dioctyl cyclohexane-1.4-dicarboxylate.

3. Positive Effect

The product obtained from this invention has excellent toxicological characteristics, good PVC compatibility, low-temperature performance, no odor, non-toxicity, and environmental friendliness. The facilities and method of this invention are applicable to industrial-scale mass production with low cost and high efficiency.

4.2.23 *The Esterification Reaction Protocols for DOTP Plasticizer (2015)*

1. Technical Background

Plasticizers are substances that, when added, increase the plasticity, mobility, and flexibility of polymeric materials such as rubber, plastics, coatings, etc. They are indispensable additives for the plastics industry. In 2015, the annual production capacity of plasticizers in China reached 2.8 million tons. China has become the first plasticizer production and consumption country in the world, which still keeps increasing. Among them, phthalates and terephthalates are >90% of the total plasticizers. However, phthalate plasticizers have potential carcinogenic risks. Green plasticizers such as epoxidized oils and citrates suffer from low production volume and are difficult to satisfy the large market need. Therefore, in the near future terephthalate plasticizers are still the major products in the market. But the traditional intermittent production method for dioctyl terephthalate (DOTP) has drawbacks of low production rate, high energy consumption, low asset utilization, and unstable product quality.

2. Main Content

The DOTP production facilities described in this invention include the esterification kettle, buffer tank, esterification tower, condenser, alcohol-water separation tank, drain solenoid valve, secondary sedimentation separation tank, and neutralization and water washing tank. The esterification kettle is connected to the feeding port via the feeding valve at its top. The esterification tower, U-shaped tube, and alcohol-water separation tank are connected successively. An interface indicator is installed on the alcohol-water separation tank.

The DOTP plasticizer esterification reaction protocol is described below:

(1) Terephthalic acid, octanol, and titanate catalyst were added to the esterification kettle and heated under stirring. Inert gas was introduced. The reaction temperature and pressure were carefully controlled at 250–265 °C and 0.2–0.7 MPa. The alcohol/acid mass ratio was (2.4–2.8):10. The mass of titanate catalyst was 2–5% of the total mass of terephthalic acid and octanol. Axial blades were used on the stirrer in the kettle to enable better material mixing. An inner coil was used to heat the reaction mixture.
(2) After condensation, the alcohol and water vapors produced during the reaction process were guided into the alcohol-water separation tank to be separated. The alcohol overflew via the U-shaped tube to the esterification tower and counter-flew with the newly formed alcohol and water vapors. The refined alcohol entered the esterification kettle for recycled use.
(3) The water from the alcohol-water separation tank was guided to the neutralization and water washing tank for recycled use.
(4) The reaction product from the esterification kettle was sent to the relay tank and then treated in the refining section to obtain the plasticizer product. The relay tank has a volume of 70 m^3 to ensure enough space for the product to facilitate the esterification reactor ready for the next run.

3. Positive Effect

Compared to current technologies, the invention results in reduced capital cost, high operation efficiency, high conversion, low energy consumption, high asset utilization, and stable product quality.

4.2.24 A Method for Preparing Cyclohexane-1,2-Dicarboxylates (2015)

1. Technical Background

Plasticizers are one of the most produced and consumed plastics additives in the world. Among them, traditional phthalates are about 80% of the total plasticizer production, but their suspected carcinogenicity and toxicity have raised close attention globally. Affected by plasticizer pollution, toxic capsule plasticizer and food wrap film plasticizer, toxic toy incidents, environmentally friendly plasticizer is the future direction for plastics additives. China is also increasing the support for the R&D of environmentally friendly plasticizers.

Cyclohexane-1,2-dicarboxylates are a type of non-phthalate environmentally friendly plasticizers that have similar structure and properties, better performance, and more importantly, superior toxicological characteristics. However, China has almost zero production of cyclohexane-1,2-dicarboxylates and relies heavily on importation. China is the major phthalate production country. Thus, ring hydrogenation of phthalate to produce cyclohexane-1,2-dicarboxylates is of significant importance to the plasticizer upgrading, which is an ideal solution for phthalate plasticizer producers to meet the safety requirements.

The relative content of cyclohexane-1,2-dicarboxylates cis–trans isomers largely affects the product quality. Specifically, trans-structured cyclohexane-1,2-dicarboxylate plasticizers can easily migrate to form an oil film on article surfaces and decrease the product quality. Thus, the lower content of trans-structured cyclohexane-1,2-dicarboxylates, the better. The formation of trans-structured cyclohexane-1,2-dicarboxylates should be avoided during preparation.

2. Main Content

This invention provides a method for preparing cyclohexane-1,2-dicarboxylates.

(1) With the presence of the first catalyst, phthalate, solvent, and H_2 were mixed in the hydrogenation reactor to obtain the first gas–liquid mixture fluid.
(2) The gas and liquid phases in the first fluid were separated.
(3) Through the nano-sized holes, H_2 was introduced into the liquid hydrogenation reactant to obtain the second gas–liquid mixture fluid.
(4) With the presence of the second catalyst, the second mixture fluid was treated in the hydrogenation reactor at 80–95 °C and 4–7 MPa to obtain the third gas–liquid mixture fluid.
(5) The gas and liquid phases in the third fluid were separated.

The above technical protocol has the following advantages:

(1) Because of the high H_2 consumption and heat generation in phthalate hydrogenation reactions, this invention uses the following measures to rapidly dissipate the generated heat to reduce the catalyst bed temperature and thus side reactions: (i) the raw materials use the alcohol or cyclohexane-1,2-dicarboxylate solution of phthalates, *i.e.*, solvent dilution; (ii) the hydrogenation reactor is an isothermal one.
(2) Before the hydrogenation reaction, nano-sized holes are used to highly disperse and dissolve H_2 in the liquid reaction mixture to improve the gas–liquid mass transport and expedite the reaction rate.
(3) The upward flow method is used to feed the raw materials to avoid bias flow, wall flow or channel flow and make full use of catalyst activity to ensure complete phthalate hydrogenation.
(4) The two-stage hydrogenation method can decrease the intensity of reactions and increase the selectivity via controlling reaction process, dissipating generated heat, and avoiding over-concentrated heat.

(5) The addition of heavy rare earth additives can induce the lattice distortion in the Rh and/or Ru nanocrystals and increase the irregularity and dispersity of crystal structures, leading to enhanced activity for the first and second catalysts. In addition, the relatively strong electron-donating property of heavy rare earth elements can increase the surface electron density of Rh and/or Ru, promote the adsorption of phthalates, reduce the activation energy of hydrogenation reactions, and improve the activity of the first and second catalysts. Therefore, the first and second catalysts are featured with low-temperature high activity.
(6) Due to the mild hydrogenation conditions, the cis-product content is high, which is advantageous for applications of downstream products.

3. Positive Effect

Compared to current technologies, the invention can achieve 100% phthalate conversion, >99.5% cyclohexane-1,2-dicarboxylates selectivity, >95.7 wt% cis-product under lower reaction temperature, pressure and H_2/ester volume ratio.

4.2.25 A Method for Preparing Environmentally Friendly Plasticizers (2015)

1. Technical Background

Plasticizers are widely applied not only in rubber, plastic thin film, plastic articles, plastic materials, packaging material industries, but also as organic synthesis intermediates and solvents, surfactants, biochemicals, etc. They are involved in multiple interdisciplinary fields and have deep impacts on many industries. In the social needs for energy saving and environmental protection, green plasticizers are a necessary clean production option to adjust to the future technological directions, economic demands, and market development.

2. Main Content

This invention provides a renewable, stable, cost-effective, non-toxic method for preparing green, environmentally friendly plasticizers, including the following steps:

(1) Raw materials preparation: citrate 15–20 parts, bio-plasticizer 10–13 parts, glycol derivative 3–6 parts, phthalic anhydride 30–35 parts, aliphatic acid ester 2–6 parts, catalyst 3–8 parts, stabilizer 4–10 parts. The stabilizer was a 3:1 (wt) mixture of epoxy compound and sodium carbonate.
(2) Stabilizer preparation: the epoxy compound and sodium carbonate at 3:1 were well mixed and maintained at 120 °C for 1.5 h with intermediate-speed stirring.
(3) The mixture of citrate, bio-plasticizer, glycol derivative, phthalic anhydride, aliphatic acid ester, and stabilizer was introduced in a flask equipped with a thermometer, addition funnel, and water trap. The mixture was stirred at a low speed and heated to 60–70 °C. The catalyst was added via the addition funnel.

The mixture was then heated to 120–140 °C and transferred to a closed rubber mixing machine for 1.5 h.

(4) The dough mixture was cooled naturally to room temperature and then placed at 0–4 °C for 20–40 min to obtain the green plasticizer.

The citrate was tributyl citrate, tripentyl citrate, trinonyl citrate, tributyl acetyl citrate, or trihexyl acetyl citrate. The stabilizer was one or several of ZnO, MgO, and CaO. The glycol derivative was bioglycerin or sodium triglyceride. The aliphatic acid ester was sucrose fatty acid polyester. The catalyst was TiO_2. The epoxy compound was epoxidized linseed oil or ESO. The low-speed stir rate was 200–300 r/min and the intermediate-speed stir rate was 600–800 r/min.

3. Positive Effect

Green plasticizers prepared by the invention have the advantages of facile operations, high industrialization level, strong replicability, renewability, stable quality, low cost, and non-toxicity. They have high plasticization performance with strong practical significance, promising to meet the future development trend.

4.2.26 A Method for Preparing Hexahydrophthalic Anhydride (2016)

1. Technical Background

Hexahydrophthalic anhydride (HHPA) is an indispensable raw material used for producing polyester-based high-end coatings. Using HHPA to substitute aromatic polycarboxylates in producing amino alkyd paint, significantly improves the coating brightness and gloss, decreases the viscosity, increases solid content, and particularly enhances the weather-resistant property. HHPA is also a raw material used for producing plasticizers, pesticides, corrosion inhibitors, etc. With the fast development of the IT industry, there have been increasing quality and quantity demands for semiconductor molding and potting materials. HHPA has been widely used in this field as an epoxy curing agent.

2. Main Content

This invention provides a method for preparing HHPA to make reasonable use of phthalic anhydride. To address the issues such as the easy formation of phthalide and complicated hydrogenation products, efficient yolk-shell PdRu catalysts were developed to achieve benzene ring hydrogenation. The resource utilization of phthalic anhydride was achieved through combined solvent recycling and refining to convert it to a high value-added chemical HHPA. In addition, the yolk-shell Pd–Ru catalysts improved the product yield. The technical protocol is described below.

The preheated (50 °C) phthalic anhydride was added to the dissolution tower and dissolved with a suitable solvent at 50–70 °C. The pressurized phthalic anhydride solution was added to the selective hydrogenation tower loaded with the catalyst. The hydrogenation product was guided into the solvent recycling tower. The dioxane evaporated at the top of the solvent recycling tower was recycled as the solvent. The material at the tower bottom was transferred to the HHPA refining tower. Pure HHPA was obtained at the tower bottom. A mixture of phthalic anhydride, 2-methylbenzoic acid, and 2-methyl cyclohexane carboxylate was obtained from the tower top, which was recrystallized to acquire phthalic anhydride.

Yolk-shell Pd–Ru catalyst was used to hydrogenate the benzene ring in phthalic anhydride. The catalyst support was one or more of Al_2O_3, MgO, ZnO, TiO_2, prepared via the rapid precipitation method. The mass fractions of Pd and Ru were 0.01–0.1% and 0.8–1.5%, respectively.

The selective hydrogenation tower was a constant pressure reaction kettle or fixed-bed reactor. When the former was used, hydrogenation conditions were: 80–140 °C, 6–12 MPa H_2 pressure, reaction time 2–10 h, HHPA/catalyst mass ratio (2–50):1. When the latter was used, hydrogenation conditions were: 120–180 °C, 6–18 MPa H_2 pressure, space velocity 0.2–0.6 h^{-1}.

After hydrogenation, the materials entered the solvent recycling tower with the tower top temperature 100–110 °C, 0.1 MPa pressure, theoretical plate number 9, reflux ratio 0.03.

The HHPA refining tower had the tower top temperature 260–290 °C, 0.1 MPa pressure, theoretical plate number 200, reflux ratio 2.6. The mixture of phthalic anhydride, 2-methylbenzoic acid and 2-methyl cyclohexane carboxylate obtained from the tower top was recrystallized twice to acquire phthalic anhydride at 0 °C.

The solvent was one or more of dioxane, THF, cyclohexane, and decahydronaphthalene. The phthalic anhydride concentration was 5–20%.

3. Positive Effect

This invention has the advantages of the high value-added product and broad applications, thus having good economic benefits and promising industrial applications.

4.2.27 A Method for Preparing Dioctyl Cyclohexane-1,4-Dicarboxylate (2016)

1. Technical Background

Because of their unique performance, DOP and DOTP plasticizers play an important role in manufacturing PVC articles and applications. However, phthalate plasticizers have inferior heat-, UV- and cold-resistant properties. Numerous animal tests show that long-time contact can induce peroxisome hyperplasia and liver tumors. Many countries and regions have restricted or prohibited their uses in direct human contact plastic articles such as medical supplies, food, children's toys.

Cyclohexane dicarboxylates, such as dioctyl cyclohexane-1,4-dicarboxylate, have demonstrated better PVC plasticization performance; also, no toxicity or genotoxicity has been observed. Thus, they are promisingly applicable not only to sensitive soft PVC products such as food wrap films, food sealing gaskets, and sports and leisure products, but also to toys for children under three, domestic food contact gloves, environmental wallpaper, decorative sheet materials, intravenous infusion tubing, medical blood bags. Therefore, the synthesis of cyclohexane dicarboxylates has received wide attention.

2. Main Content

The goal of this invention is to simultaneously improve the reaction conversion and product selectivity through using acidic molecular sieves or acidic salt catalysts at suitable reaction temperatures.

This invention provides a method for preparing dioctyl cyclohexane-1,4-dicarboxylate via the esterification reaction between cyclohexane-1,4-dicarboxylic acid and octanol with the acidic salt catalyst at 120–200 °C. The octanol can be n-octanol or isooctanol and the acidic salt can be sodium hydrogen sulfate or potassium hydrogen sulfate. In a specific example, the acid/alcohol mass ratio was 1:(1.6–10), preferably 1:(1.7–3). The acidic salt catalyst was at the 0.1–15.0 wt%, preferably 1–10 wt%, more preferably 1–2 wt%, of cyclohexane-1,4-dicarboxylic acid. Too high catalyst loading had minimal effects on the reaction, but too low loading had an effect on the conversion ratio. The acidic salt catalyst-based esterification reaction was operated at 130–180 °C, preferably 140–160 °C.

This invention provides an additional method for preparing dioctyl cyclohexane-1,4-dicarboxylate via the esterification reaction between cyclohexane-1,4-dicarboxylic acid and octanol with acidic molecular sieves catalyst at 120–200 °C. The octanol can be n-octanol or isooctanol. The molecular sieves can be Y-type, β-type, or ZSM-5 zeolites.

When the acidic molecular sieves catalyst is used, the acid/alcohol mass ratio was 1:(1.6–10), preferably 1:(1.7–3). The acidic salt catalyst was at the 0.1–15.0 wt%, preferably 1–10 wt%, more preferably 1–2 wt%, of cyclohexane-1,4-dicarboxylic acid. In a specific example, the reaction temperature was 130–180 °C, preferably 140–160 °C.

This invention uses special catalysts to catalyze the esterification reactions under special temperatures to achieve desirable yields of dioctyl cyclohexane-1,4-dicarboxylate. In addition, the acidic molecular sieves and acidic salt catalysts are heterogeneous catalysts and can be directed filtered to separate and recycle without a need of other chemical treatments, making the process green, environmentally friendly, safe, and energy saving. Therefore, the method described in this invention has advantages of fast reaction rate, high reactant conversion, good product selectivity, recyclability of catalysts, environmental friendliness, etc.

3. Positive Effect

The reactions are fast with simple protocols. The catalysts are cost-effective and readily accessible with excellent recyclability, leading to greatly decreased capital cost. The post-treatments are simple with reduced wastewater discharge.

4.2.28 A Method for Preparing the Catalysts used for Preparing Cyclohexane Dicarboxylates (2016)

1. Technical Background

Cyclohexane dicarboxylates have wide applications including uses as polymer modification materials and environmentally friendly plasticizers. Dimethyl cyclohexane-1,2-dicarboxylate is an important organic chemical industry raw material for coating production and polymer modification, as an intermediate to prepare cyclohexane-1,2-dimethanol, and as fabric or resin modification materials to afford excellent thermal stability, thermoplasticity, and mechanical strength. Diisononyl cyclohexane-1,2-dicarboxylate (DINCH) is an environmentally friendly plasticizer. Common plasticizers are phthalates, but cyclohexane dicarboxylates have similar plasticization performance and better toxicological characteristics with no reproductive toxicity or peroxisome hyperplasia. Thus, they can be used in sensitive materials such as food packaging, medical equipment, and toys.

Currently, the preparation of cyclohexane dicarboxylates is through hydrogenation of phthalates using Pd-based catalysts enabling selective benzene ring hydrogenation yet with relatively low catalytic activity, harsh reaction conditions, and complicated catalyst preparation protocols.

2. Main Content

The goal of this invention is to design a simple method for preparing the catalysts for use to synthesize cyclohexane dicarboxylates. The obtained catalysts have high activity and stability and can enable mild reaction conditions and high selectivity.

The method for preparing the catalysts includes the following steps:

(1) Preparation of noble metal precursor solution: $RhCl_3$ and $PdCl_2$ at a molar ratio of 1:(0.1–10) were dissolved in an alcoholic solvent at a noble metal concentration of 0.003–0.015 mol/L. The solvent was one or two of ethanol, ethylene glycol or glycol.

(2) Microwave reduction: The noble metal precursor solution was added to activated carbon support. The loading of activated carbon support was 1–10wt%. The microwave reduction time was 10–120 s at the microwave frequency 2450 MHz and power 800 W.

(3) Cooling and drying: After the microwave reduction, the mixture was cooled to room temperature, washed with deionized water, and dried in a vacuum oven to obtain the catalysts.

The cyclohexane dicarboxylate preparation method includes the following steps:

To a PTFE lining were added phthalate, alcoholic solvent, and catalyst. The phthalate/alcohol mass ratio was 1:(0–15). The noble metal/phthalate mass ratio was 1:(40–200). Then the PTFE lining was placed in a high-pressure autoclave and the gas tightness was checked. H_2 was used to replace the air inside the autoclave for 3–5 times and then was pressurized to 1.0–4.0 MPa. The autoclave was heated to 50–160 °C and maintained at this temperature for 1–8 h with the stir rate of 800–1500 r/min and 4–5 MPa H_2 pressure. After cooling to room temperature, the mixture was filtered to obtain the cyclohexane dicarboxylate.

The phthalate was one of dimethyl terephthalate, dimethyl phthalate or diisononyl phthalate. The corresponding product was dimethyl cyclohexane-1,4-dicarboxylate, dimethyl cyclohexane-1,2-dicarboxylate or diisononyl cyclohexane-1,2-dicarboxylate.

Here is an example. To a PTFE lining were added dimethyl phthalate and catalyst. The noble metal/phthalate mass ratio was 1:150. Then the PTFE lining was placed in a high-pressure autoclave and the gas tightness was checked. H_2 was used to replace the air inside the autoclave for 3–5 times and then was pressurized to 2.0 MPa. The autoclave was heated to 50 °C and maintained at this temperature for 1 h with a stir rate of 1500 r/min. After cooling to room temperature, the mixture was filtered to obtain dimethyl cyclohexane-1,2-dicarboxylate. Table 4.15 lists the Experiments 1~ 16 test results of the activity and selectivity of catalysts.

3. Positive Effect

The microwave reduction method for preparing supported double metal catalysts has simple protocols, technical stability, and repeatability. When used to prepare cyclohexane dicarboxylates, the catalysts enable mild reaction conditions, short reaction time, high reactant conversion and product selectivity, and hundred times of repeated uses.

Table 4.15 Experiments 1~ 16 catalyst activity and selectivity test results

#	Reactant conversion/%	Product selectivity/%	#	Reactant conversion/%	Product selectivity/%
1	98.8	98.5	9	97.6	97.1
2	100	98.7	10	99.5	98.6
3	96.6	97.4	11	96.3	98.5
4	95.4	98.2	12	96.6	97.5
5	96.3	98.5	13	95.1	97.2
6	100	98.6	14	99.6	98.2
7	95.0	96.4	15	94.6	98.1
8	95.3	97.5	16	94.0	97.0

Note 1~8: dimethyl terephthalate reactant and dimethyl cyclohexane-1,4-dicarboxylate product; 9–12: DMP reactant and DMCH product; 13~ 16: DINP reactant and DINCH product

4.3 Conclusion

1. Foreign Inventions and Patents

In 1998, BASF (Germany) filed a patent in China titled "Method for hydrogenating benzene polycarboxylic acids or derivatives thereof by using a catalyst containing macropores" [1]. This invention used supported Ru catalysts in a vertical steel high-pressure tubular fixed-bed reactor under conditions of average temperature 125 °C, 20 MPa H_2 pressure, phthalate flow rate 0.45 kg/h, 20% excess H_2. A single hydrogenation reaction achieved 99.5% conversion and 99.2% selectivity. After the second hydrogenation (99.5% cyclohexane dicarboxylate flow rate 1 kg/h), no phthalate was detected with 99% product selectivity. The 1% low-boiling-point side products were removed by steam distillation at 170 °C and 5 kPa pressure to obtain 99.7% pure product.

During 2001–2011, USA etc. filed a number of patents, among which were the hydrogenation methods including US6284917 [2], US6888021 [3], WO2004/046078 [4], US7893295 [5], US7319161 [5], US7361714 [6], US7355084 [7], US7632961 [8], US7786201 [9]. The hydrogenation methods had high yields and good quality, but suffer from high capital and operational costs. Only BASF achieved industrialization.

The esterification methods included US6310235 [10], which used cyclohexane dicarboxylic acid and alcohol as raw materials and Ti-, Zr- or Sn-based metal catalysts to obtain the products via single esterification at 120–160 °C, second esterification at 250 °C, neutralization, dealcoholization, drying, filtration, etc.

2. Domestic Inventions and Patents

In 2001, Peizi Lin et al. from Sinopec Tianjin Branch published a patent titled "A catalyst for hydrogenating dimethyl terephthalates to dimethyl cyclohexane-1,4-dicarboxylates and the preparation thereof" [11]. The catalyst was composed of the active composition, additive, and support. The active composition was metallic Pd. The additive was from Group IIA metals (Ca, Ba, Mg, etc.) and Group VIII metals (Rh, Ru, Ir, Pt, etc.). The support was Al_2O_3. The additives improved the catalyst performance and obviously decreased the reaction pressure.

In 2006, Songyue Xie et al. from Nan Ya Plastics Corporation published a patent titled "A plasticizer containing cyclohexane multi-carboxylate and its preparation method" [12]. Cyclohexane multi-carboxylates were synthesized via single-step high-temperature esterification reactions between cyclohexane multi-carboxylic acids or derivatives and linear or branched alcohols having 4–13 carbon atoms. The cyclohexane multi-carboxylic acid or derivative used 1 mol and the alcohol uses 2.2–3.8 mol under conditions of 200–250 °C, reaction time 5–8 h, reaction pressure 0.5–76 kPa, and with the catalyst. The formed water and unreacted alcohol were removed via azeotropic distillation at 90–180 °C. The acid value of the reaction mixture was measured as 1 mg KOH/g. 9–16wt% NaOH solution was used to neutralize the mixture until the acid value became 0.08 mg KOH/g. Then, the mixture was distilled till the alcohol content reached below 300 ppm, dried, filtered, and purified below 100 °C to obtain 99.8% pure cyclohexane multi-carboxylates.

In 2008, Yunjie Ding et al. from the Chinese Academy of Science, Dalian Institute of Chemical Physics published two patents:

(1) A catalyst for preparing cyclohexane-1,2-dicarbolxylates [13]
(2) A method for preparing cyclohexane-1,2-dicarboxylates [14]

The esterification reaction used Pd-Ru-ZrO_2/Al_2O_3 catalyst and trickle bed reactor under 200 °C, 5.0 MPa, H_2/ester molar ratio 150, DINP mass space velocity 0.5 h^{-1}, 12 h to achieve 100% DINP conversion and DINCH selectivity.

In 2009 and 2011, Jiping Shang et al. from Hangzhou Lin'an Shangtong Plasticizer Co., Ltd. published two patents:

(1) A method for preparing environmentally friendly di(2-ethylhexyl) cyclohexane-1,2-dicarboxylate (DEHCH) (2009) [15]. Preparation of DEHCH used the esterification reaction between cyclohexane-1,2-dicarboxylic anhydride and 2-ethylhexanol with the catalyst at 185–220 °C for 3–4 h. The reaction mixture was dealcoholized to remove the excess alcohol, neutralized with aqueous NaOH solution at 85–95 °C, washed with deionized water at 90–95 °C, steam-distillated for 3–4 h, and finally decolorized with added activated carbon for 20–30 min to obtain >99.5% pure DEHCH.
(2) A method for preparing environmentally friendly di(2-ethylhexyl) cyclohexane-1,2-dicarboxylate (DEHCH) plasticizer (2011) [16]. A mixture of 1 mol hexahydrophthalic anhydride (HHPA), 2.0–2.2 mol 2-ethylhexanol, and 0.4–0.6wt% solid superacid SO_4^{2-}/TiO_2-Al_2O_3 catalyst was esterified at 210 °C for 1 h, steam-distilled under reduced pressure for 3–4 h, and decolorized with activated carbon for 20–30 min to obtain the purified di(2-ethylhexyl) hexahydrophthalate.

In 2010 and 2012, Pingping Jiang et al. from Jiangnan University published two patents:

(1) A method for preparing cyclohexane-1,2-dicarboxylates using organic acid catalysts (2010) [17]. Esterification reactions were performed using hexahydrophthalic anhydride (HHPA) and C_8 or C_9 alcohol at a moar ratio of 1:3, together with methanesulfonic acid or sulfamic acid at 0.05–0.5wt% of the total raw materials, at 160–230 °C for 3–7 h. The acid value was lower than 2 mg KOH/g. The mixture was decolorized with 5wt% activated carbon or hydrogen peroxide at 70 °C for 1 h and neutralized with NaOH solution until weakly alkaline (pH = 8–9) to remove the unreacted HHPA and formed cyclohexane monoacid monoester. The organic layer was washed with water 2–3 times to neutral and then dried to obtain cyclohexane-1,2-dicarboxylates in >99.5% purity.
(2) A method for preparing cyclohexane-1,2-dicarboxylates using ionic liquid catalysts (2010) [18]. The esterification reaction proceeded with cyclohexane-1,2-dicarboxylic acid and n-butanol as raw materials and ionic liquid (0.5–2wt%) as the catalyst in the N_2 atmosphere at 110–150 °C for 5–8 h until the acid value dropped below 3 mg KOH/g. The reaction mixture was neutralized with 5wt% NaOH solution at 70 °C until weakly alkaline pH (8–9)

to remove the unreacted HHPA and form cyclohexane monoacid monoester, washed with water 2–3 times to neutral, vacuum distilled to remove n-butanol and water, and further vacuum distilled at 230–250 °C and 0.09 MPa to obtain the product distillate, followed by removal of colored substances, to obtain dibutyl cyclohexane-1,2-dicarboxylate with a chromaticity (Pt–Co method) of 20–100Hazen, purity of 99.6% and yield of >98%.

In 2012, Xuekuan Li et al. from the Chinese Academy of Science, Shanxi Institute of Coal Chemistry published two patents:

(1) A type of catalysts for hydrogenating benzenecarboxylates to cyclohexanecarboxylates and the preparation method thereof (2012) [19].
(2) A type of benzenecarboxylate hydrogenation catalyst and the preparation and application thereof (2012) [21].

Under conditions of 40% Ni-1% Ca-10%Mg/SiO_2 catalyst, fixed-bed reactor, 180 °C, 4 MPa H_2 pressure, phthalate space velocity 4 h^{-1}, and H_2/ester molar ratio 60, the phthalate conversion and cyclohexane-1,2-dicarboxylate selectivity reached 100%. Under conditions of 40% Ni-1% Ca-10%Mg/SiO_2 catalyst at 9% loading, high-pressure autoclave reactor, 160 °C, 2 MPa H_2 pressure, and 15 h, the phthalate conversion and cyclohexane-1,2-dicarboxylate selectivity reached as high as 100%.

In 2012, Jigang Luo et al. from Fushun Xinrui Catalyst Co., Ltd. published a patent titled "Catalysts for hydrogenating benzenecarboxylate to cyclohexanecarboxylates and the preparation and application thereof" [20]. Under conditions of supported Ni catalyst, continuous fixed-bed reactor, 150–200 °C, 3.0–15.0 MPa H_2 pressure, volume space velocity 0.1–0.5 h^{-1}, and H_2/feedstock volume ratio (300–1000):1, the phthalate conversion and cyclohexane-1,2-dicarboxylate selectivity reached 100%.

In 2012, Zhigang Qiu et al. from Jiangsu Kangheng Chemical Co., Ltd. published a patent titled "A method for preparing diisooctyl cyclohexane-1,2-dicarboxylate plasticizer" [21]. Under conditions of Ru/C catalyst at 5wt% loading, 1L high-pressure autoclave, diisooctyl phthalate as raw material and isooctanol as the solvent with a mass ratio of 1:2.5, 170–180 °C, 6–7 MPa H_2 pressure, 6–8 h, the selectivity reached >98%. Then, the product was distilled under high vacuum to obtain diisooctyl cyclohexane-1,2-dicarboxylate with 96% yield and >99% purity.

In 2012–2015, Xiaoguo Li, Haibin Yu, Peng Gao et al. from China National Offshore Oil Corporation, Tianjin Chemical Research and Design Institute published 7 patents:

(1) A method for preparing the catalysts for hydrogenation-based cyclohexane-1,2-dicarboxylates (2012) [22]. Under conditions of supported Ru and/or Pd catalyst, 20 mL fixed-bed reactor, 130 °C, 3 MPa, volume space velocity 0.5 h^{-1}, and H_2/oil volume ratio 1000:1, the phthalate conversion and cyclohexane-1,2-dicarboxylate selectivity reached >99.9%.
(2) 4.2.14 A method for preparing diisooctyl cyclohexane-1,2-dicarboxylates via hydrogenation (2012) [23]. Under conditions of Pd-Ru/Al_2O_3 catalyst, trickle bed reactor, 175 °C, 15 MPa, phthalate volume space velocity 0.4 h^{-1}, and

H_2/ester molar ratio 600, and 24 h, the phthalate conversion and cyclohexane-1,2-dicarboxylate selectivity reached >99.9%.

(3) A method for preparing cyclohexane-1,2-dicarboxylates via hydrogenation (2012) [24]. Under conditions of supported Ru and/or Pd catalyst, Al_2O_3 support, fixed-bed reactor, 120–220 °C, 6–16 MPa H_2 pressure, phthalate volume space velocity 0.2–0.8 h^{-1}, and H_2/phthalate volume ratio (1000–400):1, the phthalate conversion and cyclohexane-1,2-dicarboxylate selectivity reached >99.9%.

(4) A method for preparing environmentally friendly cyclohexane-1,2-dicarboxylates (2013) [25]. Under conditions of supported Ru and/or Pd catalyst, Al_2O_3 or SiO_2-Al_2O_3 support, 20 mL fixed-bed reactor, 130 °C, 3 MPa, volume space velocity 0.5 h^{-1}, and H_2/oil volume ratio 1000:1, the conversion ratio, selectivity and yield reached >99.1%, >99.5%, and >99.4%, respectively.

(5) A method for preparing phthalate selective hydrogenation catalysts (2013) [26]. Under conditions of supported Pd-Ru/SiO_2-Al_2O_3 catalyst, fixed-bed reactor, 140 °C, 15 MPa, H_2/ester molar ratio 15, volume space velocity 0.4 h^{-1}, and 24 h, the phthalate conversion, and cyclohexane-1,2-dicarboxylate selectivity reached >99.9%.

(6) A method for preparing phthalate hydrogenation catalysts (2015)[27]. Under conditions of Ru catalyst on amorphous SiO_2 or Al_2O_3 or both, and fixed-bed reactor, the phthalate conversion, and cyclohexane-1,2-dicarboxylate selectivity reached >99%.

(7) An integrated method for preparing cyclohexane-1,2-dicarboxylates (2015). The integrated method includes esterification, hydrogenation, high-pressure separation, crude dealcoholization, base wash, water wash, steam distillation, and drying and decolorization. (i) Esterification: with non-acid catalysts, preferably tetraisopropyl titanate or tetrabutyl titanate, phthalic anhydride, and isooctanol as raw materials. The esterification product had an acid value below 0.5 mg KOH/g, which, after water washing, entered the hydrogenation reactor; (ii) Hydrogenation: with Ru/Pd catalyst, 150 °C, 15 MPa H_2 pressure, volume space velocity 0.3 h^{-1}, and H_2/phthalate ratio 600; (iii) High-pressure separation: the hydrogenation products containing diisooctyl cyclohexane-1,2-dicarboxylate, isooctanol, isooctyl (methylcyclohexane) carboxylate were separated into the gas phase and liquid phase by the high-pressure splitter; (iv) Crude dealcoholization: the liquid phase was dealcoholized in the dealcoholization tower until the isooctanol content was below 1wt% and was then guided into the base wash kettle; (v) Base wash: with 3% Na_2CO_3 solution, aqueous/oil mass ratio 1:10, 90 °C; (vi) Water wash: water/oil mass ratio 2:10, 90 °C; (vii) Steam distillation: raw material temperature 140 °C, tower bottom temperature 155 °C, tower top temperature 135 °C, steam/oil mass ratio 1:20. The upper product in the steam distillation tower was separated from the organic phase, merged with the crude dealcoholization tower top product, and flowed into the esterification reactor. The distillate at the bottom of the steam distillation tower was moved to the drying tower; (viii) Drying and decolorization: 150 °C. The dried material was decolorized in the decolorization kettle and filtered to

obtain the final product. The physico-chemical properties of the product were measured.

In 2013, Jinxian Shen et al. from Zhejiang Weibo Chemical Technology Co., Ltd. published a patent titled "A type of efficient plasticizer hydrogenation catalysts and their preparation method". The supported catalyst used the active composition of Fe and/or Ni, the additive of Na, K, Ca, Cu and their oxides, the support of N-doped carbon nanotubes or macroporous Al_2O_3 modified with N-doped carbon nanotubes. Under conditions of a fixed-bed reactor, 200 °C, 7.0 MPa H_2 pressure, H_2/ester molar ratio 150, mass space velocity 0.5 h^{-1}, and 20 h, the DEHP conversion and DEHCH selectivity was 97.2% and 98.0%, respectively; the DBP conversion and DBCH selectivity were 96.8% and 95.6%, respectively; the DINP conversion and DINCH selectivity were 89.6%–97.6% and 94.1%–98.5%, respectively.

In 2013, Zhipeng Lin et al. from Foshan Shunde Tiansheng Trade Co., Ltd. published a patent titled "A type of non-linear plasticizers and the preparation and applications thereof". Non-linear plasticizers were synthesized using non-linear alcohols (branched-chain alcohols, such as isoheptanol, isooctanol, isononanol, isodecanol) as raw materials. Under conditions of single or mixed tetraisopropyl titanate (TPT), sulfuric acid, organotin (tin oxalate), or triphenylphosphine catalyst, alcohol/acid molar ratio (2–3):1, 120–200 °C, and 4–5 h, the end acid value was $\leq$ 0.3 mg KOH /g. The product contains zero phthalate or benzene ring, is non-toxic and environmentally friendly, and have excellent evaporation resistance (for toys, home products), aging resistance (for wires, cables, car decorations, water-proof cloth, rubber tapes, etc.), and good solubility, dispersity and adhesion (for dyes, inks or coatings).

In 2013, Hongxi Fang et al. from Ningbo Donglai Chemical Co., Ltd. published a patent titled "A type of apparatus for preparing dioctyl cyclohexane-1,2-dicarboxylate". The facilities include the esterification kettle, refining kettle, filtration tank, and hydrogenation reaction unit. The hydrogenation reaction proceeded in a fixed-bed reactor at 200 °C and 12.9 MPa. The excellent toxicological characteristics make it suitable for use in sensitive soft PVC products. It has good PVC compatibility, processibility, low volatility, low-temperature performance, no odor, environmental friendliness, and non-toxicity. During granulation or plasticization, it has remarkable PVC compatibility and low viscosity, making it suitable for special injection molds even with delicate designs. It has very low migration and can be combined with ABS, SAN, PC resins to ensure long durability and calendar life.

In 2015, Guangquan Xing et al. from Shandong Qilu Plasticizer Co., Ltd. published a patent titled "The esterification reaction protocols for DOTP plasticizer". Compared to current technologies, the invention results in reduced capital cost, high operation efficiency, high conversion, low energy consumption, high asset utilization, and stable product quality.

In 2015, Xichun She et al. from Hunan Changling Petrochemical Technology Co., Ltd. published a patent titled "A method for preparing cyclohexane-1,2-dicarboxylates". Under conditions of Rh and/or Ru catalyst with heavy rare earth element additive, 80–95 °C, 4–7 MPa H_2 pressure, the phthalate conversion was

100% and the cyclohexane-1,2-dicarboxylate selectivity was >99.5% with >95.7 wt% cis-structure content.

In 2015, Bo Zhou from Suzhou Goodtime Technology Development Co., Ltd. published a patent titled "A method for preparing environmentally friendly plasticizers". The esterification reaction used citrates, bio-plasticizer, glycol derivative, phthalic anhydride, aliphatic acid ester as raw materials, TiO2 as the catalyst, and epoxy compound and sodium carbonate as the stabilizer, and proceeded at 60–70 °C. The reaction mixture was dehydrated at 90 °C to remove the formed water, refined at 120–140 °C for 1.5 h, and settled at 0–4 °C for 20–40 min to obtain the green, environmentally friendly plasticizers.

In 2016, Changhai Liang et al. from Dalian University of Technology published a patent titled "A method for preparing hexahydrophthalic anhydride". Yolk-shell Pd–Ru catalyst was used with one or several of Al_2O_3, MgO, ZnO, TiO_2 as the support. Using single or mixed dioxane, THF, cyclohexane, and decahydronaphthalene as the solvent, the phthalic anhydride was selectively hydrogenated to form hexahydrophthalic anhydride at 50–70 °C. When a constant pressure reaction kettle was used, hydrogenation conditions were: 80–140 °C, 6–12 MPa H_2 pressure, reaction time 2–10 h, HHPA/catalyst mass ratio (2–50):1. When a fixed-bed reactor was used, hydrogenation conditions were: 120–180 °C, 6–18 MPa H_2 pressure, space velocity 0.2–0.6 h^{-1}. After hydrogenation, the materials entered the solvent recycling tower with the tower top temperature 100–110 °C, 0.1 MPa pressure, theoretical plate number 9, reflux ratio 0.03. The HHPA refining tower had the tower top temperature 260–290 °C, 0.1 MPa pressure, theoretical plate number 200, reflux ratio 2.6. The mixture of phthalic anhydride, 2-methylbenzoic acid, and 2-methyl cyclohexane carboxylate obtained from the tower top was recrystallized twice at 0 °C to acquire high purity phthalic anhydride. Esterification of HHPA can prepare cyclohexane-1,2-dicarboxylate plasticizers.

In 2016, Qun Wang et al. from Jiangsu Qingquan Chemical Co., Ltd. published a patent titled "A method for preparing dioctyl cyclohexane-1,4-dicarboxylate". Using acidic molecular sieves or acidic salt catalyst, cyclohexane-1,4-dicarboxylic acid and octanol were esterified at 140–160 °C for 3 h. The yield of dioctyl cyclohexane-1,4-dicarboxylate reached an acceptable value.

In 2016, Qiang Bai et al. from Qingdao University of Science and Technology published a patent titled "A method for preparing the catalysts used for preparing cyclohexane dicarboxylates". Under conditions of the supported catalyst with Ru–Pd active composition on an activated carbon support, high-pressure autoclave with PTFE lining, phthalate as the raw material with a noble metal/phthalate mass ratio of 1:(40–200), alcohol as the solvent with phthalate/alcohol mass ratio of 1:(0–15), 50–180 °C, 4–5 MPa H_2 pressure, 800–1500 r/min, and 1–8 h, the phthalate conversion and cyclohexane-1,2-dicarboxylate selectivity were >95% and >97%, respectively.

In summary, among these patents, environmentally friendly plasticizers include bio-based and cyclohexane dicarboxylate plasticizers. Bio-based plasticizers include the citrates, glycol derivatives, aliphatic acid esters invented by Suzhou Goodtime Technology Development Co., Ltd.

The preparation of cyclohexane dicarboxylate plasticizers can follow the esterification method and the hydrogenation method. Agencies using the esterification method include Nan Ya Plastics Corporation, Hangzhou Lin'an Shangtong Plasticizer Co., Ltd., Jiangnan University, Foshan Shunde Tiansheng Trade Co., Ltd., Jiangsu Qingquan Chemical Co, Ltd., etc. The hydrogenation method can use fixed-bed reactors or autoclave reactors. Agencies using fixed-bed reactors include German BASF, Chinese Academy of Science-Dalian Institute of Chemical Physics, Chinese Academy of Science-Shanxi Institute of Coal Chemistry, Fushun Xinrui Catalyst Co., Ltd, China National Offshore Oil Corporation-Tianjin Chemical Research and Design Institute, Zhejiang Weibo Chemical Technology Co., Ltd, Hunan Changling Petrochemical Technology Co., Ltd., etc. Agencies using autoclave reactors include Jiangsu Kangheng Chemical Co., Ltd., Qingdao University of Science and Technology, etc. In addition, some agencies used an integrated esterification–hydrogenation method including China National Offshore Oil Corporation-Tianjin Chemical Research and Design Institute, Ningbo Donglai Chemical Co., Ltd., etc.

References

1. Staples CA, Peterson DR, Parkerton TF, et al. The environmental fate of phthalate esters: a literature review. Chemosphere. 1997;35:667–749.
2. Onorato TM, Brown PW, Morris P. Mono-(2-ethylhexyl)phthalate increase spermatocyte mitochondrial peroxiredoxin and cyclooxygenase. J Androl. 2008;29:293–303.
3. Erkekoglu P, Rachidi W, Favier A, et al. Evaluation of cytotoxicity and oxidative DNA damaging effects of di(2-ethylhexyl)-phthalate (DEHP) and mono (2-ethylhexyl)-phthalate (MEHP) on MA-10 leyding cells and protection by selenium. Toxicol Appl Pharmacol. 2010;248(1):52–62.
4. Hao YL, Zheng GY, Li QZ, et al. Combined toxic effects of DBP and DEHP on sperm in male mice. In: Du WJ (eds) Informatics and management science. London: Springer; 2013, vol 204. p. 729-34.
5. He SF. Discussion on quality and control of epoxy soybean oil. Guangzhou Chem Ind. 2005;33(3):81–2.
6. Tang YF. Development and application of green environmental protection polyester plasticizer. Nanjing: Nanjing Publishing Media Group; 2016.
7. Gosse C, Larson TM, Legrand PJP, et al. Plasticized polyvinyl chloride. USA, US 7297738. 2007-11-20.
8. He YC. Discussion on plasticizer in food packaging materials. Grand View Weekly. 2012;51(5):141–2.
9. Frados J. Plastics engineering handbook of the society of the plastics industry. New York: Van Nostrand Reinhold; 1976.
10. Li CC, Jiang PP, Dong YM. Synthesis of environmental protection plasticizer cyclohexane 1, 2-diisooctyl dicarboxylate. Fine Chem. 2011;28(11):1124–8.
11. Jiang PP, Shen ML, Dong YM. Homogeneous catalytic synthesis of polyhexahydrophthalic anhydride diethylene glycol ester by titanium complex. Chem Prog. 2011;30(3):634–7.
12. Zhao J, Xue M, Huang Y, et al. Hydrogenation of dioctyl phthalate over supported Ni catalysts. Catal Commun. 2011;16(1):30–4.
13. Sabatier P. How I have been led to the direct hydrogenation method by metallic catalysts. Ind Eng Chem. 1926;18(10):1005–8.
14. Maitlis PM. (Pentamethylcyclopentadienyl) rhodium and-iridium complexes: approaches to new types of homogeneous catalysts. Acc Chem Res. 1978;11(8):301–7.

15. Moyes RB, Wells PB. The chemisorption of benzene. Adv Catal. 1973;23:121–56.
16. Garnett JL. π-Complex intermediates in homogeneous and heterogeneous catalytic exchange reactions of hydrocarbons and derivatives with metals. Catal Rev. 1972;5(1):229–67.
17. Mccusker-Orth JE. Process for a cyclohexanedimethanol using raney metal catalysts. USA, US 6919489. 2005-07-19.
18. Sumner J, Gustafson BL. Hydrogenation of phthalic acids to cyclohexane-dicarboxylic acid. USA, US 6291706. 2001-09-18.
19. Gustafson BL, Kuo YJ, Ptice TW, et al. Low pressure process for the hydrogenation of dimethyl benzenedicarboxylates to the corresponding dimethyl cyclohexanedicarboxlates. USA, US 5286898. 1994-02-15.
20. Bohnen H, Klein T, Bergrath K. Process for preparing cyclohexanedicarboxylic esters. USA, US 6740773. 2004-05-25.
21. Li TX, Zhang XF. Study on amorphous Ultrafine Nickel alloy catalyst II kinetics of hydrogenation of benzene over Ni-B catalyst. Chin J Catal 1995;16(5):399–02.
22. Hileman B. Panel concludes phthalate is safe. Chem Eng News. 2001;79(26):12–5.
23. Hileman B. FDA suggests replacing DEHP in plastics. Chem Eng News. 2002;80(37):6–9.
24. Wang B, Wang KZ, Gong YL. Research progress of environmental protection plasticizer. China Plast Ind. 2013;41(5):12–5.
25. Ventrice P, Ventrice D, Russo E, et al. Phthalates: European regulation, chemistry, pharmacokinetic and related toxicity. Environ Toxicol Pharmacol. 2013;36(1):88–96.
26. Sang L. Current situation and development of plasticizer industry. Sci Technol Inf Develop Econ. 2009;19(27):124–5.
27. Shi WC, Zhao CY. Toxicity of plasticizers and related regulations. Plast Addit. 2007;2:46–51.

Chapter 5
Catalytic Ring Hydrogenation of Phthalate Plasticizers

This chapter summarizes the catalytic ring hydrogenation of the most consumed DOP and DBP, as well as commonly used DIBP, DINP, DPHP, DIDP, etc., including the following topics: the effects of solvent, temperature, and pressure; determination of suitable reaction conditions; qualitative analysis and purification of products; repeatability of catalysts; high activity and selectivity of catalysts; multi-phase reaction kinetics of catalysis reactions with the kinetic order and activation energy; characterizations of catalysts to reveal the relationship between their structure, texture, and surface functionality with their preparation and performance; ring hydrogenation reaction mechanisms; characterizations of plasticized PVC materials in terms of mechanical, thermal, and photonic properties; comparison of the plasticization performance between cyclohexane-1,2-dicarboxylate and phthalate; inspection of phthalate ring hydrogenation products by the third-party. These investigations have built a solid foundation for pilot-scale and industrialization.

5.1 Ring Hydrogenation of Phthalates

DEHP, DNOP, and DIOP all have 8 carbons on the alkyl chain, and therefore, are generally called DOP. The ring hydrogenation products are generally called DEHCH.

DEHP, di(2-ethylhexyl) phthalate, has 6 main-chain carbons and totally 8 alkyl carbons. The molecular structure and ring hydrogenation product structure is

Z. Liu, *Green Catalytic Hydrogenation of Phthalate Plasticizers*,
https://doi.org/10.1007/978-981-16-9789-0_5

DNOP, di(n-octyl) phthalate, has totally 8 alkyl carbons. The molecular structure and ring hydrogenation product structure is

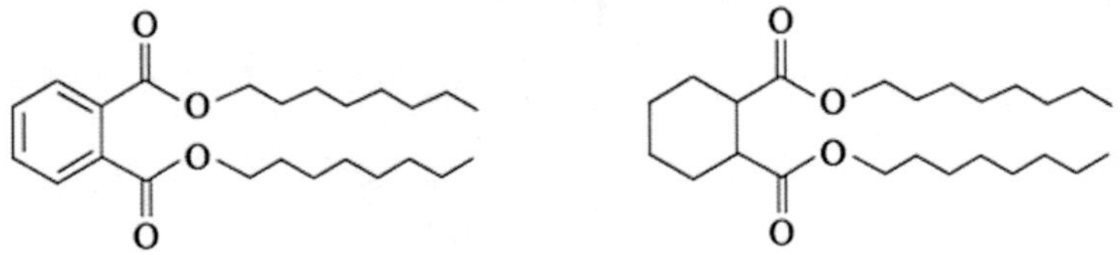

DIOP, di(isooctyl) phthalate, has 7 main-chain carbons and totally 8 alkyl carbons. The molecular structure and ring hydrogenation product structure is

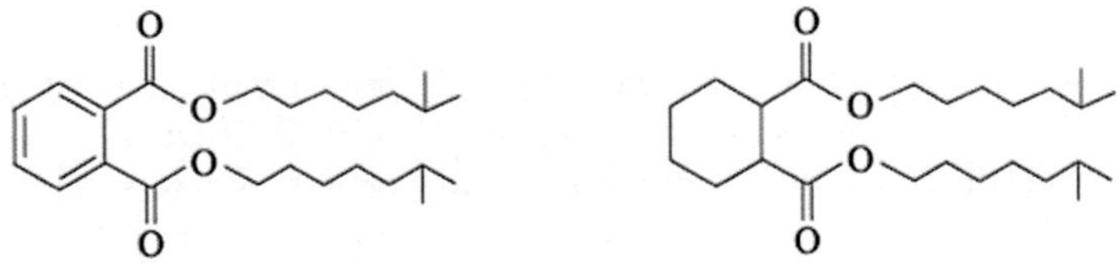

DBP, dibutyl phthalate, has totally 4 alkyl carbons. The molecular structure and ring hydrogenation product structure is

DIBP, di(iso-butyl) phthalate, has 3 main-chain carbons and totally 4 alkyl carbons. The molecular structure and ring hydrogenation product structure is

DINP, di(iso-nonyl) phthalate, has 8 main-chain carbons and totally 9 alkyl carbons. The molecular structure and ring hydrogenation product structure is

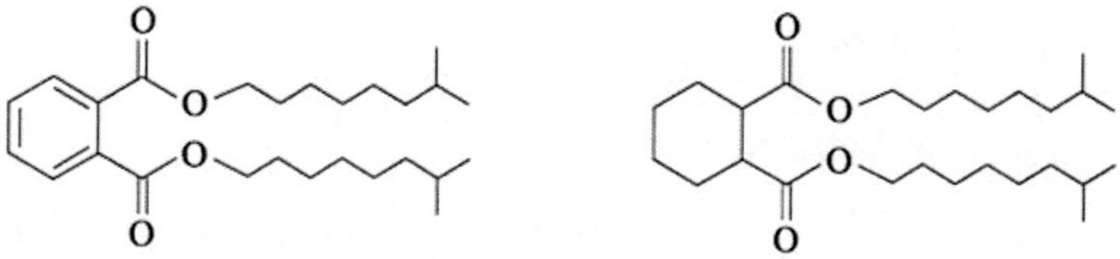

DPHP, di(2-propylheptyl) phthalate, has 4 main-chain carbons and totally 7 alkyl carbons. The molecular structure and ring hydrogenation product structure is

DIDP, di(iso-decyl) phthalate, has 9 main-chain carbons and totally 10 alkyl carbons. The molecular structure and ring hydrogenation product structure is

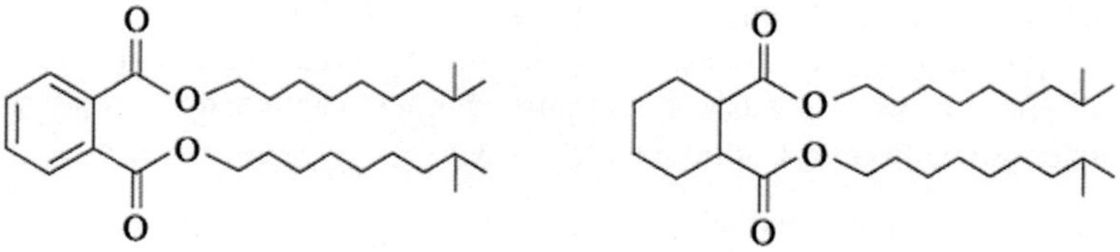

Because DOP and DBP are the major types of phthalate plasticizers used in China but are prohibited by foreign laws, their ring hydrogenations were first investigated. Their catalytic technologies were then extended to other phthalates including DIBP, DINP, DPHP, and DIDP to satisfy the need for cyclohexane-based plasticizers under different circumstances.

The core of catalytic ring hydrogenation is the catalyst, with literature reports of supported and non-supported, noble metal and non-noble metal catalysts. Different catalysts are suitable for use with different catalysis systems and reactions. The core of catalyst is the selection of active composition and additive, preparation method, and determining factors for activity and selectivity. To achieve commercialization, the production process, special equipment, and facilities must be established to match the catalysts, leading to complete sets of catalytic technologies.

5.1.1 Experimental

1. Preparation of catalysts

Supported catalysts: the active composition was one metal among Ru, Rh, Pt, Pd, and Ni; the additive was one metal among Zn, La, Co, Mn, and Ce; the support was Al_2O_3, ZrO_2 or diatomite. The precursors of the active compositions were $RuCl_3$, $RhCl_3$, $PtCl_4$, $PdCl_2$, and $NiSO_4$, respectively. The precursors of the additives were $ZnCl_2$, $La(NO_3)_3$, $CoCl_2$, $MnCl_2$, and $Ce(NO_3)_3$, respectively. The precipitator was NaOH or KOH.

To Al_2O_3, ZrO_2 or diatomite pre-mixed with water were added the precursors of both the active composition and the additive under vigorous stirring. The active composition was reduced to metal using H_2 or $NaBH_4$ under specified temperature and pressure. After a process of washing, filtration, drying, and calcination, the supported catalyst was obtained.

Non-supported catalysts: the active composition, additive, their precursors, and precipitator were the same as the supported catalysts. The sol–gel method was used to prepare the nanostructured catalysts. An aqueous solution of the hydroxide and surfactant was added rapidly to a mixed solution of active composition and additive under stirring, forming a highly dispersed system. With H_2 or $NaBH_4$ under specified temperature and pressure, the active composition was reduced to nano-scale metal crystals. After a process of washing and purification, the obtained catalyst was stored in an aqueous solution or in solid form at 60 °C.

2. Catalyst evaluation method

Evaluation of catalysts mainly includes the determination of the activity, selectivity, stability, and lifetime of the catalysts. The activity is indicated by the conversion ratio of the reactant. The selectivity is indicated by the percentage of the target product in the overall products. The stability and lifetime are indicated by the repeatable use.

The hydrogenation reaction was carried out using three autoclaves having different volumes, including 100 mL Parr Autoclave (USA), 300 mL Shijisenlang Autoclave (Beijing), and 1000 mL GS-1 Autoclave. The autoclaves were all composed of the reactor, jacket, stirrer, temperature controller, pressure controller, temperature sensor, pressure sensor, etc. A high-pressure liquid sample extracting tube was equipped with the autoclave. The inner design, material, stirring method, and heating method were different for these autoclaves. The Parr and Shijisenlang autoclaves used stainless steel material and electrical heating, while the GS-1 autoclave adopted Hastelloy alloy or PTFE liner and oil bath heating. An exploratory hydrogenation reaction was performed in the 100 mL autoclave, followed by reactions in the 300 mL and 1000 mL autoclaves. The products were analyzed using gas chromatography. The goal is to obtain repeatable and reliable results.

The reaction mixture and catalyst were added to the autoclave and sealed. The air inside was replaced by hydrogen gas. Under a suitable stirring rate, the mixture was heated to the preset temperature, followed by the introduction of pressurized hydrogen. The sample aliquots were extracted at specified time intervals to analyze the activity and selectivity of the catalyst as well as possible side products. After separation via precipitation from the products, the catalyst was subjected to repeated use to examine its lifetime.

3. Quantitative analysis of products

Agilent gas chromatographer with SE30 column (30 m × 0.32 mm × 0.5 μm) and H_2 flame ionization detector (FID). Column temperature 160–260 °C with programmed heating: initial temperature 160 °C for 1.2 min, then heating to 260 °C at 10 °C/min, then 260 °C for 5 min. The vaporization chamber temperature 80 °C, the detector temperature 290 °C, the carrier gas high purity nitrogen, sampling 0.2 μL. The samples were 4–5 times diluted by anhydrous ethanol.

The area normalization method was used to calculate the relative contents of DEHCH and phthalate, the conversion ratios at different time intervals, and the selectivity and yield of DEHCH. Taking DOP as an example, the conversion ratio of DOP (C_{DOP}), the selectivity (S_{DEHCH}), and the yield (Y_{DEHCH}) of DEHCH is defined as:

$$C_{\mathrm{DOP}} = \frac{n^0_{\mathrm{DOP}} - n_{\mathrm{DOP}}}{n^0_{\mathrm{DOP}}} \times 100\%; \quad S_{\mathrm{DEHCH}} = \frac{n_{\mathrm{DEHCH}}}{n^0_{\mathrm{DOP}} - n_{\mathrm{DOP}}} \times 100\%$$

$$Y_{\mathrm{DEHCH}} = \frac{n_{\mathrm{DEHCH}}}{n^0_{\mathrm{DOP}}} \times 100\%; \quad Y_{\mathrm{DEHCH}} = C_{\mathrm{DOP}} \times S_{\mathrm{DEHCH}}$$

where $n^0{}_{\mathrm{DOP}}$ and n_{DOP} are the amounts of DOP before and after the reaction; n_{DEHCH} is the content of DEHCH. Such definitions are also suitable for the hydrogenation of DBP and other phthalates.

4. Qualitative analysis of products

Agilent GC-MS spectrometer. Column conditions: HP-5MS capillary column (30 m × 0.25 mm × 0.25 μm), He as the carrier gas, eluent speed 1.0 mL/min, front sample inlet temperature 280 °C, sample volume 1 μL, split ratio 20:1. The heating programing: initial temperature 60 °C for 1 min, then heating to 220 °C at 20 °C/min, then heating to 290 °C at 5 °C/min, then kept at 290 °C for 3 min. The GC spectra were used to determine the relative contents of products. The MS and related databases were used to identify the molecular weights and possible structures of products.

5. Characterizations of catalysts

The structure of the catalyst was analyzed with an X'Pert PRO X-ray diffractor (Pan Analytical, Japan) with Cu Kα radiation ($\lambda = 0.15418$ nm), tube voltage 40 kV, tube current 40 mA, scan range 5–80° at 4°/min. The catalyst grain size was calculated according to Scherrer Equation $D = 0.89\lambda/\beta\cos\theta$, where D is the grain size, λ is the wavelength of the X-ray, θ is the diffraction angle, and β is the half-height width.

The texture of the catalyst was analyzed using NOVA 1000e physical adsorption analyser (Quantachrome). The sample was placed in vacuum at 150 °C for 1.5 h. N_2 was the adsorbate and catalyst was the adsorbent. At the boiling point of liquid N_2, the multi-layer physisorption and desorption of the catalyst produced the adsorption–desorption isotherm. The data was processed via BET equation and BJH equivalent cylinder model to obtain the surface area S_{BET}, average pore size D_{pore}, and overall pore volume $V_{overall}$.

Surface morphology and elemental analysis: TEM imaging and EDS analysis was performed on JEOL JEM2011 with an acceleration voltage of 200 kV. FESEM imaging was performed on JEOL JSM-7500F with an acceleration voltage of 5 kV. The S content in the catalyst was measured on BRUKE S4 Pioneer XRF.

Electronic states of catalyst surface elements were determined using XPS on PHI Quantera SXM. The activation source was Al$K\alpha$ radiation (1486.6 eV). The binding energy of C1s peak was used for calibration.

5.1.2 *Effects of Various Factors*

In the reports of catalytic hydrogenation of phthalate plasticizers, supported catalysts and fixed-bed reactors are mostly used. According to the fundamentals of multi-phase catalysis, through analysis of the characteristics of reactants and properties of reaction systems, it is determined that non-supported catalysts and slurry beds are more advantageous. Based on this argument, small-scale autoclaves were used to perform the hydrogenation reactions and investigate the effects of various factors such as solvent, temperature, pressure, etc. First, the performances of non-supported and supported catalysts were compared in terms of conversion ratio of DOP (C_{DOP}), selectivity (S_{DEHCH}), and yield (Y_{DEHCH}) of DEHCH. The results are shown in Table 5.1.

To avoid its spontaneous combustion in air, Raney Ni was in a wet state. The Ru contents in the supported and non-supported catalysts were identical. The Ru loading in the supported catalyst was 20% and the total weight of the catalyst was 0.25 g. The total weight of the non-supported catalyst was 0.05 g. When Raney Ni was used, no hydrogen was consumed at 130–150 °C; thus the temperature was raised to 170 °C.

From Table 5.1, the catalytic performance of Raney Ni is inferior compared to Ru-based catalyst. A similar phenomenon is reported in ref. [1]. Under the same

Table 5.1 DOP hydrogenation results on non-supported and supported catalysts

Catalyst	C_{DOP}/%	S_{DEHCH}/%	Y_{DEHCH}/%
Non-supported Ru-based catalyst	98.56	96.67	95.28
Raney Ni	52.04	88.78	46.20
Ru/diatomite	32.47	92.09	29.79
Ru/ZrO_2	58.05	91.25	52.97
Ru/Al_2O_3	38.90	97.48	37.92

Note Reaction conditions: 20 mL DOP, 20 mL water, 130–150 °C, 3.0 MPa H_2 pressure, 800 r/min and 3 h reaction time. Raney Ni 2.0 g, non-supported Ru catalyst 0.05 g, Ru/diatomite, Ru/ZrO_2, and Ru/Al_2O_3 0.25 g

Table 5.2 C_{DOP}, S_{DEHCH} and Y_{DEHCH} at different catalyst loadings

Mass of catalyst/g	$(m_{cat.}/m_{DOP})$/%	C_{DOP}/%	S_{DEHCH}/%	Y_{DEHCH}/%
0.0135	0.068	15.91	93.71	14.91
0.0251	0.125	60.83	93.68	56.99
0.0500	0.25	99.43	95.60	95.45
0.1000	0.5	99.27	95.97	95.27
0.2000	1.0	99.21	94.01	93.27

Note Reaction conditions: 100 mL Parr autoclave, 130 °C, 800 r/min, 3.0 MPa H_2 pressure, 40 mL water, and 20 mL DOP

conditions, non-supported Ru catalyst performs better than supported Ru/diatomite, Ru/ZrO_2, and Ru/Al_2O_3 catalysts with higher C_{DOP} and S_{DEHCH}. For liquid-phase hydrogenation of phthalates, non-supported Ru catalysts have obvious advantages.

Then, the suitable loading of the catalyst was determined. Table 5.2 lists the C_{DOP}, S_{DEHCH}, and Y_{DEHCH} at different loadings of catalyst.

From Table 5.2, under the conditions of 100 mL Parr autoclave, 130 °C, 800 r/min, 3.0 MPa hydrogen pressure, 40 mL water, and 20 mL DOP, the C_{DOP} and S_{DEHCH} increased with the amount of catalyst used. When the catalyst loading was 0.25% of the DOP mass, the C_{DOP}, S_{DEHCH}, and Y_{DEHCH} were 99.43%, 95.60%, and 95.45%, respectively. When the loading of catalyst increased above 0.25%, there was no obvious change in the C_{DOP} but a decrease in the S_{DEHCH}. Therefore, the optimal catalyst loading was 0.25% of the DOP mass.

Based on this result, a 100 mL autoclave and 0.25 wt% loading of non-supported Ru catalyst were used to investigate the effects of solvent, temperature, hydrogen pressure, stirring speed, and reaction time on C_{DOP}, S_{DEHCH}, and Y_{DEHCH}. Especially, the solvent effect study was repeated in a 1 L autoclave. 100 mL and 300 mL autoclaves were used to study the effects of reactor volume and structure.

1. Solvent

Table 5.3 compares the effects of water and ethanol solvents on C_{DOP}, S_{DEHCH}, and Y_{DEHCH}.

Under the conditions of 0.20 g Ru catalyst, 20 mL DOP, 2.0 MPa hydrogen pressure, 150 °C, 800 r/min, and reaction time of 3 h, when 40 mL water is used as the solvent, the C_{DOP}, S_{DEHCH}, and Y_{DEHCH} are 38.31%, 86.63%, and 33.15%, respectively, when 40 mL ethanol is used as the solvent, the C_{DOP}, S_{DEHCH}, and

Table 5.3 Effects of water and ethanol on the C_{DOP}, S_{DEHCH}, and Y_{DEHCH}

Solvent	C_{DOP}/%	S_{DEHCH}/%	Y_{DEHCH}/%
H_2O	38.31	86.63	33.19
C_2H_5OH	12.96	89.04	11.54

Note Reaction conditions: Ru-based catalyst 0.20 g, 20 mL DOP, 40 mL solvent, 2.0 MPa H_2 pressure, 150 °C, 800 r/min, and 3 h reaction time

Y_{DEHCH} are 12.96%, 89.04%, and 11.54%, respectively. Water affords higher C_{DOP} and DEHCH yield, but lower selectivity.

The origin of the low S_{DEHCH} lie in the tri-phase catalysis system, in which the reaction occurs at the surface of the solid catalyst. The reaction proceeds to the product via a series of elementary steps including dissolution, diffusion, adsorption, surface reaction, etc. Both water and ethanol decrease the viscosity of the reaction mixture and promote the contact between the reactant and solid catalyst, leading to an enhanced reaction rate. But water is an inorganic solvent and ethanol is an organic solvent with different polarities. This has resulted in different dissolution, diffusion, and adsorption/desorption behaviors for DOP and DEHCH. Specifically, when water is used, the product is more difficult to be desorbed from the catalyst surface, which may be the primary reason for the lower selectivity.

According to ref. [2], the solubility of hydrogen in different solvents is different. Hydrogenation is closely associated with the dissolution and transport of hydrogen. The low C_{DOP} in the case of ethanol solvent is a consequence of the low solubility of hydrogen in ethanol. Ref. [3] also shows that alkane-type solvents are inhibitive to hydrogenation reactions.

To decrease the commercialization cost, the C_{DOP}, $S_{DEHCH,}$ and Y_{DEHCH} in the case of water solvent or solvent-free conditions are shown in Table 5.4.

When water is used, the C_{DOP} and S_{DEHCH} were independent of water volume. The C_{DOP} and S_{DEHCH} are higher under solvent-free conditions. Because this finding is critically important for commercialization, this experiment was repeated in a 1 L autoclave. The result is indicated in Fig. 5.1.

From Fig. 5.1, the less water used, the higher C_{DOP}. The water-free system has the highest reaction rate and the best C_{DOP}.

It was found that, during the separation of the products after hydrogenation, the catalyst was distributed mainly in the DEHCH organic phase and precipitated out rapidly. This indicates that the catalyst surface has a high affinity to organic materials, i.e., good lipophilicity, and also the repeated use of catalyst is possible. The faster reaction rate for the water-free system is a result of exothermal reaction and temperature increase. When water is present, it absorbs the heat and slows down the

Table 5.4 C_{DOP}, S_{DEHCH} and Y_{DEHCH} with water solvent or with no solvent

Volume of wate/mL	C_{DOP}/%	S_{DEHCH}/%	Y_{DEHCH}/%
0[a]	98.40	97.10	95.54
10	98.62	96.79	95.45
20	98.32	96.88	95.25
40	98.86	96.29	95.19

Note Reaction conditions: Ru-based catalyst 0.25%, 20 mL DOP, 3.0 MPa H_2 pressure, 150 °C, 800 r/min, and 3 h reaction time
[a]When no solvent was used, due to the small volume of DOP, the liquid surface was too low to afford uniform stirring. Thus, the amounts of catalyst and DOP were doubled

Fig. 5.1 Effects of water/DOP ratio on DOP conversion ratio. Reaction conditions: catalyst loading 0.25 wt% of DOP mass, 800 r/min, total water and DOP volume 400 mL, DOP/water volume ratios: 4:0, 3:1, 1:1, 1:3, 2.5–3.5 MPa H_2 pressure, 130 °C

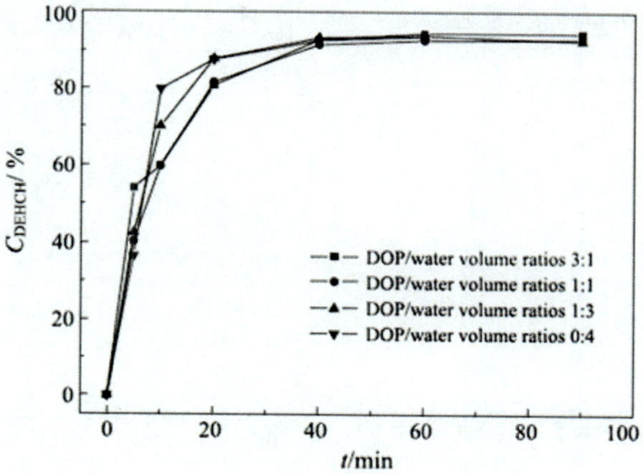

temperature increase, leading to the slowest reaction rate. But the C_{DOP} still reaches above 98% after 90 min reaction time.

2. Reaction temperature

Table 5.5 shows the effects of reaction temperature on the catalyst activity and selectivity.

At 100–120 °C, the C_{DOP} and S_{DEHCH} increase with temperature. At 120–160 °C, the S_{DEHCH} decreased due to increased side products. 130 °C is the optimal hydrogenation temperature because of the best C_{DOP} at this temperature.

3. Hydrogen pressure

Table 5.6 shows the catalyst activity and selectivity at different hydrogen pressures.

The C_{DOP} and S_{DEHCH} increases with hydrogen pressure below 3.0 MPa. When the hydrogen pressure reaches 3.0 MPa, the C_{DOP} and S_{DEHCH} are approximately the highest values. As the hydrogen pressure increases above 3.0 MPa, the C_{DOP}

Table 5.5 Effect of temperature on catalyst activity and selectivity

Temperature/°C	C_{DOP}/%	S_{DEHCH}/%	Y_{DEHCH}/%
100	57.28	99.28	56.87
110	67.68	93.76	63.46
120	98.33	96.04	94.44
130	99.21	94.01	93.27
140	91.74	91.33	83.79
150	91.48	94.47	86.43
160	97.45	91.00	88.68

Note Reaction conditions: Ru-based catalyst 0.200 g (0.25 wt% of DOP mass), 3.0 MPa H_2 pressure, 40 mL water, 20 mL DOP, 800 r/min, and 3 h reaction time

Table 5.6 The catalyst activity and selectivity under different H_2 pressures

H_2 pressure/MPa	C_{DOP}/%	S_{DEHCH}/%	Y_{DEHCH}/%
1.0	24.65	74.20	18.29
2.0	38.31	86.64	33.19
3.0	97.43	91.56	89.21
4.0	96.18	91.96	88.45
5.0	93.97	91.83	86.30

Note Reaction conditions: Ru-based catalyst 0.200 g (0.25 wt% of DOP mass), 150 °C, 40 mL water, 20 mL DOP, 800 r/min, and 3 h reaction time

Table 5.7 Effect of stir rate on catalyst performance

Stir rate/(r/min)	C_{DOP}/%	S_{DEHCH}/%	Y_{DEHCH}/%
600	33.23	97.11	32.27
800	98.32	96.97	95.25
1000	98.32	96.87	95.24
1200	98.95	97.24	96.22

Note Reaction conditions: Ru-based catalyst 0.050 g (0.25 wt% of DOP mass), 150 °C, 3.0 MPa H_2 pressure, 20 mL water, 20 mL DOP, and 3 h reaction time

does not obviously change and the S_{DEHCH} slightly decreases. Therefore, the optimal hydrogen pressure is 3.0 MPa.

4. Stir rate

Table 5.7 shows the effects of the stir rate on catalyst performance.

The C_{DOP} and S_{DEHCH} increase with the stir rate when lower than 800 r/min. Under such low stir rates, the mass transport is slower than the surface reaction and is thus the rate-determining factor. When the stir rate is higher than 800 r/min, the mass transport between phases is faster than the hydrogenation reaction rate; the latter becomes the rate-determining step. Considering too high stir rates can result in catalyst abrasion and even adhesion to the inner wall of the reactor, the optimal stir rate is kept at 800 r/min.

5. Reaction time

Table 5.8 shows the effect of reaction time on catalyst performance.

The DEHCH yield increases with reaction time, reaching 51.66% and 81.99% at 1 h and 2 h, respectively. The C_{DOP} and S_{DEHCH} become stable at 3 h. Therefore, 3 h is the optimal reaction time for hydrogenation reaction.

In summary, the optimal conditions for DOP hydrogenation include solvent-free, catalyst loading 0.25% of DOP mass, 130 °C, 3.0 MPa hydrogen pressure, stir rate 800r/min, and 3 h reaction time. The C_{DOP}, S_{DEHCH} and Y_{DEHCH} reach $> 99\%$, $> 96\%$ and $> 95\%$, respectively.

Table 5.8 Effect of reaction time on catalyst performance

Reaction time/h	C_{DOP}/%	S_{DEHCH}/%	Y_{DEHCH}/%
1	55.55	93.00	51.66
2	87.15	94.08	81.99
3	93.97	91.84	86.30
4	95.51	91.25	87.15

Note Reaction conditions: Ru-based catalyst 0.200 g, 150 °C, 5.0 MPa H_2 pressure, 20 mL water, 40 mL DOP, and 800 r/min

Table 5.9 Effects of autoclave volume and structure on catalyst performance. 100 mL Parr[a], 300 mL Shijisenlang autoclave[b]

Autoclave	Amount of water/mL	Stir rate/(r/min)	C_{DOP}/%	S_{DEHCH}/%	Y_{DEHCH}/%
100 mL Parr autoclave[a]	10	800	98.62	96.78	95.45
	20	800	98.56	96.67	95.28
	20	1000	98.32	96.88	95.25
300 mL Shijisenlang autoclave[b]	30	800	97.84	96.84	94.75
	60	800	98.70	96.29	95.04
	60	1000	98.97	94.85	93.87

Note
[a]Reaction conditions: Ru-based catalyst 0.050 g (0.25 wt% of phthalate mass), 130 °C, 3.0 MPa H_2 pressure, 20 mL water, 20 mL DOP, 800 r/min and 1000 r/min
[b]Reaction conditions: Ru-based catalyst 0.150 g (0.25 wt% of phthalate mass), 130 °C, 3.0 MPa H_2 pressure, 20 mL water, 60 mL DOP, 800 r/min and 1000 r/min

6. Autoclave volume and structure

Table 5.9 shows the effect of autoclave volume and structure on catalyst performance.

Combining Table 5.9 and Fig. 5.1, we can conclude that the autoclave volume and structure have certain but unobvious effects on C_{DOP} and S_{DEHCH}. As long as the reactants have sufficient contact with the catalyst and the mass transport is faster than the hydrogenation rate, the catalyst performance is basically independent of the autoclave volume and structure. This finding indicates the facile scalability from the lab to industrial level.

5.1.3 Effects of S Content in DOP

S is poisonous to most catalysts. In Sect. 5.1.2, the DOP used is in analytical purity and contains no S. Industrial grade DOP is produced via esterification of phthalic anhydride with corresponding alcohols. Phthalic anhydride is produced via either oxidation of naphthalene or oxidation of o-xylene [4, 5], as shown in Figs. 5.2 and 5.3.

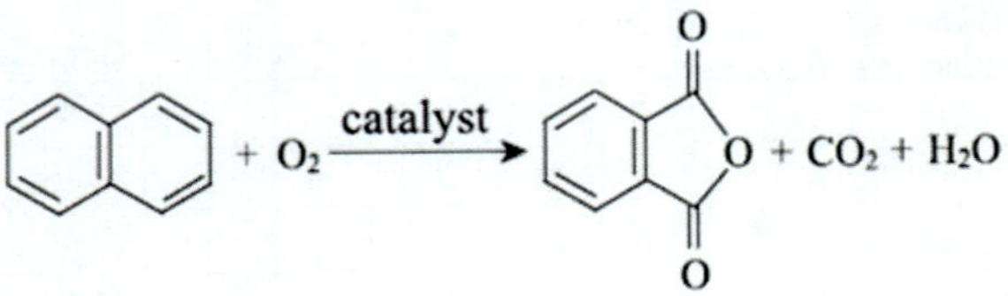

Fig. 5.2 Preparation of phthalic anhydride via oxidation of naphthalene (the naphthalene method)

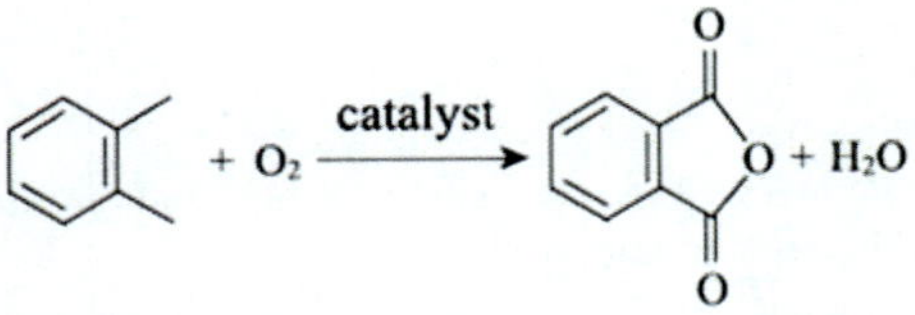

Fig. 5.3 Preparation of phthalic anhydride via oxidation of o-xylene (the o-xylene method)

Naphthalene comes from coal tar, in which a trace amount of S residue exists in the form of benzothiophene in phthalates, leading to the poisoning of catalysts. The effects of S on catalyst performance in hydrogenation reactions are investigated by using DOP from different sources. The result is shown in Fig. 5.4.

When DOP from oxidation of o-xylene is used, the C_{DOP} at 60 min is 100%. But when DOP from oxidation of naphthalene is used, the C_{DOP} at 120 min is still lower than 60%. Then, the hydrogenation results for o-xylene-derived DOP from different companies (A, B, and C) are shown in Tables 5.10, 5.11, and 5.12.

Samples were extracted every 10 min. At 60 min, when the reaction is completed, the C_{DOP} is close to 100% and the DEHCH yield is >99%. Three times of repeated use of the catalyst indicate excellent repeatability and stability. The results suggest no S content in the DOP.

No catalyst poisoning occurs in the catalytic hydrogenation of phthalates for the o-xylene-derived DOP from different vendors.

Repeated hydrogenation experiments were performed to o-xylene-derived DOP. Table 5.13 shows the results of 20 repeated hydrogenation reactions.

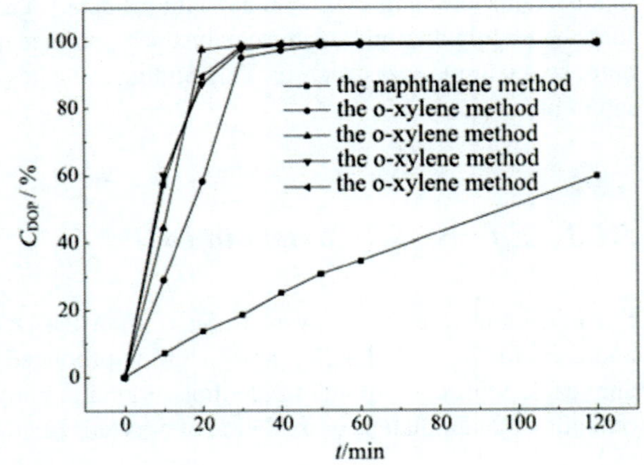

Fig. 5.4 Hydrogenation results using DOP prepared from different methods

Table 5.10 Catalytic hydrogenation results of DOP prepared via the o-xylene method from Company A

Times of repeated catalyst use	Y_{DEHCH} at different reaction times/%						C_{DOP} /%
	10 min	20 min	30 min	40 min	50 min	60 min	60 min
1	44.36	97.48	98.74	98.93	99.32	99.34	100
2	42.76	59.47	96.24	99.12	99.09	99.22	99.75
3	48.66	59.71	85.66	98.73	98.96	99.21	99.76

Note reaction conditions: 130 °C, 4.0 MPa H_2 pressure, 800 r/min, 120 mL DOP, catalyst 0.25 wt%

Table 5.11 Catalytic hydrogenation results of DOP prepared via the o-xylene method from Company B

Times of repeated catalyst use	Y_{DEHCH} at different reaction times/%						C_{DOP} /%
	10 min	20 min	30 min	40 min	50 min	60 min	60 min
1	60.14	87.30	97.52	98.75	99.10	99.02	99.73
2	58.75	86.85	97.62	98.67	99.30	98.50	99.44

Note Reaction conditions: 130 °C, 4.0 MPa H_2 pressure, 800 r/min, 120 mL DOP, catalyst 0.25 wt%

Table 5.12 Catalytic hydrogenation results of DOP prepared via the o-xylene method from Company C

Times of repeated catalyst use	Y_{DEHCH} at different reaction times/%						C_{DOP} /%
	10 min	20 min	30 min	40 min	50 min	60 min	60 min
1	57.47	89.48	98.34	99.14	99.42	99.14	99.81
2	61.16	87.47	97.49	99.30	99.30	99.21	99.81

Note Reaction conditions: 130 °C, 4.0 MPa H_2 pressure, 800 r/min, 120 mL DOP, catalyst 0.25 wt%

From Table 5.13, the C_{DOP} is as high as 100% and the Y_{DEHCH} is above 99% at different time intervals and at 60 min of hydrogenation of the o-xylene-derived DOP.

Figure 5.5 shows the results of 5 repeated uses of the catalyst in hydrogenation of naphthalene-derived DOP.

The catalyst works normally for the first hydrogenation with nearly 100% C_{DOP}. But the catalyst activity decreases obviously with the second use and even more seriously in the following hydrogenation because of the accumulation of S on the catalyst. In the fifth use, the catalyst almost loses all its activity. XRF analysis of the deactivated catalyst detects S.

In summary, o-xylene-derived phthalates should be used in industrial protocols. Before the volume production, hydrogenation tests should be performed to determine whether S is present in phthalate raw materials to avoid any loss possibly caused by S poisoning.

Table 5.13 Repeated hydrogenation results of DOP prepared via the o-xylene method

Times of repeated catalyst use	Y_{DEHCH} at different reaction times/%						C_{DOP} /%
	10 min	20 min	30 min	40 min	50 min	60 min	60 min
1	25.2	65.8	90.4	97.9	98.1	98.7	100
2	44.8	88.6	98.1	98.9	99.0	99.1	100
3	43.7	82.5	96.6	97.8	99.2	98.7	100
4	46.0	74.8	99.0	99.1	99.2	99.4	100
5	33.3	86.6	96.9	99.0	98.7	99.4	100
6	37.5	77.6	97.0	99.1	99.2	99.1	100
7	37.2	48.6	84.0	97.9	99.3	99.2	100
8	38.0	70.7	96.3	98.7	98.8	99.1	100
9	42.4	78.0	98.0	98.7	99.1	99.4	100
10	40.4	88.7	97.0	99.2	99.2	99.3	100
11	43.9	83.2	98.5	98.8	99.3	99.3	100
12	35.7	77.6	98.1	99.1	99.3	99.3	100
13	36.7	55.1	94.2	99.2	99.2	99.3	100
14	39.7	89.2	97.9	98.4	99.1	99.3	100
15	38.6	93.3	99.0	99.5	99.2	99.3	100
16	29.4	57.3	88.0	98.1	98.6	99.3	100
17	43.3	70.0	91.4	96.8	98.6	99.1	100
18	32.5	62.1	89.1	98.5	99.3	99.3	100
19	48.4	61.0	77.6	95.1	98.2	99.4	100
20	30.1	50.3	90.9	97.8	97.9	99.2	100

Note Reaction conditions: 130 °C, 3.0 MPa H_2 pressure, 800 r/min, 120 mL DOP, catalyst 0.25 wt%

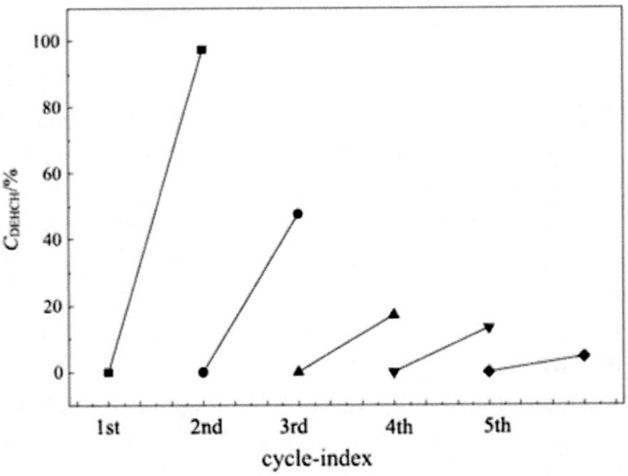

Fig. 5.5 Five repeated hydrogenation results of DOP prepared via the naphthalene method. Reaction time was 60 min

5.1.4 Qualitative Analysis of Products

The products of hydrogenation of DOP and DBP were analyzed using GC-MS. Figure 5.6 and Table 5.14 show the GC spectrum and MS analysis of DOP hydrogenation products.

In Table 5.14, cyclohexane-1,2-dicarboxylates are the main product of DOP hydrogenation, as well as other low-molecular-weight linear and branched alkanes, substituted cyclohexanes, cyclohexylethanol, etc.

Figure 5.7 and Table 5.15 show the GC spectrum and MS analysis of DBP hydrogenation products.

In Table 5.15, cyclohexane-1,2-dicarboxylates are the only products of DBP hydrogenation. The products at the retention times of 10.647 min and 10.732 min are isomers. No low-molecular-weight linear and branched alkanes, substituted cyclohexanes or cyclohexylethanol are detected.

Refs. [6–9] reported that, during hydrogenation, phthalates may undergo intramolecular elimination reactions to form 2-(alkoxycarbonyl)benzoic acid and alcohol or alkene, or directly produce phthalic anhydride and corresponding alkenes at high temperature. The mechanisms of parasitic side reactions of DOP hydrogenation include:

(1) Upon heating, DOP decomposes into phthalic anhydride and 2-ethylhexanol. The phthalic anhydride is hydrogenated and decarboxylated to form 2-methylcyclohexylmethanol that is further reduced to yield methyl cyclohexane that then produces heptane via ring opening reaction. 2-Ethylhexanol is dehydrated to yield 2-ethyl hexene.
(2) One of the two ester bonds in DOP breaks to produce 2-(2'-ethylhexoxycarbonyl)benzoic acid and 2-ethylhexene; the latter is hydrogenated to 3-methyl heptane.

The generation of low-boiling-point materials seriously deteriorate the product quality not only decreasing the flash point, viscosity, and density, but also causing

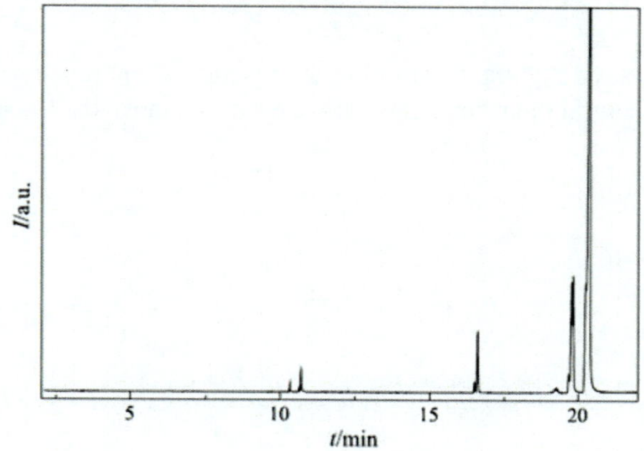

Fig. 5.6 GC spectrum of DOP hydrogenation products

Table 5.14 DOP hydrogenation products

Retention time/min	Product	Retention time/min	Product
2.157		10.715	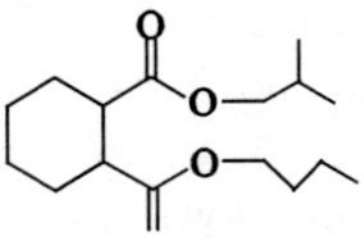
2.338	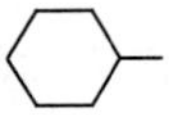	16.518	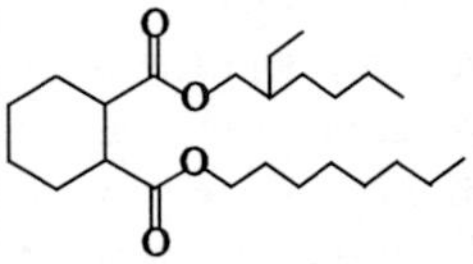
2.589	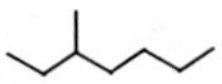	19.659	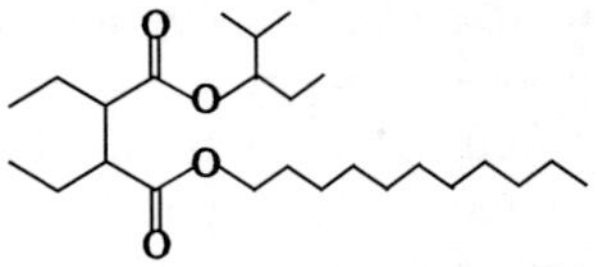
3.083	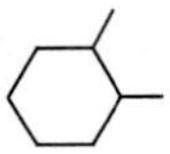	19.750	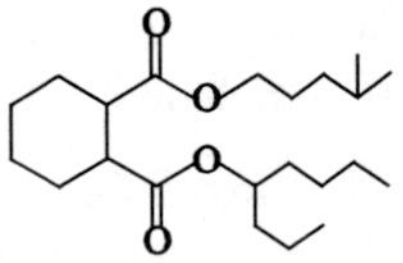
4.554	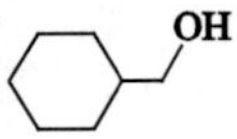	20.248	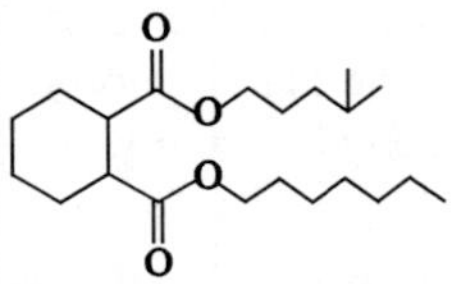
10.028	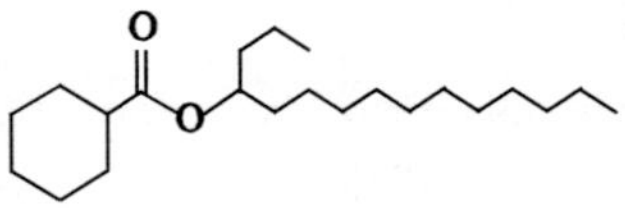	20.346	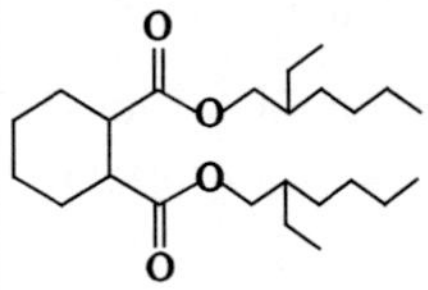

strong odors. Controlling the reaction temperature can decrease the side products. Further purification is also needed to remove the low-boiling-point materials.

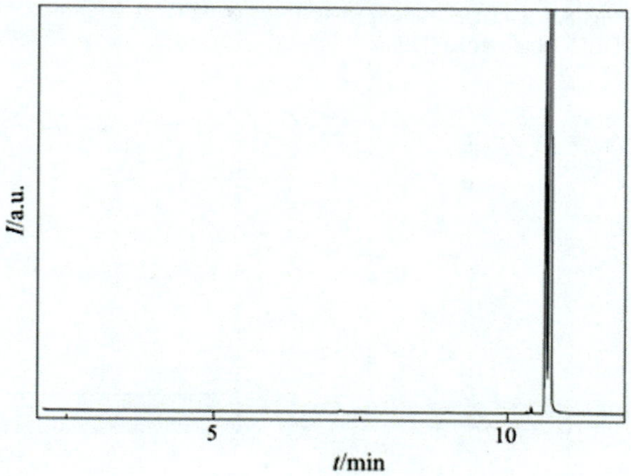

Fig. 5.7 GC spectrum of DBP hydrogenation products

Table 5.15 DBP hydrogenation product/min

Retention time/min	Product	Retention time /min	Product
10.327	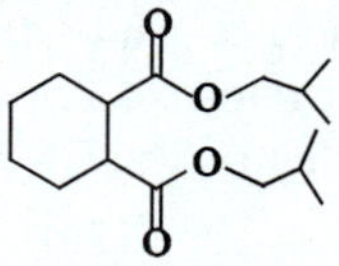	10.647	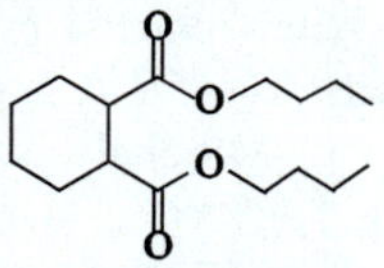
10.412	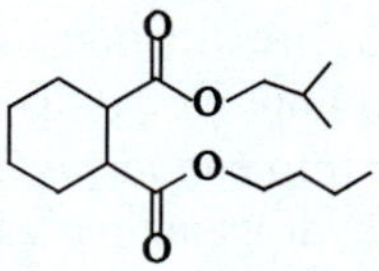	10.732	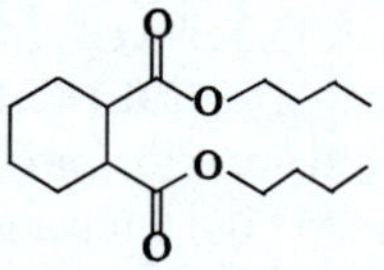

5.1.5 *Separation and Purification of Products*

From Table 5.14, catalytic hydrogenation of DOP may also generate side products including heptane, methyl cyclohexane, 3-methyloctane, 1,2-dimethyl cyclohexane, etc. Therefore, combined reduced pressure distillation and nitrogen extraction methods are used to remove these low-boiling-point substances and improve the quality of cyclohexane-based plasticizers.

Under 160–220 °C and −0.01 to −0.098 MPa, reduced pressure distillation and nitrogen extraction were performed to the hydrogenation products of DOP (DEHCH), DINP (DINCH), DBP (DBCH), DPHP (DPHCH), and DIDP (DIDCH), respectively. Among them, the optimal conditions for DOP are 170 °C and −0.1 MPa, under which the side products can be completely removed.

GC-MS analysis was performed to the product after reduced pressure distillation (170 °C and −0.1 MPa). The GC spectrum is shown in Fig. 5.8.

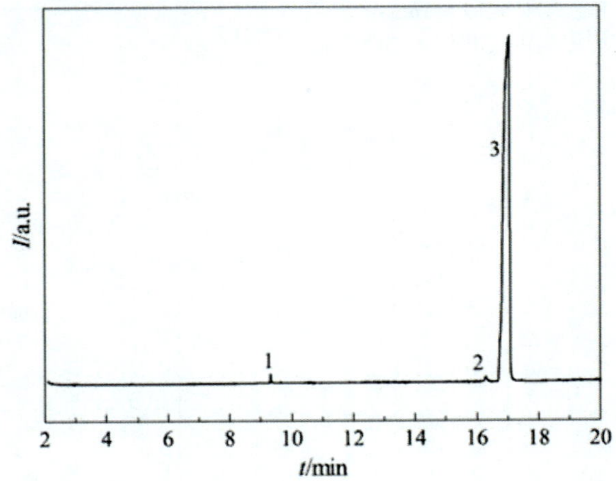

Fig. 5.8 GC spectrum of DEHCH after distillation

The MS results of the products corresponding to the chromatographic peaks in Fig. 5.8 are shown in Table 5.16.

After reduced pressure distillation under 170 °C and −0.1 MPa, the purified DEHCH is 99.78% pure with 0.22% dimethyl cyclohexane.

GC-MS analysis was performed to the distillates. The GC spectrum is shown in Fig. 5.9.

The MS results for the peaks in Fig. 5.9 are shown in Table 5.17.

After reduced pressure distillation and nitrogen extraction, the volatile alkanes and alcohols in DEHCH are removed, leading to improve product quality. Among the distillates, side products of Peaks 3,4,5,6 are formed most likely during distillation due to high temperature (170 °C), not in the hydrogenation reaction (130 °C).

Table 5.16 GC-MS results of DEHCH after distillation

Peak #	Retention time/min	Content/%	Structure
1	9.289	0.221	
2	16.184	0.821	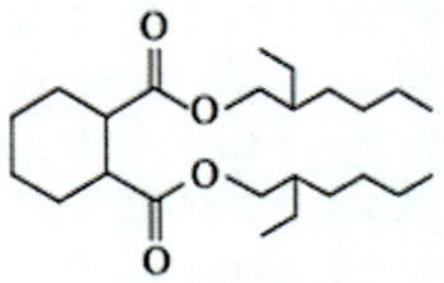
3	16.852	98.958	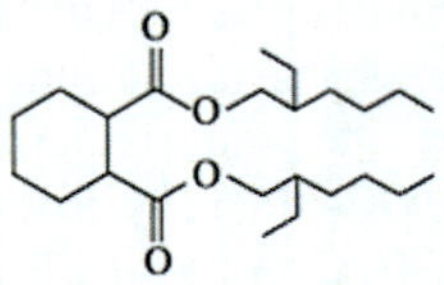

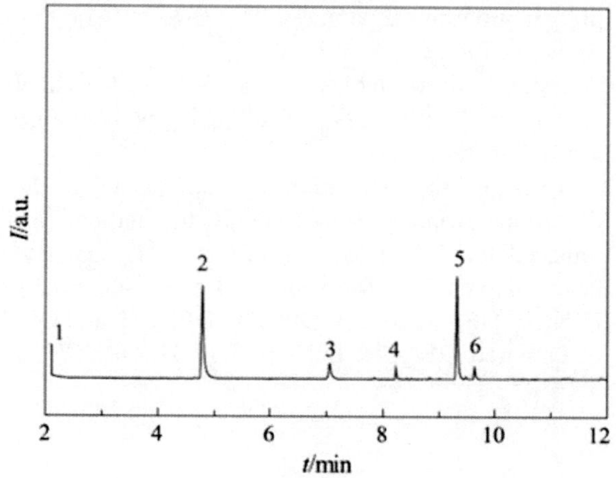

Fig. 5.9 GC spectrum of the distillate

Table 5.17 GC-MS results of the distillate

Peak #	Retention time/min	Content/%	Structure
1	2.104	3.711	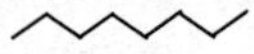
2	4.801	44.740	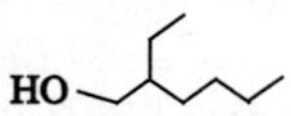
3	7.073	8.360	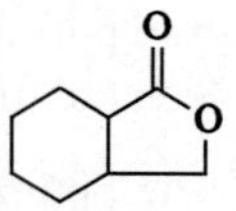
4	8.232	2.953	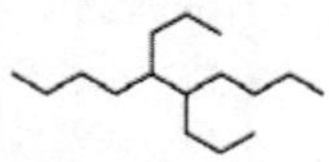
5	9.321	36.428	
6	9.642	3.808	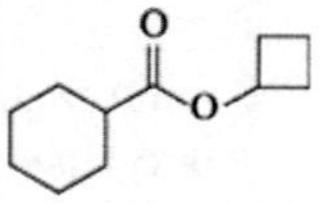

5.1.6 *Repeatability of Catalytic Hydrogenation*

Based on the above results, further modification was carried out to the catalytic hydrogenation of phthalates. In February 10–20 and April 2–18, 2017, two batches of DOP hydrogenation to prepare DEHCH were performed in in 1L autoclaves (0.75L DOP each batch). In each batch, the catalyst was repeatedly used for 70 times after precipitation and separation. The C_{DOP}, S_{DEHCH}, and Y_{DEHCH} for each batch are shown in Tables 5.18 and 5.19.

From these two tables, the catalyst still has excellent repeatability even after 70 recycles with 100% C_{DOP}. The S_{DEHCH} is maintained at >99% in Table 5.18 and at >99.7% in Table 5.19.

To expand the applications of cyclohexane-1,2-dicarboxylates and meet the needs of various property measurements, the authors carried out more experiments and replaced the 1 L autoclave with a 10 L one. By the end of 2017, the authors have completed the following hydrogenation reactions: DOP to DEHCH, DINP to DINCH, DBP to DBCH, DPHP to DPHCH, and DIDP to DIDCH. The total amount of prepared DEHCH, DINCH, DBCH, DPHCH, and DIDCH has reached about 2000 kg.

5.1.7 Conclusions

Catalytic ring hydrogenation of phthalate compounds including DOP and DBP to prepare cyclohexane-1,2-dicarboxylates has been successfully achieved. The effects of catalyst, solvent, temperature, pressure, stir rate, reaction time, and autoclave volume and structure on catalyst performance have been determined. Parasitic products have been identified via quantitative and qualitative analyses. The product separation and purification conditions have been optimized. The repeatability of catalyst have been confirmed. The major concludes are given below:

(1) The optimal hydrogenation conditions for DOP: no solvent, catalyst loading at 0.25% of DOP mass, 130 °C, 3.0 MPa, stir rate of 800 r/min, and reaction time of 3 h, under which the C_{DOP}, S_{DEHCH}, and Y_{DEHCH} reach >99%, >96%, and >95%, respectively. The autoclave uses 316 L stainless steel material that can satisfy the hydrogenation requirements.

(2) Through comparing the hydrogenation performances, naphthalene-derived DOP causes catalyst poisoning and deactivation due to the presence of S residue. For o-xylene-derived DOP, the catalyst can be repeatedly used for more than 20 times with well-maintained catalytic performance. Therefore, o-xylene-derived phthalate plasticizers should be used as the raw material to avoid catalyst poisoning and extend the lifetime.

(3) The main side products are 2-(alkoxycarbonyl)benzoic acid, 2-ethylhexanol and 3-methyl heptane. The parasitic mechanisms include thermal decomposition of DOP, cleavage of an ester bond, carbon chain rearrangements, and ring opening of phthalic anhydride. A well-controlled reaction temperature can reduce the side products that also need to be separated from the main product.

(4) Reduced pressure distillation and nitrogen extraction were performed at 160–220 °C and −0.01 to −0.098 MPa. The optimal conditions for DOP is 170 °C and −0.1 MPa, under which the side products are completely removed.

(5) The catalyst exhibits high repeatability with 100% C_{DOP} and >99.7% S_{DEHCH}.

Table 5.18 C_{DOP}, S_{DEHCH}, and Y_{DEHCH} (the first batch). Times of repeated catalyst use

Times of repeated catalyst use	C_{DOP}/%	S_{DEHCH}/%	Y_{DEHCH}/%	Times of repeated catalyst use	C_{DOP}/%	S_{DEHCH}/%	Y_{DEHCH}/%
1	100	99.21	99.21	14	100	99.30	99.30
2	100	99.22	99.22	15	100	99.24	99.24
3	100	99.32	99.32	16	100	99.27	99.27
4	100	99.34	99.34	17	100	99.21	99.21
5	100	99.12	99.12	18	100	99.25	99.25
6	100	99.31	99.31	19	100	99.20	99.20
7	100	99.25	99.25	20	100	99.27	99.27
8	100	99.28	99.28	21	100	99.23	99.23
9	100	99.25	99.25	22	100	99.29	99.29
10	100	99.28	99.28	23	100	99.24	99.24
11	100	99.27	99.27	24	100	99.21	99.21
12	100	99.31	99.31	25	100	99.35	99.35
13	100	99.32	99.32	26	100	99.31	99.31
27	100	99.26	99.26	49	100	99.31	99.31
28	100	99.31	99.31	50	100	99.29	99.29
29	100	99.28	99.28	51	100	99.35	99.35
30	100	99.19	99.19	52	100	99.24	99.24
31	100	99.25	99.25	53	100	99.26	99.26
32	100	99.28	99.28	54	100	99.24	99.24
33	100	99.22	99.22	55	100	99.26	99.26
34	100	99.31	99.31	56	100	99.32	99.32
35	100	99.26	99.26	57	100	99.28	99.28
36	100	99.19	99.19	58	100	99.33	99.33
37	100	99.19	99.19	59	100	99.26	99.26
38	100	99.28	99.28	60	100	99.30	99.30
39	100	99.31	99.31	61	100	99.29	99.29
40	100	99.26	99.26	62	100	99.24	99.24
41	100	99.6	99.26	63	100	99.26	99.26
42	100	99.31	99.31	64	100	99.19	99.19
43	100	99.29	99.29	65	100	99.23	99.23
44	100	99.32	99.32	66	100	99.24	99.24
45	100	99.33	99.33	67	100	99.18	99.18
46	100	99.31	99.31	68	100	99.19	99.19
47	100	99.31	99.31	69	100	99.30	99.30

(continued)

Table 5.18 (continued)

Times of repeated catalyst use	C_{DOP}/%	S_{DEHCH}/%	Y_{DEHCH}/%	Times of repeated catalyst use	C_{DOP}/%	S_{DEHCH}/%	Y_{DEHCH}/%
48	100	99.26	99.26	70	100	99.22	99.22

Note Reaction conditions: 130 °C, 4.0 MPa H_2 pressure, 800 r/min, 750 mL DOP, catalyst 0.25 wt%

5.2 Multi-phase Catalysis Kinetics of DOP Hydrogenation

Chemical kinetics is "the media language of quantitative chemistry" and is expressed as mathematical functions that are associated with optimization of operation conditions and also provide theoretical basis for reaction mechanism studies. Multi-phase catalysis kinetics involves elementary steps of diffusion, adsorption, surface reaction, and desorption, which are closely associated with the active center and surface properties of the catalysts, and shed light on the scientific nature of the interactions between the active center and the reaction system.

Because of the unsaturated coordination, the surface atoms of solid catalysts generally receive uneven interactions from adjacent atoms and are in unbalanced force fields. Thus, they tend to adsorb external gas and liquid molecules forming active centers.

The physical adsorption of the active centers to adsorbates is mainly non-selective van der Waals forces without electron transfer or chemical bond cleavage/formation, but can decrease the activation energy of the ensuing chemical adsorption. Chemical adsorption is similar to chemical reaction mainly in the form of chemical bonding forces that can activate the reactant molecules to jump from ground state to excited state and induce chemical reaction.

The most common type of multi-phase catalysis bimolecular reactions is the L–H (Langmuir-Hinshewood) process, i.e., the reaction between two molecules activated by chemical adsorption, as shown in Fig. 5.10.

For the catalytic hydrogenation of phthalate plasticizers, the L–H process is followed to analyze the reaction mechanism, in which A is phthalate molecule and B is H_2. The benzene ring in phthalate structure is adsorbed to the active center and the alkyl groups are further away from the catalyst surface. The two H atoms in H_2 dissociate with each other and are adsorbed to other active centers. The activated H atom and benzene ring undergo chemical reaction to form cyclohexane-1,2-dicarboxylate plasticizers. The reaction rate is: $r = k_2\theta_A\theta_B$, where θ_A is the coverage of phthalate molecules and θ_B is the coverage of dissociated H atoms. Therefore, the catalytic hydrogenation of phthalate plasticizers should follow the bimolecular L–H process.

The other mechanism is the Rideal process, i.e., the reaction between activated phthalate molecules on catalyst surface with gas phase molecules, as shown in Fig. 5.11.

Table 5.19 C_{DOP}, S_{DEHCH}, and Y_{DEHCH} (the second batch)

Times of repeated catalyst use	C_{DOP}/%	S_{DEHCH}/%	Y_{DEHCH}/%	Times of repeated catalyst use	C_{DOP}/%	S_{DEHCH}/%	Y_{DEHCH}/%
1	100	99.76	99.76	10	100	99.89	99.89
2	100	99.81	99.81	11	100	99.87	99.87
3	100	99.86	99.86	12	100	99.89	99.89
4	100	99.76	99.76	13	100	99.76	99.76
5	100	99.82	99.82	14	100	99.81	99.81
6	100	99.68	99.68	15	100	99.82	99.82
7	100	99.72	99.72	16	100	99.80	99.80
8	100	99.81	99.81	17	100	99.79	99.79
9	100	99.90	99.90	18	100	99.91	99.91
19	100	99.88	99.88	45	100	99.88	99.88
20	100	99.86	99.86	46	100	99.82	99.82
21	100	99.88	99.88	47	100	99.85	99.85
22	100	99.79	99.79	48	100	99.88	99.88
23	100	99.81	99.81	49	100	99.86	99.86
24	100	99.80	99.80	50	100	99.87	99.87
25	100	99.79	99.79	51	100	99.87	99.87
26	100	99.88	99.88	52	100	99.86	99.86
27	100	99.90	99.90	53	100	99.87	99.87
28	100	99.80	99.80	54	100	99.85	99.85
29	100	99.86	99.86	55	100	99.84	99.84
30	100	99.87	99.87	56	100	99.82	99.82
31	100	99.79	99.79	57	100	99.86	99.86
32	100	99.84	99.84	58	100	99.75	99.75
33	100	99.82	99.82	59	100	99.82	99.82
34	100	99.83	99.83	60	100	99.81	99.81
35	100	99.88	99.88	61	100	99.87	99.87
36	100	99.88	99.88	62	100	99.89	99.89
37	100	99.79	99.79	63	100	99.84	99.84
38	100	99.86	99.86	64	100	99.86	99.86
39	100	99.85	99.85	65	100	99.87	99.87
40	100	99.86	99.86	66	100	99.85	99.85
41	100	99.82	99.82	67	100	99.87	99.87
42	100	99.86	99.86	68	100	99.86	99.86
43	100	99.82	99.82	69	100	99.87	99.87

(continued)

Table 5.19 (continued)

Times of repeated catalyst use	C_{DOP}/%	S_{DEHCH}/%	Y_{DEHCH}/%	Times of repeated catalyst use	C_{DOP}/%	S_{DEHCH}/%	Y_{DEHCH}/%
44	100	99.87	99.87	70	100	99.82	99.82

Note Reaction conditions: 130 °C, 4.0 MPa H_2 pressure, 800 r/min, 750 mL DOP, catalyst 0.25 wt%

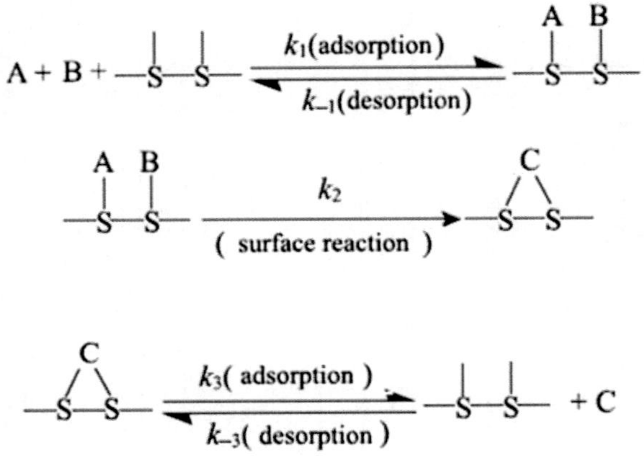

Fig. 5.10 The bimolecular L–H process

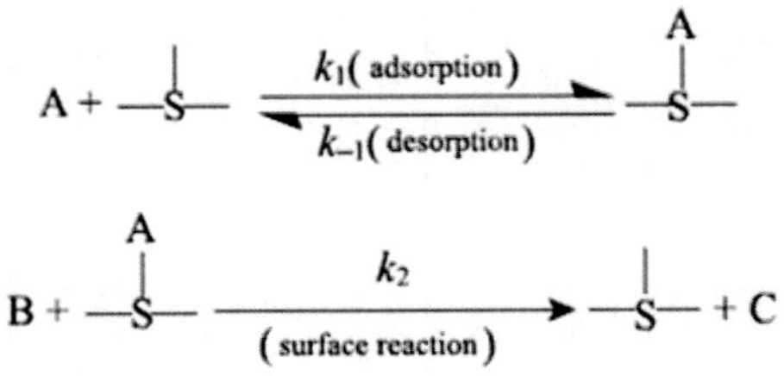

Fig. 5.11 The bimolecular Rideal process

For the catalytic hydrogenation of phthalate plasticizers, if the Rideal process is true, A is phthalate molecule with the benzene ring adsorbed to active centers and B is H_2 in gas or liquid phase. However, the molecular H_2 is difficult to have chemcial reactions with the activated benzene ring. Therefore, the catalytic hydrogenation of phthalate plasticizers is unlikely a Rideal process.

The hazards of phthalates to human, animals, and plants are primarily ascribed to the presence of a benzene ring that is stable and toxic. In phthalate structures, the large 10-atom π bonds conjugated among the benzene ring and adjacent carbonyls are certainly more stable. Therefore, phthalate plasticizers are non-degradable, tend to accumulate in human body, and have concerns of mutagenicity, carcinogenicity and teratogenicity. Long-time contact may bring damages to the liver and kidney of human and animals.

In 2002, BASF in Germany realized the structural modification of DINP via catalytic hydrogenation of the benzene ring to obtain DINCH. In 2008, BASF registered the Hexamoll® DINCH product in China as plasticizers for PVC and other polar polymers used in toxicologically sensitive applications.

For the adsorption of benzene on catalyst surface, early reports considered the Balandin theory of multilplets: benzene is aligned with metal atoms forming a planar hexaplet adsorption on metal surface. The core theme of this theory is that the activity of a catalyst largely depends on whether the atomic space lattice can aggregate reactant and product molecules. Taking hydrogenation of benzene and dehydrogenation of cyclohexane as examples, only those metals with a hexagonal arrangement of atoms and an inter-atomic spacing of 0.24–0.28 nm can have the desired catalytic activity. Pt, Pd, and Ni satisfy this requirement and are good catalysts; in this sense, Fe, Th, and Ca do not have any activity. Later, more research detected metal surface rearrangements after gas adsorption and in the course of catalytic reactions. The catalyst activity reflects the dynamic processes in the period of the reaction, while the lattice spacing only shows the geometric parameters needed for the catalyst system.

Currently, the active center theory has been widely accepted. The core is that the solid surface is not uniform and the active sites are only a small portion of catalyst surface that are called active centers. Only when a reactant is adsorbed to the active center can it be activated and undergo a chemical reaction. Different surface locations can have different active centers and exhibit different selectivities. There is a great deal of evidence to support this theory. For example, the heat of adsorption decreases with catalyst surface coverage, indicating that the adsorption with the highest thermal effect occurs only at the active centers. If a trace amount of impurity is adsorbed to the active centers, the catalyst is poisoned and loses the activity. The catalyst activity tends to be damaged by heat; before the catalyst is calcinated, the surface area does not change much but the active centers are damaged. The poisoned catalyst may lose its activity to certain reactions but still can catalyze other different reactions. The active centers can be atomic clusters or single atoms; especially the latter has currently become a hot research area.

Regarding the adsorption of H_2 on catalyst surface, metals that can enable H_2–D_2 exchange reaction are generally considered capable of dissociating H–H bond and forming adsorbed H species [10]. Fig. 5.12 shows the adsorption model for H_2 on Ru and Table 5.20 lists the molar energy of M-H bonds formed on various metals.

Figure 5.12 represents the physical snapshots of the transition state of H_2 from physisorption to chemisorption. At physisorption, the covalent H–H bond length remains unchanged. At the transition state, the H–H bond becomes loose and longer, meanwhile loose H–Ru bonds appear. At chemisorption, the H–H bond is completely broken and strong H–Ru chemical bond is formed; a H atom is bonded to one or two Ru atoms.

Major adsorption-determining factors include temperature, pressure, and characteristics of the adsorbent and adsorbate. The adsorption is an exothermal process. When temperature increases, the adsorption capacity decreases, while the adsorption rate increases. From Table 5.20, as well as the hydrogenation activity order, the bond energies of Ir, Rh, Ru, Pt, and Pd at chemisorption are neither too strong nor

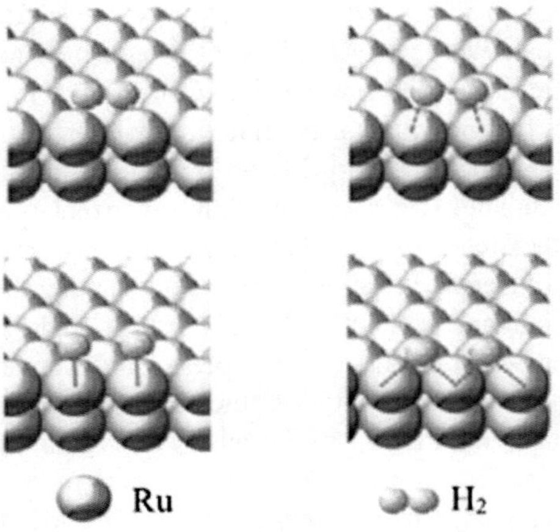

Fig. 5.12 The hydrogen adsorption model on Ru metal

Table 5.20 Molar bond energies of M–H bonds

Metal	Molar bond energy $E_{M—H}$/(kJ/mol)
Co	266
Fe	287
Ni	280
Ir, Rh, Ru	270
Pt, Pd	275

too weak; these metals usually exhibit excellent hydrogenation activity. Too strong bonding makes the reaction difficult to occur and even poison the catalyst, while too weak bonding brings no activating effects.

Hydrogenation of DOP is even more complicated, because it needs to hydrogenate the benzene ring while keeping the plasticizing ester groups. This requires adsorption of benzene ring to the catalyst but the esters kept away from the catalyst surface. This feature guides the design and development of the catalyst, but is also the challenge. Only nanocrystal catalysts can satisfy such a requirement. The kinetics studies will unravel the effects of concentration and temperature on the reaction rate, which can offer clues to understand the reaction mechanisms of multi-phase catalysis of phthalates and provide scientific basis for the development of highly active and selective catalysts.

5.2.1 *Effects of Concentration on DOP Hydrogenation Rate*

The effects of concentration on reaction rate are expressed in the forms of differential functions and integral functions. The differential functions indicate the relationship between reaction rate and reactant concentration (or pressure) at specified temperature and often have simple kinetic orders (zero or positive integers). The integral functions indicate the relationship between reactant concentration and reaction time. The rate constant stands for the rate under unit concentration and unit time, and thus can be used to compare fast or slow reactions.

The key to determine the rate law is to identify the reaction order that is the fundamental data for reactor design. Strictly speaking, the reaction order depends on the reaction mechanism, but most reaction mechanisms are not well understood. To figure out the reaction mechanism, the first thing is to look at the rate law. Therefore, the reaction order can offer clues on reaction mechanism.

The reaction order is usually experimentally determined via measuring the relationship between reactant/product concentration c and time t, i.e., the c–t data that can be processed using either integral or differential methods.

The integral method is based on the integral equation of the rate law and uses the linear relationship to determine the reaction order. For the reaction

$$\mathrm{A} \rightarrow \mathrm{P}$$

The integral equation of the rate law of a zero order reaction is

$$c_A^0 - c_A = k_0 t \tag{5.1}$$

where c_A^0 is the initial concentration of reactant A, c_A is the concentration of A at time t, k_0 is the rate constant.

c_A is plotted with respect to t to obtain a linear curve. The negative of the slope is the rate constant k_0.

The integral equation of the rate law of a first order reaction is

$$\mathrm{ln}c_A^0 - \mathrm{ln}c_A = k_1 t \tag{5.2}$$

From Eq. 5.2, $\mathrm{ln}c_\mathrm{A}$–t is a linear relationship. The negative of the slope is the rate constant k_1.

The differential method is based on the differential equation of the rate law. Assume the rate law is

$$r = -\frac{d_{c_A}}{d_t} = kc_A^n \tag{5.3}$$

Taking the logarithm to obtain

$$\mathrm{lg}r = \mathrm{lg}\left[-\frac{d_{c_A}}{d_t}\right] = \mathrm{lg}k + \mathrm{lg}c_A \tag{5.4}$$

The data are processed in two steps. In the first step, a c_A–t curve is plotted. At a series of concentrations c_1, c_2, ..., the slopes of the curve are obtained, i.e., r_1, r_2,

.... In the second step, lg$r$ is plotted as a function of lgc_A. If the rate law equation is correct, a linear relationship will be obtained, the slope n of which is the reaction order.

Differential Method 1: the reaction order is determined via measurements of a single sample. Differential Method 2: the reaction order is determined through multiple samples, typically at least 5 samples. In the two different methods, the physical meanings of the obtained reaction order are different. Method 1 is called the reaction order versus time, including the effects of product on reaction rate. Method 2 reflects the effects on reactant concentration on reaction rate and stands for what a reaction order really means.

The above two methods were both used when determining the reaction order of DOP hydrogenation. The integral method is used when measuring the reaction order in terms of DOP. The differential method is used when measuring the order in terms of H_2 pressure.

The reaction rate is related with the catalyst concentration for multi-phase catalysis systems. Theoretically, the amount of catalyst affects the reaction rate but not the reaction order. In general, the reaction order can be determined using a linear relationship obtained with a fixed catalyst loading. But using multiple linear curves obtained with a variety of catalyst loadings is also a possible option.

The DOP hydrogenation reaction can be written as

$$\mathrm{DOP} + 3\mathrm{H_2} \rightarrow \mathrm{DEHCH}$$

The rate law is expressed as

$$-\frac{d_{c_{DOP}}}{d_t} = \frac{d_{c_{DEHCH}}}{d_t} = k \bullet c_{DOP}^a \bullet p_{H_2}^b \tag{5.5}$$

where k is the rate constant; c_{DOP} is the DOP concentration; p_{H2} is the hydrogen pressure; a and b are the reaction orders with respect to DOP concentration and hydrogen pressure, respectively.

The elementary steps of multi-phase catalysis include diffusion and adsorption of reactants, surface reaction, desorption and diffusion of products. The real situation is more complicated.

When the surface reaction is the rate-determining step, the rate law can be expressed as

$$r = k\theta_{\mathrm{DOP}}\theta_{\mathrm{H}}$$

where θ_{DOP} is the coverage of DOP on catalyst surface; θ_{H} is the coverage of dissociated H atoms on catalyst surface.

when the adsorption of H_2 or DOP is the rate-determining step, the rate law is expressed as

$$r = kc_{H_2}\theta_0 \quad \text{or} \quad r = kc_{DOP}\theta_0$$

where c_{H_2} and c_{DOP} are the concentrations of H_2 and DOP in liquid phase; θ_0 is the blank coverage on catalyst surface.

When the product desorption is the rate-determining step, the rate law is expressed as

$$r = k_d\theta_{DEHCH}$$

where k_d is the DEHCH desorption rate constant; θ_{DEHCH} is the coverage of DEHCH on catalyst surface.

Sufficiently high stir rates can eliminate the diffusion control of H_2 in liquid phase, thin out the retentate layer on catalyst surface, and enhance the mass transport at the liquid/solid interface. Inside the catalyst nanocrystals, the channels become less and shorter leading to decreased control by the inner diffusion. The reaction mainly occurs at the outer surface of catalysts.

The reaction order with respect to DOP concentration is determined using the graphing (or integral) method.

Assume the rate law for DOP hydrogenation is

$$-\frac{d_{c_{DOP}}}{d_t} = \frac{d_{c_{DEHCH}}}{d_t} = k \bullet c_{DOP}^a \tag{5.6}$$

where $k^{'} = kp_{H_2}^b$ is the apparent rate constant, merging the $p_{H_2}^b$ in Eq. 5.5 into k, which requires a constant hydrogen pressure. For zero order reaction, $a = 0$ in Eq. 5.6, and the indefinite integral equation is expressed as

$$c_{DEHCH} = k't + C \tag{5.7}$$

where C is the indefinite integral constant.

Telling from Eq. 5.7, the c_{DEHCH}-t relationship is linear. The slope is the apparent rate constant k'.

1. c_{DEHCH}–t curves at different catalyst loadings

Figure 5.13 shows the c_{DEHCH}-t curves at different catalyst loadings. The c_{DEHCH} were measured in the samples taken every 10 min.

Under the conditions of 130 °C, 3.0 MPa hydrogen pressure and 800 r/min, the c_{DEHCH}–t relationship is linear at different catalyst loadings, which is a typical zero order reaction. As the reaction proceeds, DOP is consumed and DEHCH is produced. As DOP and DEHCH is miscible, the DOP concentration decreases gradually. The linear c_{DEHCH}–t relationship suggests that the reaction rate is independent on DOP

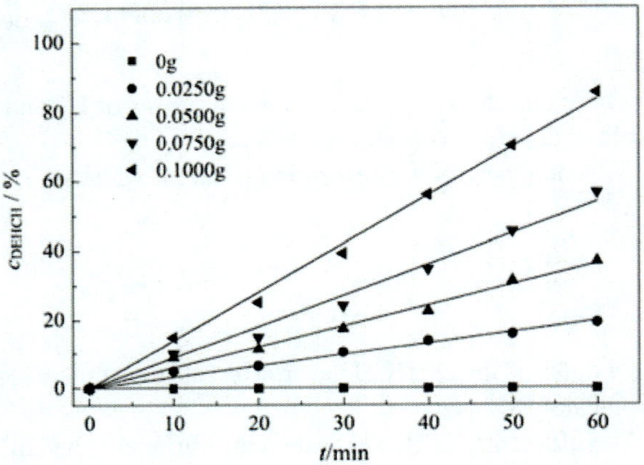

Fig. 5.13 The c_{DEHCH}–t curves at different catalyst loadings. Reaction conditions: 130 °C, 3.0 MPa H_2 pressure, 800 r/min

concentration in a zero order. As the catalyst loading increases, the slope and the rate constant k increase, indicating faster rates with higher catalyst loadings. Table 5.21 shows the rate constants of DOP hydrogenation reaction at different catalyst loadings.

The rate constant k is plotted against the catalyst loading, as shown in Fig. 5.14. The DOP hydrogenation rate increases linearly with the catalyst loading. A mechanistic explanation is that the increase in catalyst loading leads to a linear increase in the surface and the number of active centers. The catalyst nanocrystals are highly dispersed in DOP to have reactions on their outer surfaces. Theoretically, only nano-scale catalyst particles can approximately exhibit this type of relationship.

Table 5.21 The DOP hydrogenation reaction rate constants at diflerent catalyst loadings

Catalyst Loading/g	0	0.0250	0.0500	0.0750	0.1000
k/(concentration/min)	0	0.328	0.603	0.896	1.394

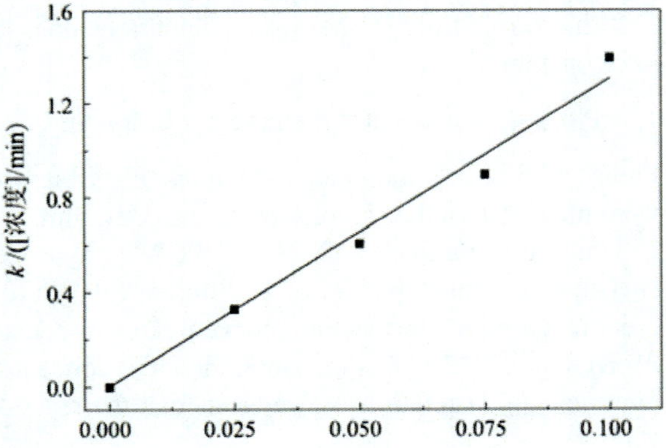

Fig. 5.14 The rate constant k as a function of catalyst loading

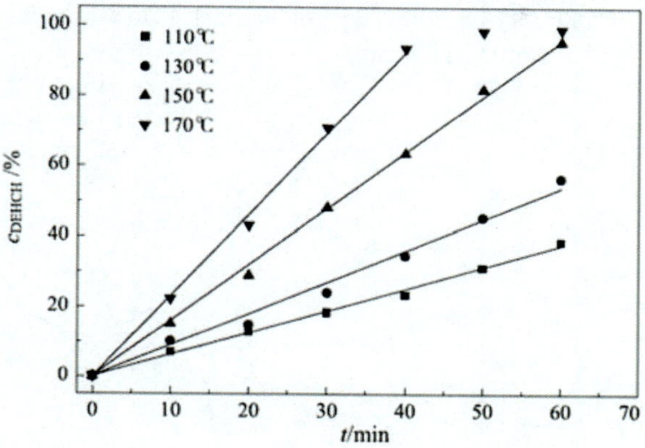

Fig. 5.15 The c_{DEHCH}–t curves at different temperatures. Reaction conditions: 3.0 MPa H_2 pressure, 800 r/min

2. c_{DEHCH}-t curves at different temperatures

Under a fixed catalyst/DOP mass ratio of 0.25%, the reaction orders with respect to DOP concentration were investigated at different temperatures.

Figure 5.15 shows the c_{DEHCH}-t curves at 110, 130, 150, and 170 °C under the fixed amount initial DOP, catalyst loading, stir rate, and hydrogen pressure.

The c_{DEHCH}–t curves are linear at different temperatures. As the temperature increases, the slope (i.e., rate constant) increases. This indicates the hydrogenation rate is zero order to DOP concentration, independent of temperature. Similar to the case of different catalyst loadings, under given temperature and catalyst, DOP is gradually consumed and its concentration decreases as the reaction proceeds.

Table 5.22 shows the rate constants of DOP hydrogenation reaction at different temperatures.

To explore the common features of the hydrogenation reactions of phthalates and terephthalates, the DBP and DOTP hydrogenation reactions were investigated to obtain the c_{DBHCH}–t and c_{DOTCH}–t curves.

The DBP hydrogenation reaction is

$$\mathrm{DBP} + 3\mathrm{H}_2 \rightarrow \mathrm{DBHCH}$$

The DOTP hydrogenation reaction is

$$\mathrm{DOTP} + 3\mathrm{H}_2 \rightarrow \mathrm{DOTP}$$

Table 5.22 The DOP hydrogenation reaction rate constants at different temperatures

Temperature/°C	110	130	150	170
Thermodynamic temperature /K	383.15	403.15	423.15	443.15
k/(concentration/min)	0.6258	0.8965	1.5977	2.3185

DOTCH is dioctyl cyclohexane-1,4-dicarboxylate. The molecular structures of DOTP and DOTCH are

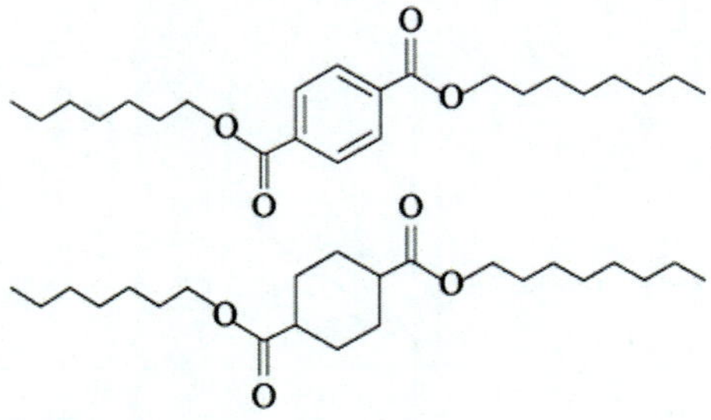

Although DBP and DOTP have different alkyl chain lengths and sites from DOP, their products share the same feature of hydrogenated benzene ring with carbonyls untouched. That is, they all require the benzene ring adsorbed to the catalyst and the alkyl groups kept away from catalyst surface. Theoretically, they should have the same rate law.

Figure 5.16 shows the c_{DBHCH}-t curves at 110, 130, 150, and 170 °C under the fixed amount initial DBP, catalyst loading, stir rate, and hydrogen pressure.

The c_{DBHCH}–t curves are linear at different temperatures. Similar to DOP, the DBP hydrogenation rate is zero order to DBP concentration.

Table 5.23 shows the rate constants of DBP hydrogenation reaction at different temperatures.

Figure 5.17 shows the c_{DOTCH}–t curves at 110, 130, 150, and 170 °C under the fixed amount initial DOTP, catalyst loading, stir rate, and hydrogen pressure.

The c_{DOTCH}–t curves are linear at different temperatures. Similar to DOP and DBP, the DOTP hydrogenation rate is zero order to DOTP concentration.

Table 5.24 shows the rate constants of DOTP hydrogenation reaction at different

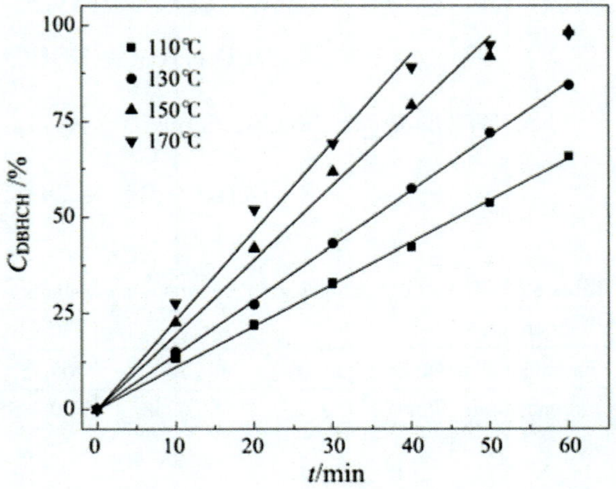

Fig. 5.16 The c_{DBHCH}–t curves at different temperatures. Reaction conditions: 3.0 MPa H_2 pressure, 800 r/min

Table 5.23 The DBP hydrogenation reaction rate constants at different temperatures

Temperature /°C	110	130	150	170
Thermodynamic temperature /K	383.15	403.15	423.15	443.15
k/(concentration/min)	1.086	1.422	1.950	2.317

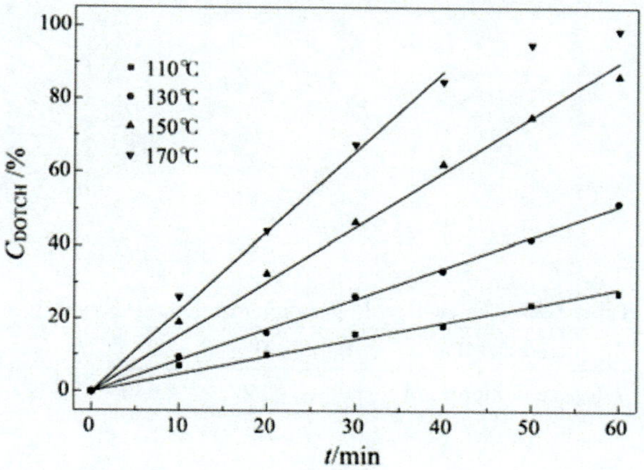

Fig. 5.17 The c_{DOTCH}–t curves of DOTP hydrogenation reaction at different temperatures

Table 5.24 The DOTP hydrogenation reaction rate constants at different temperatures

Temperature/°C	110	130	150	170
Thermodynamic temperature/K	383.15	403.15	423.15	443.15
k/(concentration/min)	0.473	0.849	1.502	2.189

temperatures.

The hydrogenation rates are all zero order to the reactant concentration for DBP, DOTP, and DOP, indicating that they have the same reaction mechanism: during the reaction the benzene ring is adsorbed to the catalyst while the alkyl group is kept away from catalyst surface.

3. c_{DEHCH}-t curves at different hydrogen pressures

Under fixed catalyst loading, reaction temperature and stir rate, the c_{DEHCH}–t curves were measured at different hydrogen pressures and are shown in Fig. 5.18.

The c_{DEHCH}–t curves are linear at different hydrogen pressures. As the hydrogen pressure increases, the slope (i.e., rate constant) increases indicating faster rates. Table 5.25 shows the rate constants of DOP hydrogenation reaction at different hydrogen pressures.

Based on Figs. 5.13, 5.14, 5.15, 5.16, 5.17, and 5.18, a conclusion can be reached that the reaction rates of catalytic hydrogenation reactions of phthalates and terephthalates are zero order to reactant concentration. The reaction order is independent on catalyst loading, temperature and hydrogen pressure.

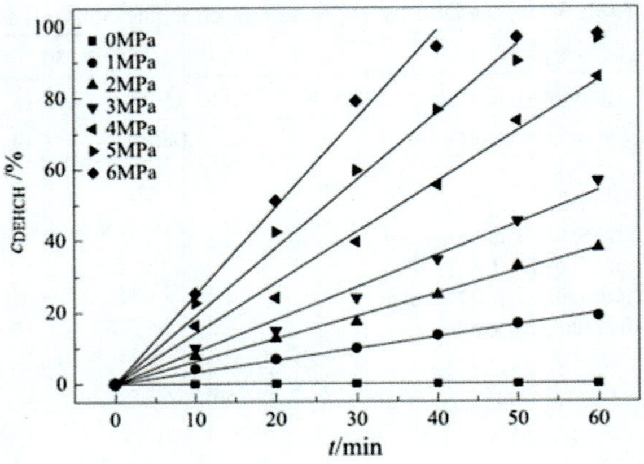

Fig. 5.18 The c_{DEHCH}-t curves at different H_2 pressures

Table 5.25 The DOP hydrogenation reaction rate constants at different H_2 pressures

H_2 pressures/MPa	0	1.0	2.0	3.0	4.0	5.0	6.0
k/(concentration/min)	0	0.328	0.630	0.897	1.412	1.895	2.470

The zero order reaction of multi-phase catalysis is often explained as follows: the reaction occurs on catalyst surface and the rate depends on reactant concentration on surface; as the reactant adsorption has reached saturation, changes of the reactant concentration in gas or liquid phase do not alter its surface concentration; the reaction rate is independent on the reactant concentration, i.e., zero order.

The zero order reaction indicates that the catalyst surface has a high affinity to benzene ring and the active centers have excellent adsorption and activation effects.

A feature of zero order reaction is the limited time for reaction completion. It is similar to the uniform linear motion in physics, in that, a limited time is needed to complete a distance at a fixed speed. No other reaction orders have this feature. For example, the time for completion of first order (decay of radiative element) and second order reactions is infinite. This feature of zero order reaction provides a theoretical basis for achieving 100% conversion of phthalates in catalytic hydrogenation. Domestic and foreign laws require that the 6P content does not exceed 0.1%, i.e., 1000 ppm. With this catalytic hydrogenation technology, phthalate residual concentrations are lower than the detection limit (10 ppm) in the inspections by different authorities. And this is theoretically possible.

Based on Table 5.25, the reaction order with respect to hydrogen pressure can be determined using the differential method.

For the rate law of DOP hydrogenation reaction in Eq. 5.5

$$-\frac{dc_{\text{DOP}}}{dt} = \frac{dc_{\text{DEHCH}}}{dt} = k \cdot c_{\text{DOP}}^{a} \cdot p_{\text{H}_2}^{b}$$

The reaction order with respect to DOP concentration is zero ($a = 0$). Then the above equation is rewritten as

$$r = -\frac{d_{c_{\text{DOP}}}}{d_t} = \frac{d_{c_{\text{DEHCH}}}}{d_t} = k \bullet p_{H_2}^b \tag{5.8}$$

Taking the logarithm at both sides

$$\lg r = \lg\left[-\frac{d_{c_{\text{DOP}}}}{d_t}\right] = \lg\left[\frac{d_{c_{\text{DEHCH}}}}{d_t}\right] = \lg k + \lg p_{\text{H}} \tag{5.9}$$

The plot of $\lg r$ as a function of lg p_{H_2} should be linear. The slope is the reaction order of the rate with respect to hydrogen pressure.

Theoretically, to derive the reaction order b, one should obtain both the p_{H_2}–t curves and the reaction rate r at time t using the differential method, followed by plotting the $\lg r$–lg p_{H_2} relationship. In reality, the experiments proceed under constant hydrogen pressure. Therefore, a different method can be used to obtain b.

From the c_{DEHCH}–t curves in Fig. 5.16, one can obtain

$$r = \frac{dc_{\text{DEHCH}}}{dt} = k$$

The $\lg r$–lg p_{H_2} plot should be linear. The slope is the reaction order b. The $\lg k$ and lg p_{H_2} values are obtained from Table 5.25 and are shown in Table 5.26.

The $\lg k$ is plotted against lg p_{H_2}, shown in Fig. 5.19.

The slope is $b \approx 1$. Thus, the DOP hydrogenation rate is approximately first order to hydrogen pressure.

In summary, the rate law of DOP hydrogenation is expressed as

$$r = -\frac{dc_{\text{DOP}}}{dt} = \frac{dc_{\text{DEHCH}}}{dt} = k \cdot p_{H_2}$$

Table 5.26 Values of $\lg k$ and $\lg p_{H_2}$

lg p_{H_2}	0	0.301	0.477	0.602	0.699	0.778
lg k	−0.484	−0.201	−0.047	−0.150	0.278	0.393

Fig. 5.19 $\lg k$–$\lg p_{H_2}$ plot

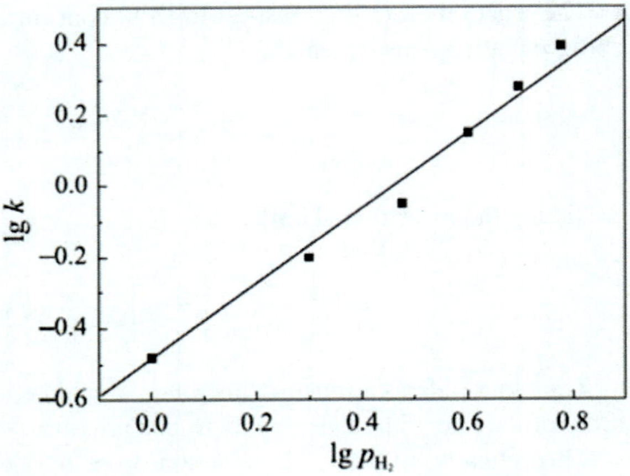

5.2.2 *Arrhenius Equation of Phthalate Hydrogenation Reaction*

To investigate the effects of temperature on reaction rate, the reactant concentration must be fixed. In fact, this is the investigation of the effects of temperature on rate constant *k* using Arrhenius equation.

From Figs. 5.15, 5.16, 5.17, 5.18, 5.19, and 5.20, the reaction rate increases with temperature for DOP, DBP, and DOTP.

Using the rate constants *k* in Tables 5.18, 5.19, and 5.20, the ln*k*–T^{-1} curves are plotted in Figs. 5.20, 5.21, and 5.22.

According to the slope in Fig. 5.20, the apparent activation energy of DOP hydrogenation is $Ea = 31.75$ kJ/mol.

According to the slope in Fig. 5.21, the apparent activation energy of DBP hydrogenation is $Ea = 18.31$ kJ/mol.

Fig. 5.20 The ln*k*–T^{-1} plot of DOP hydrogenation reaction

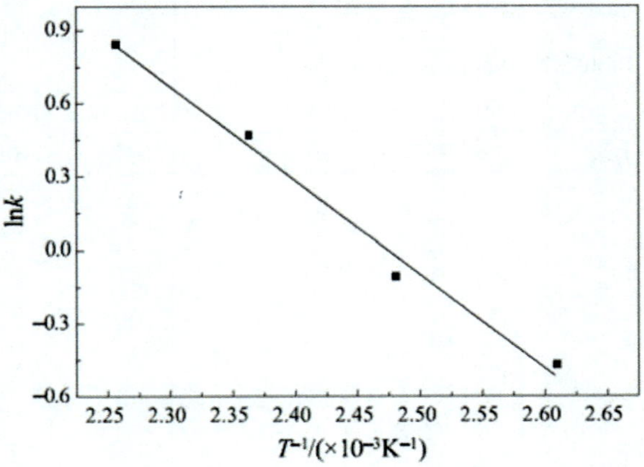

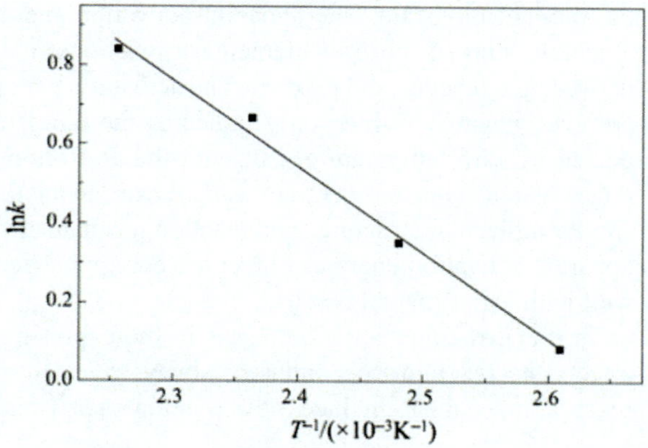

Fig. 5.21 The ln$k$–T^{-1} plot of DBP hydrogenation reaction

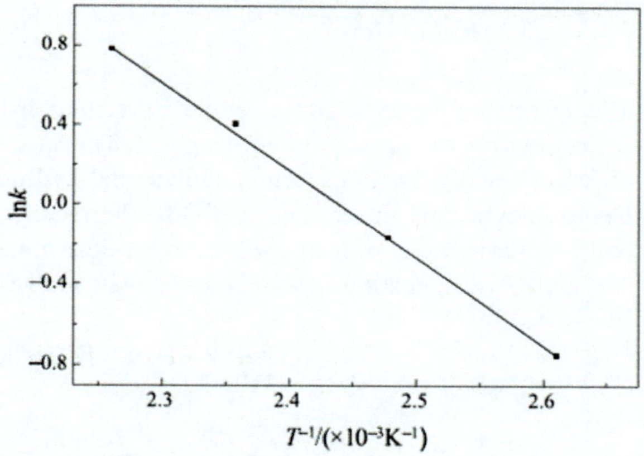

Fig. 5.22 The ln$k$–T^{-1} plot of DOTP hydrogenation reaction

According to the slope in Fig. 5.20, the apparent activation energy of DOTP hydrogenation is $E\mathrm{a} = 36.83$ kJ/mol.

Then, their Arrhenius equations are written as

$$\text{DBP } k = \mathrm{A}\exp\left(\frac{18.31}{RT}\right)$$

$$\text{DOP } k = \mathrm{A}\exp\left(\frac{31.75}{RT}\right)$$

$$\text{DOTP } k = \mathrm{A}\exp\left(\frac{36.83}{RT}\right)$$

Theoretically, only the activation energy of elementary reactions has clear physical meaning, representing the average energy of 1 mol activated molecules above regular molecules, i.e., the energy barrier that the reactant molecule must overcome to form the product molecule. The apparent activation energy of complex reactions is the algebraic sum of those of elementary reactions; it's no longer the energy barrier between the reactant and product. The activation energies of some complex reactions are even negative, which only indicates the temperature coefficient, i.e., the rate decreases with temperature. But with the activation energy, one can calculate the rate constant at one temperature using the one at another temperature. Because of the similar molecular structures and reaction mechanisms for DOP, DBP, and DOTP, the apparent activation energies reflect the energy barriers and reaction rates that agree well with experimental results.

In the Arrhenius equation, Ea is in the exponential term and thus significantly affects the reaction rate. Industry always select the optimal catalyst to tailor the reaction mechanism, reduce the activation energy and enhance the reaction rate.

5.2.3 *Conclusions*

The effects of concentration, temperature, and hydrogen pressure to phthalate hydrogenation reactions were investigated. The c_{DEHCH}–t relationship curves under different catalyst loadings, temperatures. and hydrogen pressures were measured. Using integral and differential methods, the reaction orders with respect to DOP concentration and hydrogen pressure were determined to be 0 and 1, respectively. The DOP hydrogenation reaction and rate law are written as

$$\mathrm{DOP} + 3\mathrm{H_2} \rightarrow \mathrm{DEHCH}$$

$$r = -\frac{dc_{\mathrm{DOP}}}{dt} = \frac{dc_{\mathrm{DEHCH}}}{dt} = k \cdot p_{\mathrm{H_2}}$$

The conclusions apply to catalytic hydrogenation reactions of DBP and DOTP.

Based on the ln$k$–T^{-1} plots, the apparent activation energies of DOP, DBP, and DOTP hydrogenation reactions were determined as 31.75 kJ/mol, 18.31 kJ/mol, and 36.83 kJ/mol, respectively.

The kinetics study provides the theoretical basis for 100% conversion of phthalates, offers clues on identifying reaction mechanisms, and guides the development of highly active and selective catalysts.

5.3 Characterizations of Catalysts

In order to unravel the correlation between the preparation, the performance and the structure, texture and surface properties, the Ru-based catalysts were characterized using XRD, TEM, XPS, and low-temperature N_2 physisorption.

5.3.1 XRD

XRD can offer the structural information. Powder XRD spectra of Ru/ZrO_2, Ru/Al_2O_3, Ru/diatomite and non-supported Ru catalysts are shown in Fig. 5.23.

Seen from Fig. 5.23a–c, after introduction of Ru, the characteristic peaks of the ZrO_2, Al_2O_3 and diatomite supports become weaker. The characteristic peaks of Ru in the 2θ range of 40–45° become shorter and broadened. This indicates that, in supported catalysts, the size of active composition Ru nanocrystals becomes smaller; meanwhile the presence of Ru leads to higher dispersity of the support.

Figure 5.23d shows the XRD spectra of non-supported Ru catalyst before and after the hydrogenation reaction. The peaks at 2θ angles of 38.4°, 42.2°, 44.0°, 58.2°, 69.5°, and 78.4° correspond to the (100), (002), (101), (102), (110), and (103) planes of the hexagonal Ru crystal [11–15]. According to Bragg Eq. $2d\sin\theta = n\lambda$ ($\lambda = 0.15418$ nm), the d-spacings of Ru(100), (002), (101), (102), (110), and (103) planes are obtained as 2.34 Å, 2.14 Å, 2.06 Å, 1.58 Å, 1.35 Å, and 1.22 Å, respectively. According to Scherrer equation, the Ru nanocrystal size (Å) = $85.729\theta/\beta_{1/2}$, where θ is the width (mm) of 1° in radian; $\beta_{1/2}$ is the peak width at the half height of Ru(101)

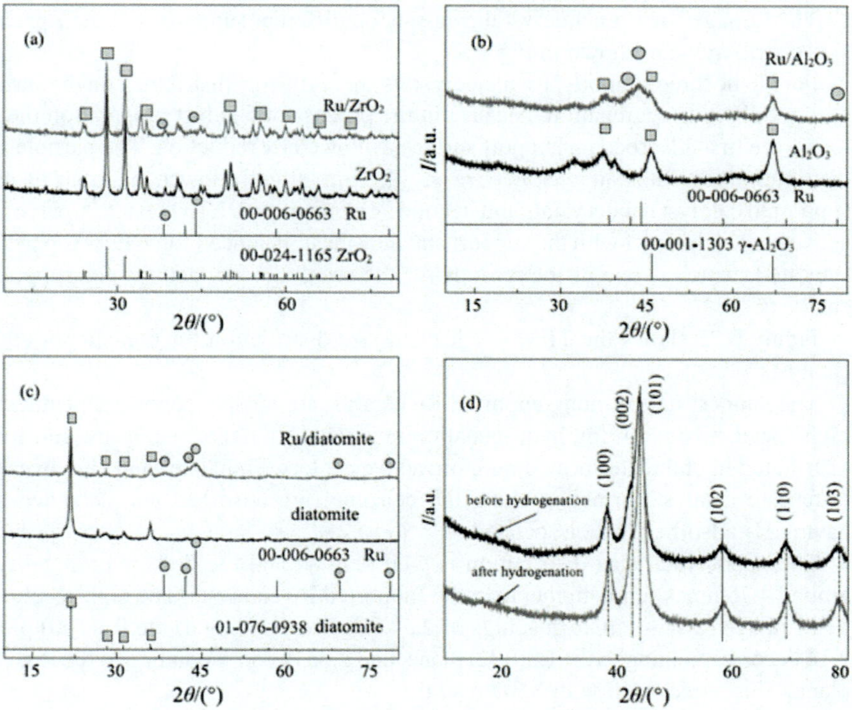

Fig. 5.23 XRD spectra of Ru/ZrO_2 (**a**), Ru/Al_2O_3 (**b**), Ru/diatomite (**c**), and non-supported Ru catalysts (**d**)

peak. The Ru nanocrystal size is 4.31 nm before hydrogenation. After 20 repeated uses, the size increases slightly to 5.02 nm.

Nanocrystals are composed of a limited number of atoms or atomic clusters with high ratios of surface to inner atoms. The surface atoms are in unbalanced force fields and usually have high chemical activity.

Preparation of cyclohexane-1,2-dicarboxylate plasticizers from phthalates features hydrogenation of the benzene ring with the ester groups unchanged, which requires adsorption of the benzene ring to catalyst surface with the ester group kept away. Only when the catalyst nanocrystal size is close to molecular size can it is possible to satisfy this requirement. Although the particle size distribution is the random combination of Ru nanocrystals, the nano-sized catalysts provide a possibility to meet the above requirement.

5.3.2 *Tem*

TEM provides the information of surface morphology and particle size for the catalysts. TEM images of supported and non-supported Ru catalysts are shown below.

TEM images of Ru nanocrystal catalysts on different supports and their particle size distribution are shown in Fig. 5.24.

For all the three supports, Ru nanocrystals are uniformly distributed on the surface without obvious aggregation, because in the process of catalyst preparation the Ru ions were first adsorbed on support surface, followed by reduction. The particle size distribution indicates an average size of ~4.2 nm, slightly lower than that of non-supported catalyst nanocrystals and in good agreement with XRD result. A part of Ru active centers interact with the support and thus the unbalanced force field is partially saturated, which is one of the reasons for the relatively low activity for supported catalysts.

Figure 5.25 shows the TEM images and size distribution for non-supported Ru catalysts.

The nanocrystals of non-supported Ru catalyst are loosely connected. Under the high-speed stirring during hydrogenation reactions, Ru nanocrystals are uniformly distributed in phthalates providing more active centers. Phthalates and H_2 can easily diffuse to catalyst surface, making the benzene ring adsorbed and activated and having H_2 adsorbed and dissociated.

The nanocrystal size distribution is well defined and the highest probability is around 4.75 nm, slightly higher than the supported Ru catalysts. The high-resolution TEM image reveals lattice spacings of 2.34 Å corresponding to the Ru(100) plane, 2.14 Å corresponding to the Ru(002) plane, and 2.05 Å corresponding to the Ru(101) plane. This agrees well with XRD result.

Since 1990s, the rapid development of nanoscience has provided a new strategy to design and develop highly active and selective catalysts, as well as a new perspective for understanding the activity and selectivity of catalysts. Nanocrystals are composed by a limited number of atoms or atomic clusters. Due to their size within

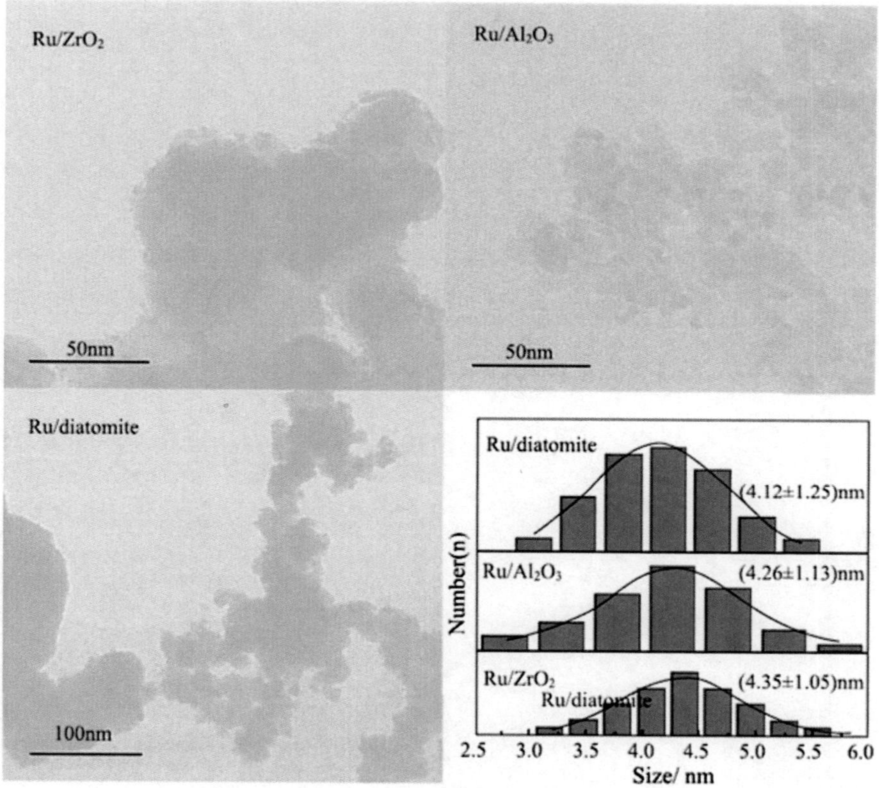

Fig. 5.24 TEM images of Ru/ZrO_2 (**a**), Ru/Al_2O_3 (**b**), Ru/diatomite (**c**), and their particle size distributions

1–100 nm, nanocrystals can easily form the so-called single-atom or atomic cluster active centers. Nanoparticles have four fundamental features: small dimension effect, surface effect, quantum size effect, and macroscopic quantum tunneling effect. These features enable a series of novel physical and chemical properties and afford high activity and selectivity for catalysts.

However, the high dispersity of nano catalysts also brings challenges to their use. Phthalate hydrogenation experiments indicate that the catalysts can be easily separated via gravimetric precipitation, especially at high temperatures with decreased product viscosity, which helps achieve the recycling of catalysts.

5.3.3 *Texture Analysis of Catalysts*

Low-temperature N_2 physisorption can provide texture information such as BET surface area, average pore size and total pore volume. The adsorption–desorption

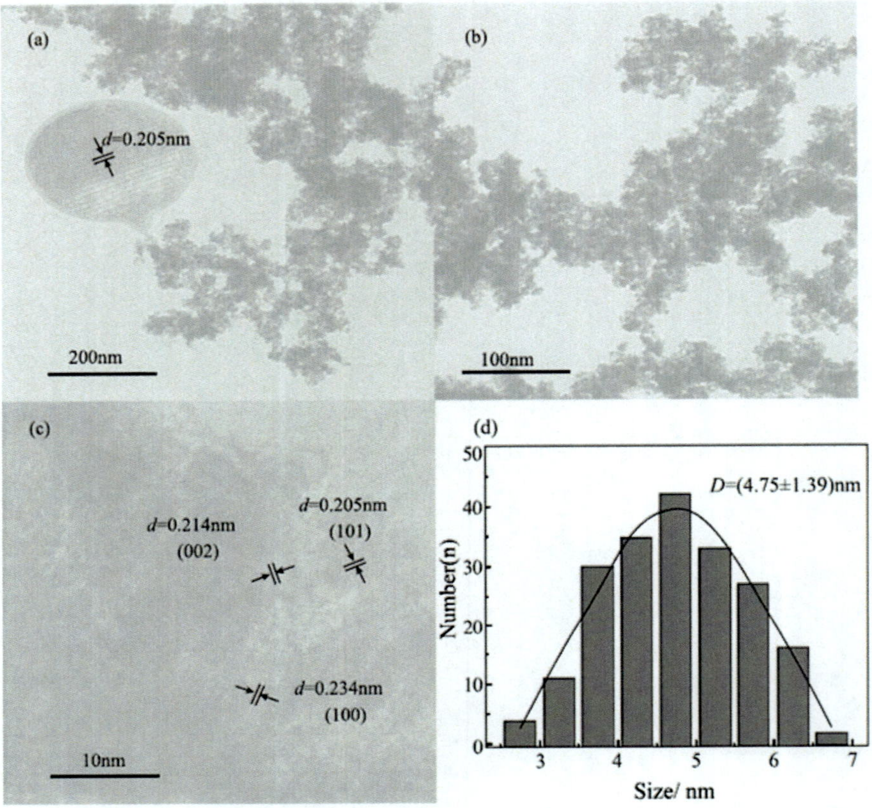

Fig. 5.25 TEM images of non-supported Ru catalyst and particle size distribution

isotherms of the catalyst were measured at liquid N_2 temperature (−195.8 °C) using N_2 as the adsorbent, as shown in Fig. 5.26.

Table 5.27 lists the surface area S_{BET}, average pore size D_{pore}, and total pore volume V_{total} of non-supported Ru catalysts obtained using BET equation and BJH effective cylinder model.

From Fig. 5.26a–c, the adsorption–desorption isotherms of Ru/ZrO_2, Ru/Al_2O_3, and Ru/diatomite catalysts are roughly identical to those of the ZrO_2, Al_2O_3, and

Table 5.27 S_{BET}, D_{pore} and V_{total} of non-supported Ru catalyst

Sample	$S_{BET}/(m^2/g)$	$D_{孔}$/nm	$V_{总}/(cm^3/g)$
Ru BH[a]	73	3.67	0.13
Ru AH[b]	70	3.67	0.13

Note
[a]Non-supported Ru catalyst before hydrogenation
[b]non-supported Ru catalyst after 20 repeated uses

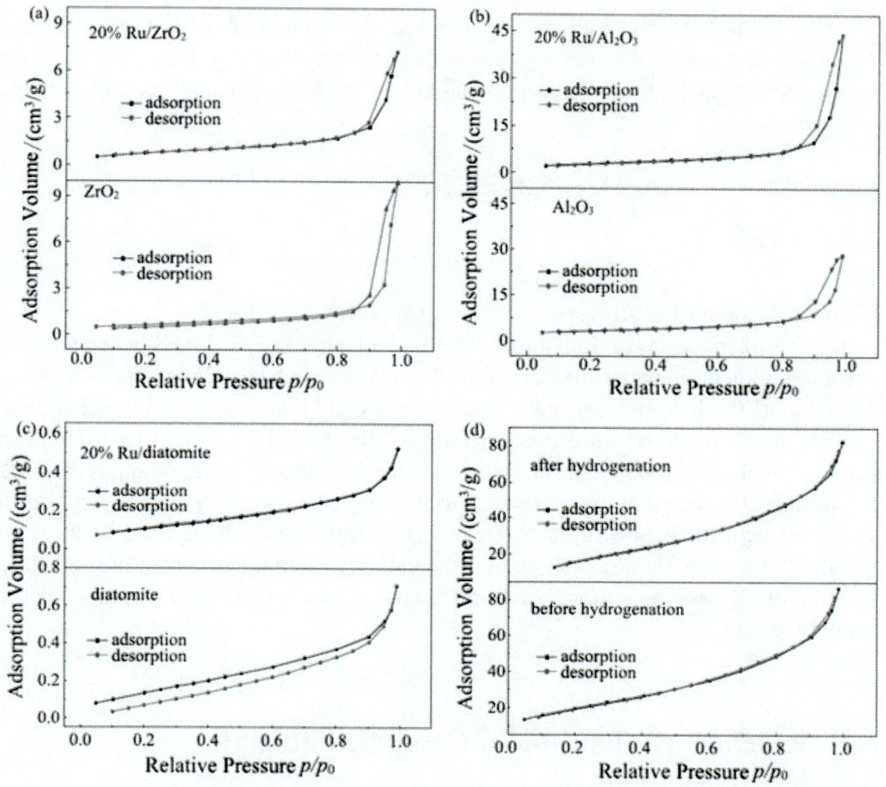

Fig. 5.26 N_2 adsorption–desorption isotherms of Ru/ZrO_2, Ru/Al_2O_3, Ru/diatomite, and non-supported Ru catalysts

diatomite supports, suggesting that the texture properties of supported catalysts are controlled by the support. The occurrence of catalytic phthalate hydrogenation reactions inside the pores is not desired because of the potential challenges for materials diffusion, adsorption, desorption, and even side reactions. The lower activity of supported than non-supported catalysts is mostly because of the texture structure of the support.

In the adsorption–desorption isotherms, the low-pressure zone (relative pressure p/p_0 <0.8) is related with single-layer adsorption. Single molecular layer adsorption is reversible and does not have hysteresis. The separation of isotherm lines occurs at p/p_0 >0.8, which is typical for mesoporous structures.

From Fig. 5.26d, the adsorption and desorption isotherm lines of non-supported catalysts are almost completely overlapped, suggesting that both adsorption and desorption mainly occur on the outer surface of catalysts. Only with very small nanocrystal size and very few inner pores can this phenomenon occur. This finding provides an experimental support to the linear increase of DOP hydrogenation rate constant with the loading of non-supported Ru catalyst (Fig. 5.14).

From Table 5.27, the catalyst surface areas before and after hydrogenation are 73 m^2/g and 70m^2/g, respectively, yet with unchanged average pore size and total pore volume. The texture properties can explain the high repeatability and stability of catalysts.

5.3.4 *XPS*

Figure 5.27 shows the XPS spectra of non-supported Ru catalysts.

In the full spectrum of non-supported Ru catalyst in Fig. 5.27a, O peaks are also observed, indicating partial oxidation of Ru on catalyst surface. From Fig. 5.27b, the $3d_{5/2}$ line is divided into two peaks. The 280 eV one belongs to metallic Ru and the 281 eV one to oxidized Ru. Ru is partially oxidized when exposed to air. This can explain the observation that the activity is relatively low in the first use and is restored to normal after repeated uses. Before the first use, the catalyst is usually stored in air for a long time or overnight. After repeated uses, the oxidized Ru is reduced to metallic Ru in the H_2 atmosphere and the catalytic activity becomes normal. It is the metallic Ru as single atoms or atomic clusters, not the oxidized Ru, that has the catalytic effect.

5.4 Reaction Mechanisms of Catalytic Phthalate Hydrogenation

The ring hydrogenation of phthalates uses a multi-phase catalysis system involving solid catalyst, liquid phthalate, and hydrogen gas. The catalyst is in the form of 3–6 nm nanocrystals composed of a number of atoms and atomic clusters of transition metals with empty *d* orbitals. On its surface exist single-atom or multi-atom

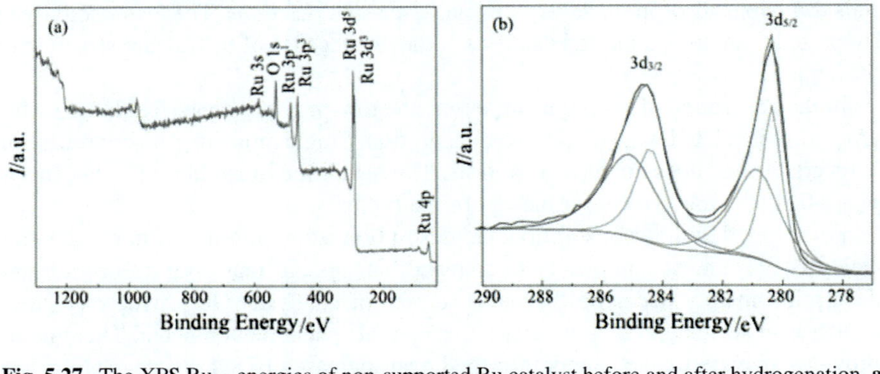

Fig. 5.27 The XPS Ru_{3d} energies of non-supported Ru catalyst before and after hydrogenation. **a** Full spectrum before hydrogenation; **b** Ru_{3d} energies after hydrogenation

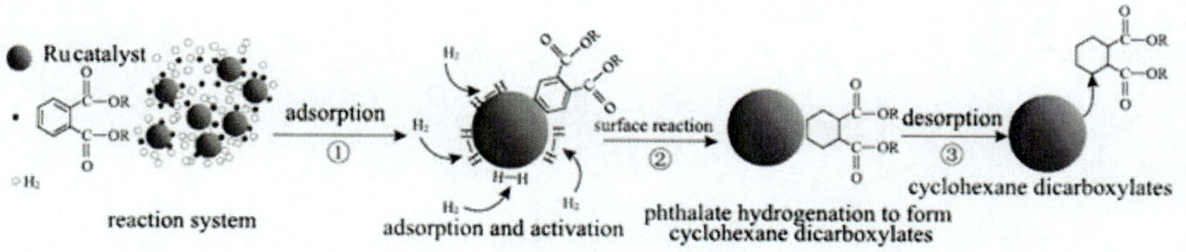

Fig. 5.28 Schematic multi-phase catalysis mechanism of catalytic hydrogenation of phthalates

active centers with high affinity to benzene ring. H_2 molecules are dissociated into H atoms at the single-atom active centers via a series of elementary steps including dissolution, diffusion, and adsorption. The benzene ring in phthalate is adsorbed at the multi-atom active centers, while the two ester groups are kept away from the active centers. Following a bimolecular L–H reaction process, the dissociated H and activated benzene ring react with each other. The benzene ring is 100% converted to cyclohexane with the unaffected ester groups to form the cyclohexane-1,2-dicarboxylate. Due to the decrease in electron density, the six-membered ring is rapidly desorbed with a low possibility of re-adsorption. Thus, the production protocol is simplified with the only need to separate out the low-boiling-point side products, leading to achievement of high selectivity and high yield for the target product under relatively mild conditions, as shown in Fig. 5.28.

Through the GC-MS analysis of the product and distillates, the mechanisms of parasitic side reactions including decarboxylation and isomerization of phthalates are illustrated in Fig. 5.29.

In Fig. 5.29, DOP is thermally decomposed into phthalic anhydride, 2-ethylhexanol and 2-ethylhexene. DOP is hydrogenated to form DEHCH. Phthalic anhydride is converted to heptane via multiple reduction and decarboxylation. 2-ethylhexene is converted to 1,4-dimethylcyclohexane via free radical reactions.

Figure 5.30 shows the possible reaction mechanisms during reduced pressure distillation.

DEHCH can be possibly converted to 5,6-dipropyldecane, 2-methxoy cyclohexanecarboxylic acid lactone, and cyclobutyl cyclohexanecarboxylate.

5.5 Characterizations of PVC Plasticized by Cyclohexane-1,2-Dicarboxylates and Phthalates

Plasticizers not only need excellent weather-resistance (light-, heat- and oxidation-stability) and durability (evaporation-, extraction- and migration-resistance), but also afford excellent mechanical, thermal, and photonic properties to the plasticized PVC materials. Therefore, DEHCH, DOP, and plasticized PVC materials were characterized for their mechanical, thermal, dynamic mechanical, rheological, photonic, anti-extraction, and anti-evaporation properties.

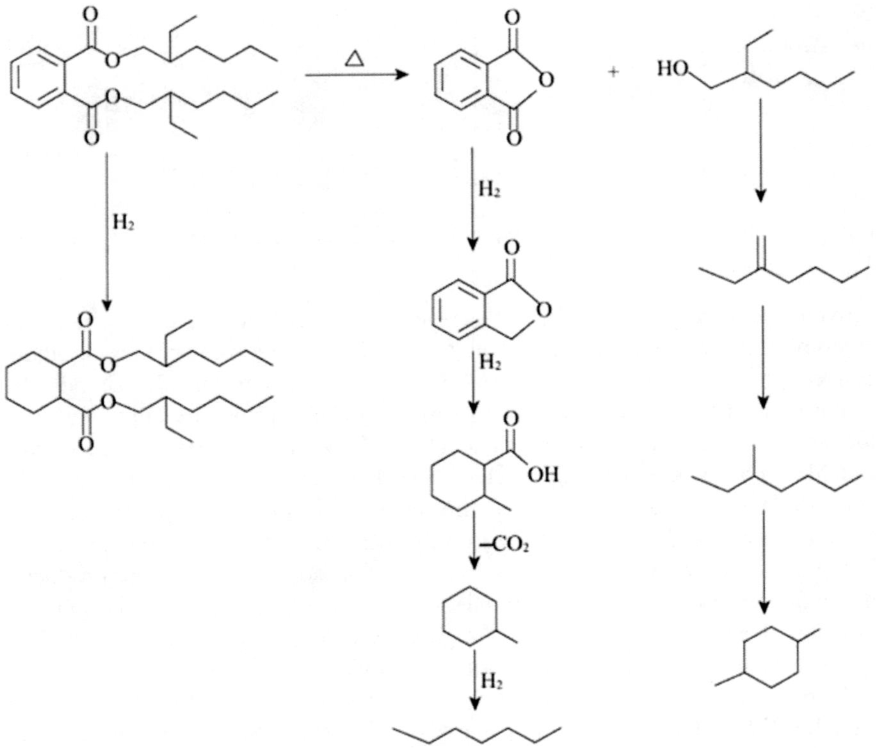

Fig. 5.29 Catalytic reaction mechanisms during the hydrogenation reaction

5.5.1 Mechanical Properties of PVC/DEHCH and PVC/DOP Composite Materials

To S-G5 PVC resin were added 2phr (parts per hundreds of resin) dibutyltin dilaurate (DBTDL, 95%, Aladdin) heat stabilizer. After drying at 80 °C for 5 h, DEHCH or DOP plasticizer at different ratios (10–70 phr) were added. The mixture was blended with a high-speed mixer and was plasticized with a double-roll mill at 160–165 °C. Then, the mixture was cooled down naturally and crushed using a crusher. Finally, the material was placed in a 1 mm-thick dumbbell mold, hot pressed for 5 min at 180 °C and 10 MPa using a flat vulcanization machine (Qingdao YaXing, 600 mm × 600 mm × 2 mm), and cold pressed for 5 min to obtain the DEHCH- or DOP-containing PVC test plates. Their mechanical and rheological properties were measured using an electronic universal testing machine (Shenzhen New SanSi, CMT 5104) and a rotational rheometer (UK Malvern Panalytical, Bohlin Gemini2). Their thermal properties were measured using thermogravimetric Analyzer (German NETZSCH, TG 209), differential scanning calorimeter (German NETZSCH, DSC 204), and dynamic mechanical analyzer (German NETZSCH, DMA 242).

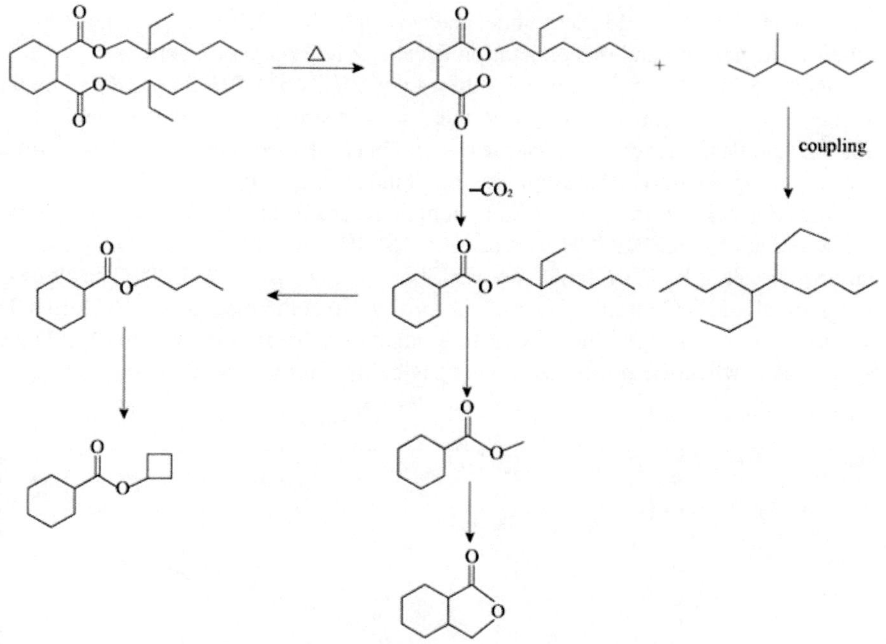

Fig. 5.30 Possible reaction mechanisms during the reduced pressure distillation

Mechanical properties include elongation at break, tensile modulus, tensile strength, and tensile yield stress.

The elongation at break refers to the ratio between increased length and initial length after breakage of the tested specimen, expressed as δ. It measures the plasticity of a material.

The tensile modulus is the elasticity in tension, the ratio of the tensile stress (force per unit length) along the central axis to the cross-sectional area of a material. Equation: tensile modulus (N/m^2) $= f/S$, where f is the force needed (N) and S is the cross-sectional area (m^2). The tensile modulus at 100% elongation is called 100% modulus.

The tensile strength is the stress at the maximum plastic deformation of a material. In the tensile testing, the maximum stress a material can withstand while being stretched before breaking is the tensile strength. Equation: $\sigma_t = p/(b \cdot d)$, where σ_t is the tensile strength (MPa), p is the maximum stress (N), b is the width of the sample and d is the thickness of the sample. The lower the tensile strength, the higher the elongation at break and the better the plasticization performance.

The tensile yield stress is the real stress a material needs to undergo plastic flow when unidirectionally stretched. If work hardening is neglected, the tensile yield stress can be considered as a constant value approximately the same as the yield limit. In the stage of plastic deformation, each data-point in the stress curve can be considered as the yield point under the corresponding deformation; this stress is called yield strength and this elongation is called yield point elongation.

Figures 5.31, 5.32, 5.33, and 5.34 shows the effects of DEHCH or DOP content on the elongation at break, tensile modulus, tensile strength and tensile yield stress of PVC samples. The tensile properties of dumbbell-shaped samples were measured at a stretching speed of 20 mm/min, using Standard GB/T 1040.2-2006.

From Fig. 5.31, as the plasticizer content increases from 10 to 70 phr, the elongation at break of PVC/DEHCH increases by 314.4% from 30.5 to 344.9%; the elongation at break of PVC/DOP increases by 300.3% from 30.9 to 331.2%. The elongation at break of PVC samples is largely improved. In comparison, the PVC/DEHCH sample has an extra 14.1% increase in the elongation at break compared to PVC/DOP. Specifically, when the plasticizer content is below 50 phr, the elongation at break

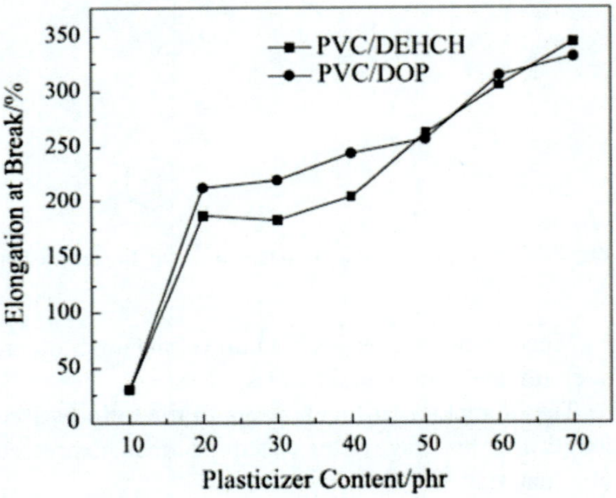

Fig. 5.31 Effect of the DEHCH or DOP content on the elongation at break of PVC samples

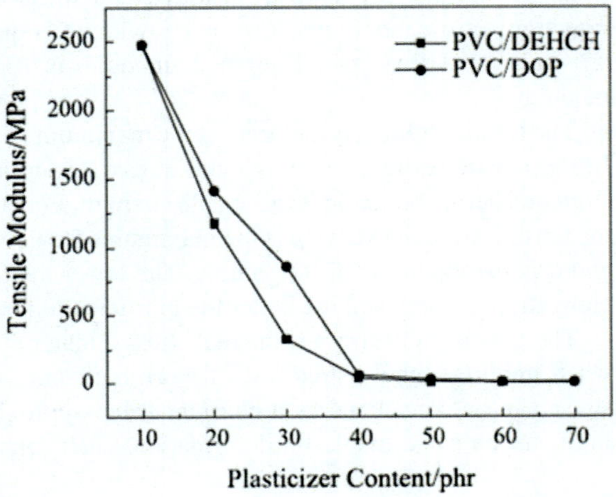

Fig. 5.32 Effect of the DEHCH or DOP content on the tensile modulus of PVC samples

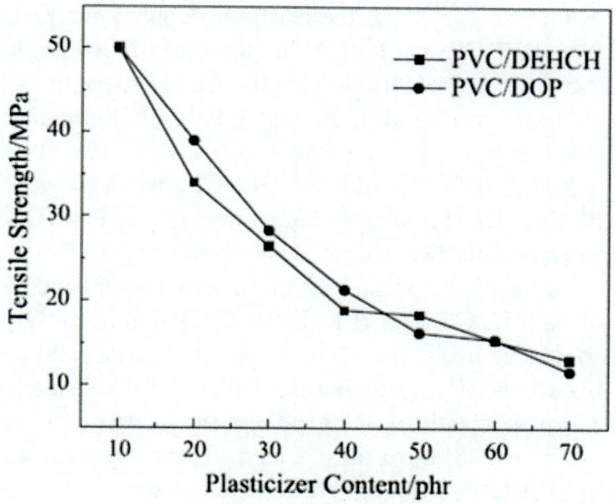

Fig. 5.33 Effect of the DEHCH or DOP content on the tensile strength of PVC samples

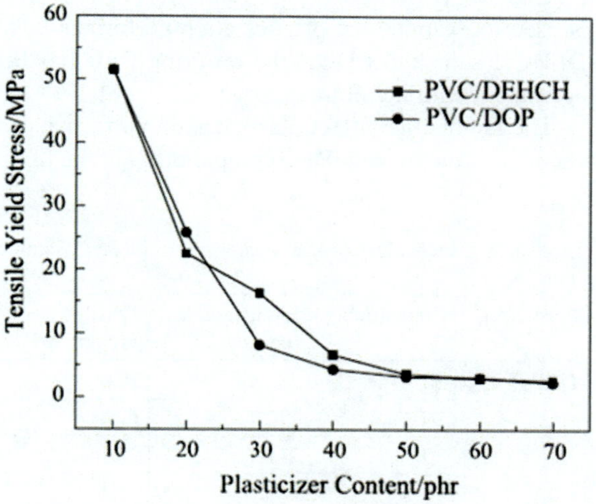

Fig. 5.34 Effect of the DEHCH or DOP content on the tensile yield stress of PVC samples

of PVC/DOP is higher than that of PVC/DEHCH; when above 50 phr, the parameters of these two samples are very close; when reaching 70 phr, the elongation at break of PVC/DEHCH is higher than that of PVC/DOP. Therefore, at low plasticizer contents, DOP has a better plasticization performance than DEHCH. When the plasticizer content is above 50 phr, the plasticization effects are very close.

From Fig. 5.32, as the plasticizer content increases from 10 to 70 phr, the tensile modulus of PVC/DEHCH decreases from 2447.3 MPa to 5.2 MPa; the tensile modulus of PVC/DOP decreases from 2478.7 MPa to 7.1 MPa. In general, DEHCH and DOP produce comparable drops to the tensile modulus of PVC, although PVC/DEHCH has a higher PVC/DOP when above 40 phr.

From Fig. 5.33, as the plasticizer content increases, the tensile strengths of both PVC/DEHCH and PVC/DOP significantly decrease. When the plasticizer content increases from 10 to 70 phr, the tensile strength of PVC/DEHCH decreases from 49.2 MPa to 13.2 MPa and that of PVC/DOP decreases from 50.0 MPa to 11.7 MPa. When the plasticizer content is below 40 phr, the tensile strength of PVC/DOP is higher than that of PVC/DEHCH; when above 40 phr, the tensile strength of PVC/DEHCH is slightly higher than that of PVC/DOP. The plasticization effects of these two plasticizers are generally very close.

From Fig. 5.34, as the plasticizer content increases, the tensile yield stresses of both PVC/DEHCH and PVC/DOP significantly decrease. When the plasticizer content increases from 10 to 70 phr, the tensile yield stress of PVC/DEHCH decreases from 51.6 MPa to 2.6 MPa and that of PVC/DOP decreases from 51.6 MPa to 2.4 MPa. The plasticization effects of these two plasticizers are very close.

Table 5.28 shows the mechanical property parameters of PVC samples plasticized by DEHCH and DOP at different contents.

From Table 5.28, at the same plasticizer content, the mechanical properties of PVC/DEHCH and PVC/DOP samples are very close. If the elongation at break is used to examine the plasticization performance, DEHCH is better than DOP. If DEHCH substitutes DOP, the resulting PVC has better mechanical properties, in agreement with literature reports.

The factors that affect plasticization performance include the molecular weight, chemical structure and PVC compatibility of the plasticizer. DOP and DEHCH have

Table 5.28 Mechanical property parameters of PVC samples with different DEHCH or DOP contents

Plasticizer	Content/phr	Elongation at break/%	Tensile strength/MPa	Tensile modulus/MPa	Tensile yield stress/MPa
DOP	10	30.9	50.0	2478.7	51.6
	20	212.7	39.0	1411.8	25.9
	30	219.7	28.3	858.0	8.2
	40	243.9	21.4	58.9	4.4
	50	256.6	16.3	26.8	3.3
	60	314.3	15.5	9.4	3.1
	70	331.2	11.7	7.1	2.4
DEHCH	10	30.5	49.2	2447.3	51.6
	20	186.7	34.0	1175.5	22.6
	30	182.3	26.5	326.9	16.3
	40	204.3	18.9	36.9	6.7
	50	261.9	18.4	11.5	3.7
	60	305.2	15.4	7.6	3.0
	70	344.9	13.2	5.2	2.6

small molecular weights. Their insertion between the PVC chains weakens the interchain interactions, leading to decreased entanglement and increased mobility of PVC chains and consequently the plasticity of PVC. DEHCH and DOP mainly occupy the amorphous regions in PVC resin, which increases the free volume between the polymers and produces certain shielding effect, resulting in stronger lubrication and slippage for polymers. Therefore, the elongation at break significantly increases with the plasticizer content, while the tensile strength, tensile modulus, and tensile yield stress significantly decrease. In addition, DOP is more polar than DEHCH. As PVC is polar, this may be a reason why the slightly different trends in the mechanical properties of PVC/DEHCH and PVC/DOP with the increase of plasticizer content. Under certain circumstances, DEHCH has better plasticization performance than DOP due to its slightly smaller molecular weight. Plasticizers are used mainly for production of soft PVC articles, with an average content of 45–50 parts in 100 parts PVC resin. At such high plasticizer contents, DEHCH has more advantageous plasticization performance than DOP.

5.5.2 *Thermal Properties of PVC/DEHCH and PVC/DOP Composite Materials*

PVC has advantages of corrosion resistance, good insulation, non-flammability, high strength, friction resistance, etc., but its molecular structure causes inferior thermal stability. When exposed to alkaline, heat or light, Cl-containing polymer chains tend to release HCl and form C=C double bonds. HCl can catalyze further thermal decomposition (i.e., synergetic induction) to release Cl on the adjacent carbons and form conjugated double bonds. With more conjugated double bonds, the material turns from light color to dark and becomes more susceptible to oxidation resulting in degradation and cross-linking. Therefore, it is very important to examine the effects of plasticizer on the thermal properties of PVC materials.

1. Thermal properties of DEHCH and DOP

Figures 5.35 and 5.36 plot the TGA and DTG curves of DEHCH and DOP. Table 5.29 lists the heated mass losses and thermal decomposition temperatures of DEHCH and DOP.

Compared to DOP, DEHCH has a very similar heated mass loss, a higher onset decomposition temperature (by 38.3 °C) and a higher maximum decomposition rate temperature (by 43.2 °C). Thus, the ring-hydrogenated DEHCH has greatly improved thermal stability, which will directly impact the thermoplasticity of PVC materials. In terms of thermal properties, DEHCH is more advantageous than DOP.

2. Thermal properties of PVC plasticized with DOP at different contents

Figures 5.37 and 5.38 plot the TGA and DTG curves of PVC samples plasticized

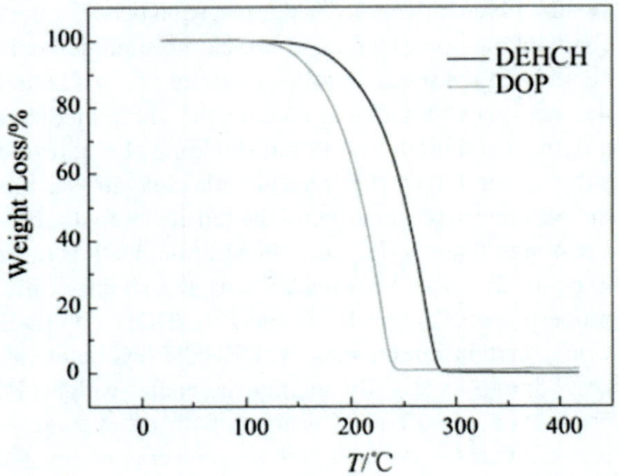

Fig. 5.35 TG curves of DEHCH and DOP

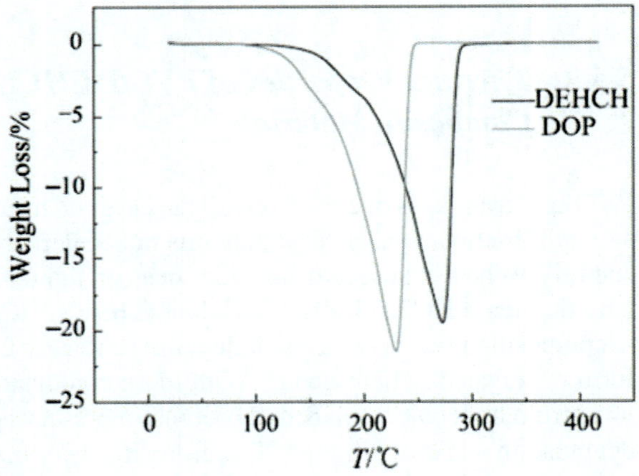

Fig. 5.36 DTG curves of DEHCH and DOP

Table 5.29 Heated mass loss and thermal decomposition temperature of DEHCH and DOP

Sample	Heated loss/%	Onset decomposition temperature T_{onset}/°C	Maximum decomposition temperature T_{max}/°C
DEHCH	99.2	228.8	272.9
DOP	98.8	190.5	229.7

Note Test conditions: sample weight 5 mg, heated in N_2 atmosphere from 25 to 450 °C at a rate of 10 °C/min

with DOP at different contents. Table 5.30 lists the heated mass losses and thermal decomposition temperatures of PVC samples plasticized with DOP at different contents.

From Fig. 5.37, the PVC samples with different DOP contents are the most rapidly decomposed at two temperature ranges: the first one between 200–300 °C and the

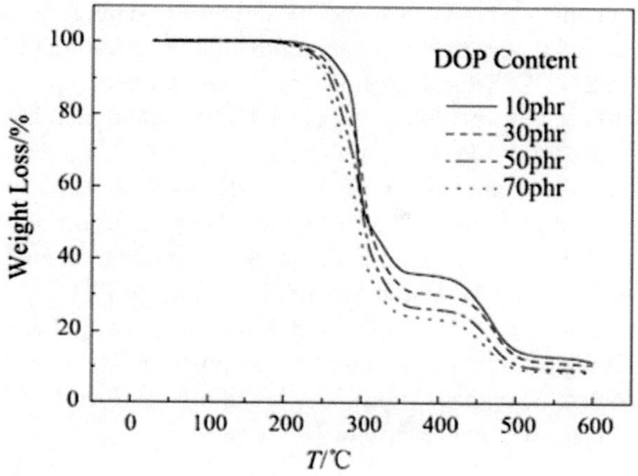

Fig. 5.37 TG curves of PVC samples with different DOP contents

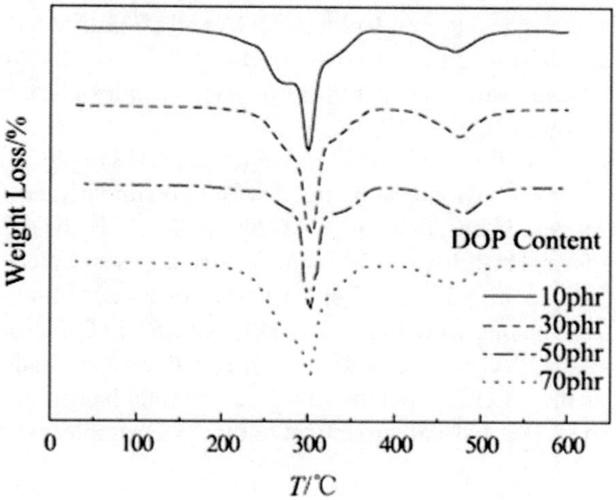

Fig. 5.38 DTG curves of PVC samples with different DOP contents

Table 5.30 Decomposition temperature and heated mass loss of PVC samples with different DOP contents

DOP content/phr	Onset decomposition temperature T_{onset}/°C	Maximum decomposition temperature in the first range $T_{1,max}$/°C	Maximum decomposition temperature in the second range $T_{2,max}$/°C	Heated loss in the first range/%	Total heated loss/%
10	281.4	293.6	467.9	64.7	87.2
30	277.1	301.6	471.8	70.2	89.2
50	266.8	302.1	466.8	74.6	90.7
70	257.1	299.9	455.4	76.2	91.6

Note Test conditions: sample weight 5 mg, heated in N_2 atmosphere from 25 to 450 °C at a rate of 10 °C/min second one between 400–500 °C. In the first range, as the DOP content increases from 10 to 70 phr, the onset decomposition temperature of PVC/DOP decreases from 281 to 257 °C. From Fig. 5.38, the maximum decomposition rate temperature is around 300 °C. From Table 5.30, as the DOP content increases from 10 to 70 phr, the heated mass loss of PVC samples in the 200–300 °C range increases from 64.7% to 76.2% and the total heated mass loss within 600 °C increases from 87.2% to 91.6%.

From Table 5.29, the onset thermal decomposition temperature of DOP is 190.5 °C, the maximum decomposition rate temperature is 229.7 °C, and the heated mass loss is 98.8%. When DOP is added to PVC, as the DOP content increases, the overall onset decomposition temperature of PVC/DOP decreases. In the range of 200–300 °C, the maximum decomposition rate temperature ($T_{1,max}$) does not change much. The heated mass loss is mainly ascribed to the cleavage of PVC chains and the consequent release of HCl gas.

3. Thermal properties of PVC plasticized with DEHCH at different contents

Figures 5.39 and 5.40 plot the TGA and DTG curves of PVC samples plasticized with DEHCH at different contents. Table 5.31 lists the heated mass losses and thermal decomposition temperatures of PVC samples plasticized with DEHCH at different contents.

From Fig. 5.39, the TGA curves of PVC samples with different DEHCH and DOP contents have the same trend with two similar thermal decomposition temperature ranges. In the first range of 200–300 °C, as the DEHCH contents increases, the thermal stability of PVC samples decreases significantly. From Table 5.31, the PVC sample with 10phr DEHCH has an onset decomposition temperature of 279.8 °C and the one with 70 phr DEHCH has 257.7 °C. These values are basically identical to the PVC samples with the same DOP content. In the second range of 400–500 °C, as the DEHCH content increases, the total heated mass loss increases from 86.86% to 92.0%, basically identical to the PVC samples with the same DOP content.

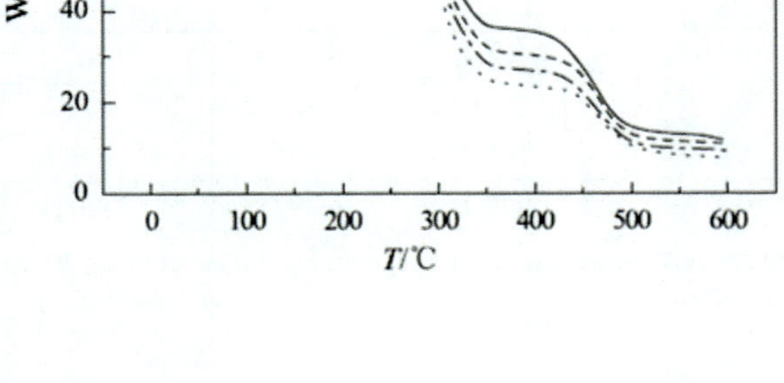

Fig. 5.39 TG curves of PVC samples with different DEHCH contents

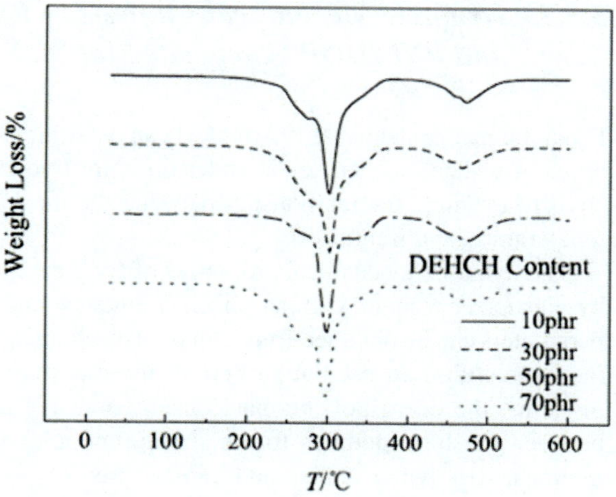

Fig. 5.40 DTG curves of PVC samples with different DEHCH contents

Table 5.31 Thermal stability parameters of PVC samples with different DEHCH contents

DEHCH content/phr	Onset decomposition temperature T_{onset}/°C	Maximum decomposition temperature in the first range $T_{1,max}$/°C	Maximum decomposition temperature in the second range $T_{2,max}$/°C	Heated loss in the first range/%	Heated loss in the second range/%
10	279.8	291.9	470.0	66.67	86.86
30	279.7	302.3	467.2	69.74	89.63
50	268.8	299.4	468.4	73.5	91.6
70	257.7	299.9	455.4	76.2	92.0

Note Test conditions: sample weight 5 mg, heated in N_2 atmosphere from 25 to 450 °C at a rate of 10 °C/min

4. Comparative thermal properties of PVC plasticized with DEHCH and DOP at different contents

According to the TGA curves of these PVC samples, the heated mass loss of PVC composite materials is small from room temperature to 200 °C without changes in thermal properties. 250–380 °C is the main decomposition temperature range of PVC; bond cleavages occur releasing HCl gas and forming polyolefin structure. In 380–520 °C, the polyolefin isomerizes to form aromatics and the residual products decompose. From the perspective of practical applications, the thermal property study is focused mainly on the HCl release step, during which the mechanical properties change significantly due to thermal decomposition. According to Tables 5.30 and 5.31, at the same DEHCH and DOP content, the onset decomposition temperature of PVC composite materials are very close.

5.5.3 Dynamic Mechanical Analysis of PVC/DEHCH and PVC/DOP Composite Materials

Dynamic mechanical analysis (DMA) can determine the dynamic mechanical properties of viscoelastic materials under different frequencies, temperatures and loads so as to examine the relationship between the mechanical performance with time, temperature, and frequency.

The dynamic mechanical properties of PVC composite materials are determined by plasticizer content, compatibilizer, frequency and temperature. Three important parameters can be obtained from dynamic mechanical measurements: (1) the storage modulus within an oscillation period, the measure of maximum energy; (2) loss modulus, the energy of thermal consumption; (3) mechanical damping, the ratio between the loss modulus to the storage modulus that is related with molecular motions of polymer chains and can be used to measure the viscoelasticity. The modulus and damping can be obtained from DMA. Upon applying a sinusoidal alternating stress, the strain under such a stress is measured. For polymeric materials, the strain lags behind the stress with a phase difference. The phase difference reflects the storage and loss moduli of the material when deformed. Therefore, through dynamic mechanical measurements, material parameters such as mechanical properties, dynamic moduli, glass transition temperature, etc., can be obtained.

PVC is a polar polymer. Because of the strong intermolecular van der Waals interactions, PVC articles usually have low softness and inferior flexibility. Thus, a large amount of plasticizers is needed in PVC processing. Plasticizer molecules can insert between PVC chains and melt in the chain segments, leading to decreased glass transition temperature of PVC.

1. Dynamic mechanical behaviors of PVC/DEHCH and PVC/DOP composite materials

Figures 5.41 and 5.42 shows the loss angle tangent–temperature ($\tan\delta$–T) curves of PVC samples containing different DOP and DEHCH contents.

Because the glass transition appears within a broad temperature range, the temperature corresponding to the internal friction peak or the loss modulus peak is usually used as the glass transition temperature T_g. Here, the temperature at the maximum internal friction peak (i.e., $\tan\delta$) in the curves is selected as T_g of PVC.

In Fig. 5.41, when the DOP content is 10–70 phr, each sample has only one internal friction peak within -50 to 100 °C, indicating the excellent compatibility between DOP and PVC. The T_g of DOP-free PVC is 87 °C, which obviously drops after addition of DOP. As the DOP content increases, the T_g shifts to the low-temperature zone. When the DOP content is 10–70 phr, the maximum values of internal friction peak (i.e., $\tan\delta$) are 80 °C, 64 °C, 50 °C, 31 °C, 16 °C, 3 °C, and − 11 °C, respectively. This indicate the excellent plasticization performance of DOP with increasing processibility of PVC materials.

In Fig. 5.42, when the DEHCH content is 10–70 phr, each sample has only one internal friction peak within -50–100 °C, indicating the excellent compatibility

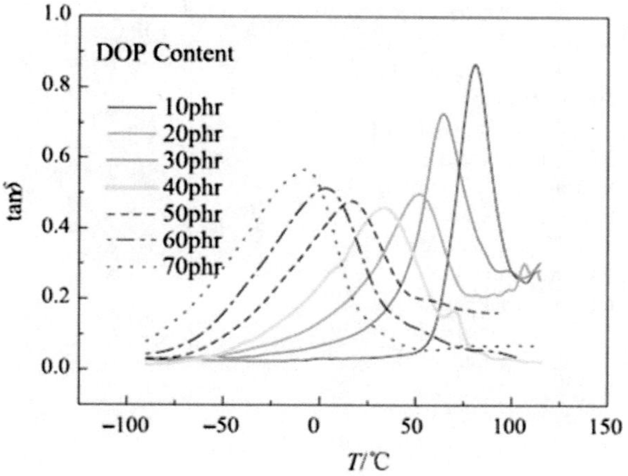

Fig. 5.41 The tanδ–T curves of PVC samples containing different DOP contents

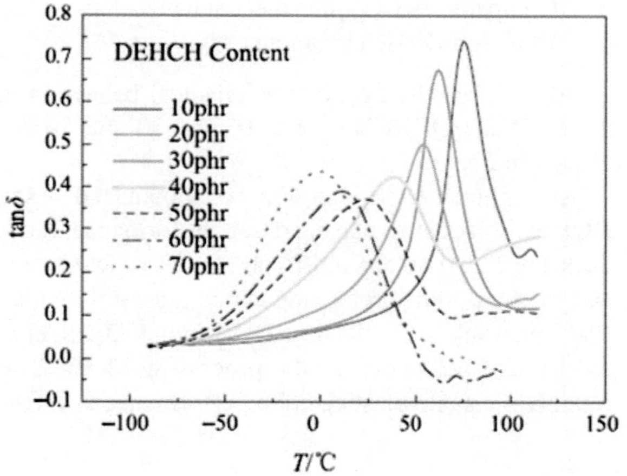

Fig. 5.42 The tanδ–T curves of PVC samples containing different DEHCH contents. Test conditions: frequency 1.0, 2.0, 5.0, 10.0, and 33.3 Hz, heating rate 1.0 °C/min, temperature scan range 100–150 °C, tensile mode. Dynamic force 2 N, additional static pressure 3.5 N, sample size 10 mm × 2 mm × 0.5 mm, clamp span length 10 mm, amplitude 1200 μm

between DEHCH and PVC. The T_g of PVC samples obviously drops after addition of DEHCH. As the DEHCH content increases, the T_g shifts to the low-temperature zone. When the DEHCH content is 10–70 phr, the maximum values of internal friction peak (i.e., tanδ) are 74 °C, 62 °C, 54 °C, 38 °C, 23 °C, 11 °C, and 0 °C, respectively. This indicate the excellent plasticization performance of DEHCH with increasing processibility of PVC materials.

DOP and DEHCH molecules penetrate into the void space between PVC chains and have coupling interactions with PVC molecules, leading to smaller chain-chain forces and larger free volume. This effectively decreases the T_g, realizes the heated melting, and affords excellent processibility of PVC.

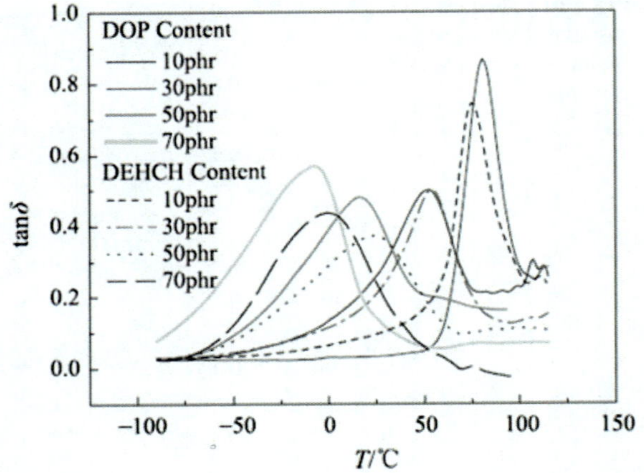

Fig. 5.43 The tanδ–T curves of PVC samples containing different DOP or DEHCH contents. Test conditions: frequency 1.0, 2.0, 5.0, 10.0, and 33.3 Hz, heating rate 1.0 °C/min, temperature scan range 100–150 °C, tensile mode. Dynamic force 2 N, additional static pressure 3.5 N, sample size 10 mm × 2 mm × 0.5 mm, clamp span length 10 mm, amplitude 1200 μm.

2. Comparative dynamic mechanical behaviors of PVC samples with different DOP and DEHCH contents

To compare the dynamic mechanical behaviors in a more straightforward way, the DOP and DEHCH data at 10 phr, 30 phr, 50 phr, and 70 phr are selected and plotted in Fig. 5.43.

At 10 phr, DEHCH is slightly better than DOP. As the plasticizer content increases, DOP exhibits better plasticization performance. At 30 phr, DEHCH and DOP have the same T_g. At 50 phr and 70 phr, the T_g of DOP are lower than that of DEHCH by 9 and 11 °C. But the processing temperatures of regular materials are above 0 °C. The PVC materials plasticized by DOP and by DEHCH have roughly the same dynamic mechanical behaviors. In the processing of PVC materials, DEHCH can achieve roughly the same plasticization performance as DOP.

5.5.4 Rheology of PVC/DOP and PVC/DEHCH Composite Materials

Rheology is the study of deformation and/or flow of substances with respect to stress, strain, temperature, and time, and is very important for polymer processing. Specifically, rheology can guide the selection of raw materials, processing facility, mold design, and technical conditions in blending processing, leading to improved product quality. For blending materials, rheological behavior tests can be used to determine whether chemical reactions between polymeric components occur in melt blending, as there are many factors that can affect the performance of blended systems. Rheological behaviors of blending materials are more complicated.

1. Rheology of PVC samples with different DOP and DEHCH contents

Figures 5.44 and 5.45 shows the elastic modulus G' and viscosity of PVC samples

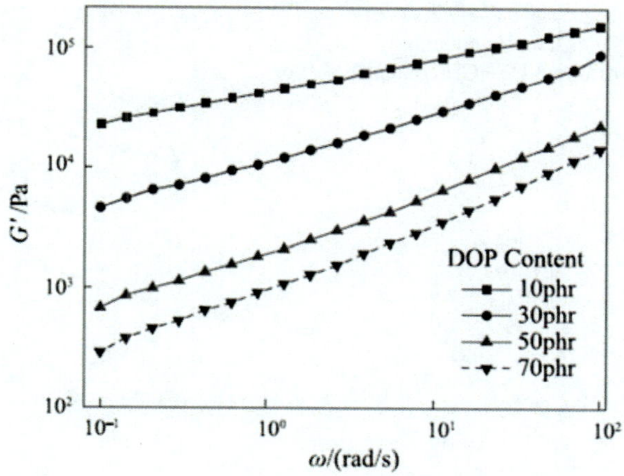

Fig. 5.44 G'–ω curves of PVC samples containing different DOP contents

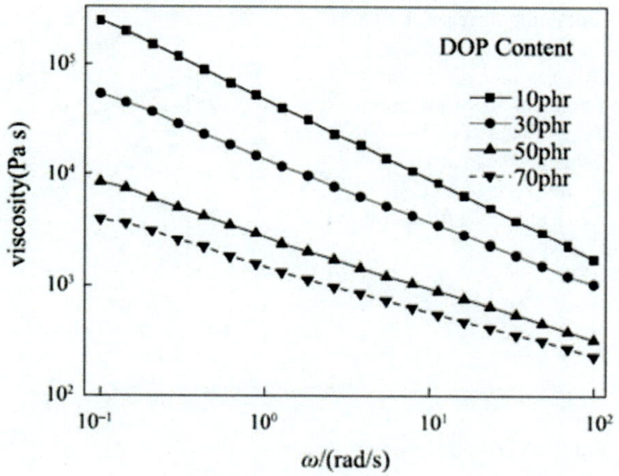

Fig. 5.45 Viscosity–ω curves of PVC samples containing different DOP contents. Test conditions: Disk samples of 25 mm diameter and 1 mm thickness were subjected to dynamic frequency scan at 190 °C in the frequency range of 0.1–100 rad/s with a strain of 5%.

containing different DOP contents as a function of frequency ω.

Figures 5.46 and 5.47 shows the elastic modulus G' and viscosity of PVC samples containing different DEHCH contents as a function of frequency ω.

G' is an elasticity-related quantity that measures the stored energy in polymer materials when deformed elastically. From Figs. 5.44, 5.45, 5.4.46, and 5.47, the G' of blends with different DOP and DEHCH contents increase while the viscosity decreases with frequency ω. They have the same trend. The mobility of PVC chains increases with the DOP or DEHCH content, leading to decreased stiffness and improved rheological properties.

2. Comparative rheological properties of PVC samples with different DOP and DEHCH contents

Figures 5.48 and 5.49 shows the elastic modulus G' and viscosity of PVC samples

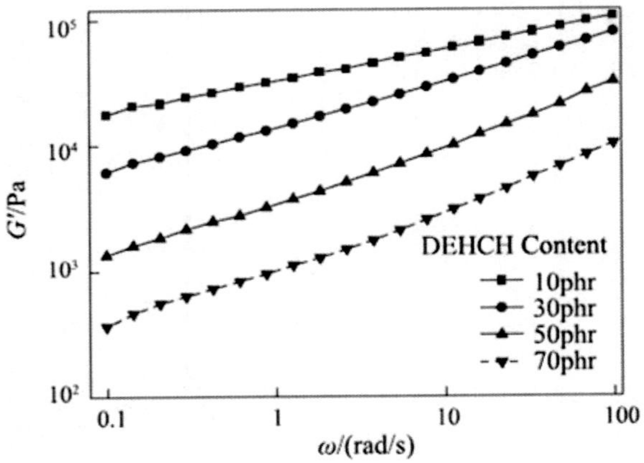

Fig. 5.46 G'–ω curves of PVC samples containing different DEHCH contents

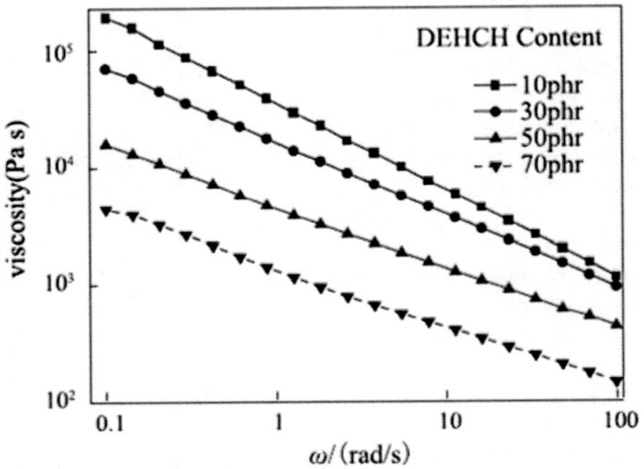

Fig. 5.47 Viscosity–ω curves of PVC samples containing different DEHCH contents. Test conditions: Disk samples of 25 mm diameter and 1 mm thickness were subjected to dynamic frequency scan at 190 °C in the frequency range of 0.1–100 rad/s with a strain of 5%.

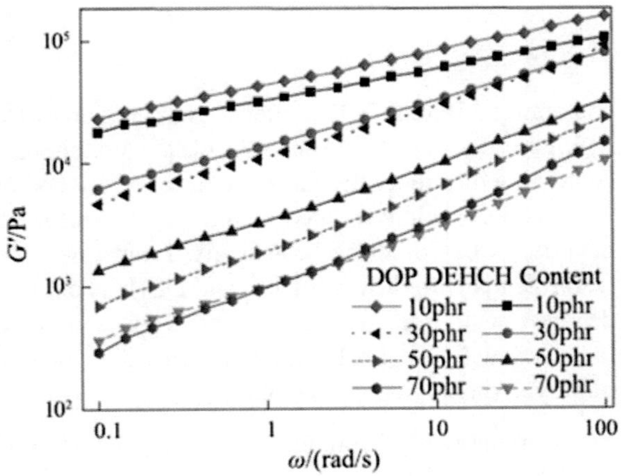

Fig. 5.48 G'–ω curves of PVC samples containing different DOP and DEHCH contents

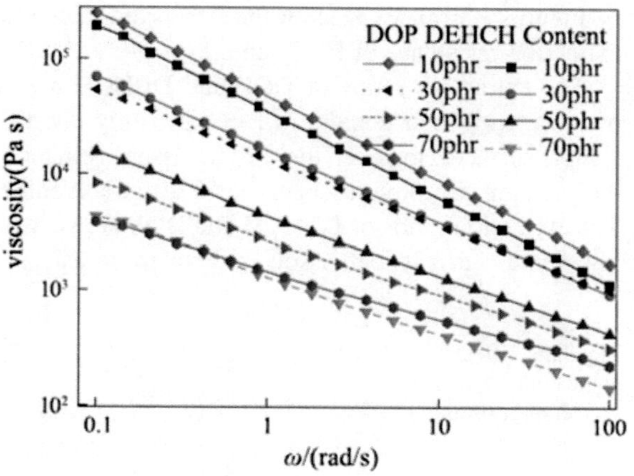

Fig. 5.49 Viscosity–ω curves of PVC samples containing different DOP and DEHCH contents

containing different DEHCH and DOP contents as a function of frequency ω at 190 °C.

At 10 phr, the elastic modulus and viscosity of DOP are slightly higher than those of DEHCH, indicating the slightly better processibility of the latter. At 30 phr, the elastic modulus and viscosity of both are very close, suggesting these two can substitute each other at this condition. At 50 phr, the elastic modulus and viscosity of DOP are slightly lower than those of DEHCH, indicating the slightly better processibility of the former. As the plasticizer content increases, the plasticity of both materials is nearly identical, indicating similar processibility.

5.5.5 Solvent Extraction Resistance of PVC Samples with Different DEHCH and DOP Contents

PVC articles are often in contact with various solvents or used as the containers for solvents. If the plasticizers are extracted by the solvent, the performance of these articles will be deteriorated and the solvents will be contaminated. Especially, PVC articles used in medical devices and food packaging have strict requirements to the solvent extraction resistance of plasticizers. The extraction ratio can measure the extent of solvent extraction of plasticizers when DOP- or DEHCH-plasticized PVC is in contact with solvents.

PVC samples with different DEHCH and DOP contents were soaked in ethanol, petroleum ether, and water for a specified period of time, followed by being dried. The weight change before and after soaking was used to calculate the extraction ratio. The tests followed Standard ISO 175-2010. The PVC samples were cut into 10 mm × 10 mm square plates that were weighed and then were soaked in 200 mL of three solvents (water, ethanol, and petroleum ether) for 48 h. The soaked plates were dried at 30 °C for 12 h and then were weighed.

Figures 5.50 and 5.51 show the effects of DEHCH and DOP content on the solvent extraction resistance of PVC samples.

The extraction ratios of DOP and DEHCH are very low in water, but slightly increase with their content in PVC, possibly due to the loss of plasticizers on the sample surface. However, the weight losses in ethanol and petroleum ether especially the latter are very obvious, because these two solvents can dissolve DOP and DEHCH. The extraction ratios of DEHCH and DOP are very close, but DEHCH extraction is lower, indicating its better solvent extraction resistance than DOP.

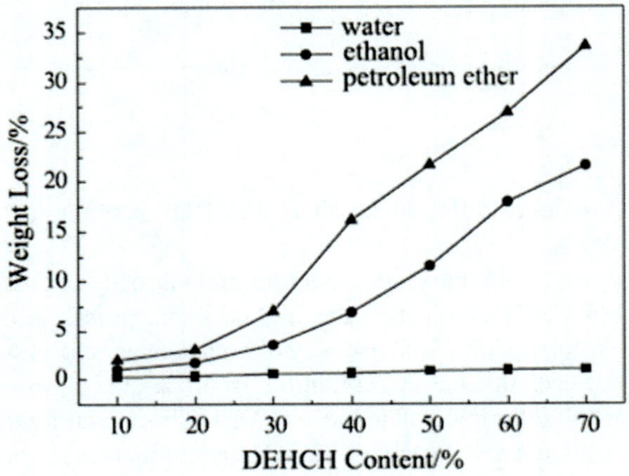

Fig. 5.50 Effect of DEHCH content on the solvent extraction resistance of PVC samples

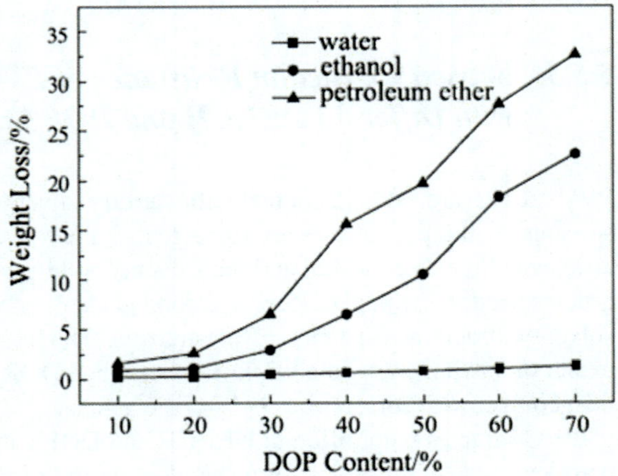

Fig. 5.51 Effect of DOP content on the solvent extraction resistance of PVC samples

5.5.6 Evaporation Resistance of PVC Composite Materials with Different DEHCH and DOP Contents

At high temperatures, DEHCH and DOP will slowly migrate from inside to surface of PVC and evaporate, which adversely affects the stability of PVC materials and causes environmental pollution.

The evaporation resistance was measured following Standard ISO176-2005. PVC samples were cut into 30 mm × 30 mm square plates and placed into a desiccator for 12 h. Then, the samples were weighed and placed in a metal container loaded with activated carbon. The test temperature was controlled at (100 ± 1) °C for 24 h. Finally, the samples were weighed to obtain the weight loss.

Figure 5.52 shows the evaporation curve of PVC samples with different DEHCH and DOP contents.

As the plasticizer content increases, the weight loss of PVC samples increases, indicating higher volatility and worse stability. The weight loss of DOP is smaller than that of DEHCH, indicating the higher stability of the former. According to ISO176-2005, low volatility is defined for <0.81% weight loss and intermediate volatility is defined for 3.83% weight loss. Therefore, these two plasticizers have intermediate volatility with little difference.

Based on Figs. 5.50, 5.51, annd 5.52, the extraction and evaporation parameters of PVC samples with different DEHCH and DOP contents are listed in Table 5.32.

The weight losses caused by extraction and evaporation are important parameters for examining plasticizer performance. The extraction ratio suggests the suitable applications and the evaporation loss indicates the temperature range the material can sustain. Therefore, solvent extraction and evaporation resistances are important index to determine the product safety to human and environment.

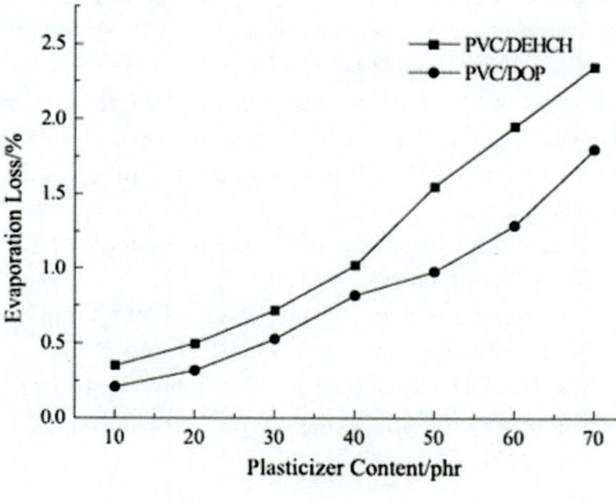

Fig. 5.52 Effect of DOP and DEHCH contents on the volatility of PVC samples

Table 5.32 The extraction resistance and volatility of PVC samples with different DOP and DEHCH contents

Plasticizer	Content/phr	Water extraction loss/%	Ethanol extraction loss/%	Petroleum ether extraction loss/%	Evaporation loss/%
DOP	10	0.15	0.82	1.55	0.21
	20	0.22	1.02	2.57	0.32
	30	0.39	2.88	6.54	0.53
	40	0.58	6.52	15.58	0.82
	50	0.72	10.53	19.64	0.98
	60	0.96	18.29	27.65	1.29
	70	1.25	22.57	32.58	0.35
DEHCH	10	0.25	1.02	1.98	0.5
	20	0.35	1.69	3.02	0.72
	30	0.53	3.55	7.02	1.02
	40	0.62	6.92	16.13	1.55
	50	0.82	11.53	21.64	1.95
	60	0.93	17.93	26.89	2.35
	70	1.02	21.57	33.58	0.35

5.5.7 Photonic Properties of PVC Composite Materials with Different DEHCH and DOP Contents

The transmittance of PVC composite materials was investigated, which is defined as the percentage of transmitted light to incident light for transparent or semitransparent materials. The type, content and internal structure of plasticizers can largely affect the transmittance. The transmittances of these two plasticizers are compared.

The transmittance was measured following Standard GB/T 2410-2008. The transmittance and haze of sample plates of 50 mm × 50 mm × 2.0 mm dimension were measured using WGT-S light transmittance and haze tester. The measurement was performed at 5 different points on the samples and the average value was used as the final result.

Figure 5.53 shows the transmittance curves of PVC samples with different DOP or DEHCH contents.

The transmittances of plasticized PVC samples increase with the plasticizer content, indicating that both DEHCH and DOP can improve the transmittance of PVC. DEHCH-plasticized PVC has obviously higher transmittance indicating its advantages in making transparent PVC articles.

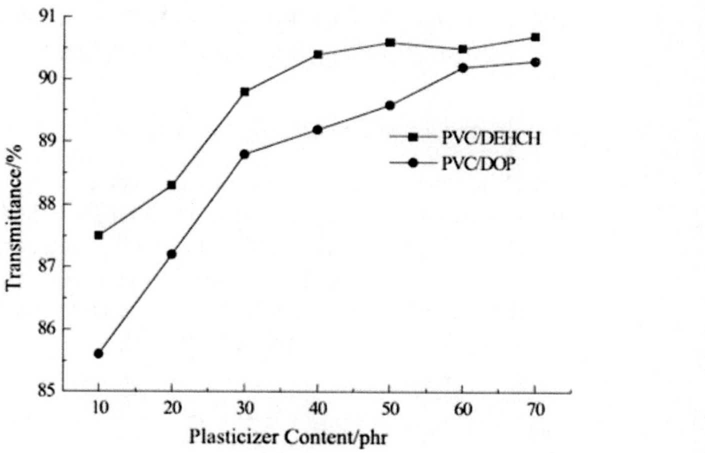

Fig. 5.53 Evolution of the transmittance of PVC samples with the DOP or DEHCH contents

5.5.8 Conclusions

Through comparing the mechanical, thermal, dynamic mechanical, rheological, photonic, solvent extraction resistant and evaporation resistant properties of DEHCH, DOP, and plasticized PVC materials, we can conclude that ring-hydrogenated phthalates have comparable and even better plasticization performance than DOP. Due to the absence of benzene ring, ring-hydrogenated phthalates exhibit no toxicity and thus can substitute traditional phthalates for use in food packaging, medical supplies, and children's toys, which is important for keeping human health and reducing environmental pollution.

5.6 Third-Party Inspections

To commercialize cyclohexane-based plasticizers for use in PVC materials, the raw materials, pellets and products all need to be inspected for their physico-chemical properties and biological toxicity by recognized third parties following national (GB/T 16886.1-2011) and industrial standards.

5.6.1 *Physico-Chemical Property Inspection for Ring-Hydrogenated Products*

1. Inspection report from Henan Analytical and Test Center

On November 19, 2016, our team entrusted Henan Analytical and Test Center to inspect the DOP hydrogenation product DEHCH including viscosity, density, chromaticity, refractive index, acid value, acidity, and contents of water, ester, DOP, Sb, As, Ba, Sn, Cr, Pb, Hg, Se, Cd, etc. The inspection results are shown in Table 5.33.

The DEHCH content is 97.83% and the DOP content is 0.01%. Other indexes all meet the requirements of related standards.

Table 5.33 DEHCH Inspection Report (No. 16110024)

Inspection items	Unit	Inspection result	Inspection standard
Viscosity (20 °C)	mPa s	6	GB/T 10247-2008
Density (20 °C)	cm^3/g	0.954	GB/T 4472-2011
Chromaticity (Pt–Co)	Hazen	<40	GB 3143-1982
Refractive index (n_D^{20})		1.460	GB/T 6488-2008
Acid value	mg/g	0.004	GB/T 1668-2008
Water content	% (wet weight)	0.18	GB/T 6283-2008
DEHCH content	% (wet weight)	97.83	GB/T 9722-2006
DOP content	% (wet weight)	0.01	
Others	% (wet weight)	1.97	
Sb	mg/kg	<0.5	DZ/T 0223-2001
As	mg/kg	<0.2	
Ba	mg/kg	<0.5	
Sn	mg/kg	<1	
Cr	mg/kg	<0.2	
Pb	mg/kg	<0.2	
Hg	mg/kg	<0.1	
Se	mg/kg	<1	
Cd	mg/kg	<0.1	

Table 5.34 DEHCH inspection report (No. UNQD-JLF20163582)

#	Inspection items	Inspection result	Note
1	Dynamic viscosity (20 °C)/(mm^2/s)	43.64	GB/T 265-1988
2	Chromaticity (Pt–Co)/Hazen	5	GB/T 3143-1982
3	Refractive index	1.4604	GB/T 6488-2008
4	Water content/%	0.04	GB/T 6283-2008
5	DOP content /(mg/kg)	Not detected (<10)	GC-MS

2. Inspection report from United Nations Quality Detection Technology Co., Ltd.

On October 15, 2016, our team entrusted United Nations Quality Detection Technology Co., Ltd. to inspect the DOP hydrogenation product DEHCH. The inspection report is shown in Table 5.34. No DOP is detected.

3. Inspection report from Shanghai Weipu Chemical Technology Service Co., Ltd.

On October 26, 2016, our team entrusted Shanghai Shanghai Weipu Chemical Technology Service Co., Ltd. to inspect the DOP hydrogenation product DEHCH. The inspection report is shown in Table 5.35.

The DEHCH content is >99%. No phthalates were detected.

4. Inspection report from Qingdao Zhengxin Inspection and Analytical Co., Ltd.

On May 9, 2017, our team entrusted Qingdao Zhengxin Inspection and Analytical Co., Ltd. to inspect the DOP hydrogenation product DEHCH. The inspection report is shown in Table 5.36.

The property and performance parameters of DEHCH as a plasticizer all meet the requirements of related standards.

5.6.2 Toxicity Test of Ring Hydrogenation Product

On May 14, 2017, our team entrusted Henan Disease Control and Prevention Center to inspect the acute oral toxicity tests for DINCH and DEHCH. The inspection report is shown below.

1. Materials and methods

(1) Animals: Kunming mice, SPF grade, body weight 18–22 g, 20 in number, 10 males and 10 females, provided by Henan Laboratory Animal Center, Certificate #: 41003100004410.
(2) Laboratory animal environments: barrier facilities, License #: SYXK (Henan) 2012-0005, temperature: 20–25 °C, relative humidity: 40–70%.
(3) Feed type and source: full nutrition mouse feed, Grade A, provided by Jiangsu Xietong Biomedical Engineering Co., Ltd., Certificate #: 120170210001.
(4) Feed disinfection method: irradiation.

Table 5.35 DEHCH inspection report (No. WP-16100424-JC-01R2)

Inspection items	Inspection result	Unit	Inspection standard
Viscosity	29	mPa·s	EN 14372: 2004
Density	0.9541	cm^3/g	
Chromaticity	5	Hazen	
Refractive index	1.4651	–	
Acid value	0.277	mg KOH/g	
Ester content	> 99.0	%	
Water content	0.1	%	
Sb	0.01	mg/kg	
As	0.005	mg/kg	
Ba	0.02	mg/kg	
Sn	0.03	mg/kg	
Cr	0.02	mg/kg	
Pb	0.03	mg/kg	
Hg	0.01	mg/kg	
Se	0.02	mg/kg	
Cd	0.01	mg/kg	
DMP	N.D. (<50)	mg/kg	EN 14372: 2004
DEP	N.D. (<50)	mg/kg	
DPRP	N.D. (<50)	mg/kg	
DIBP	N.D. (<50)	mg/kg	
DEP	N.D. (<50)	mg/kg	
DPP	N.D. (<50)	mg/kg	
DHP	N.D. (<50)	mg/kg	
BBP	N.D. (<50)	mg/kg	
DCHP	N.D. (<50)	mg/kg	
DIOP	N.D. (<50)	mg/kg	
DNOP	N.D. (<50)	mg/kg	
DINP	N.D. (<50)	mg/kg	
DIDP	N.D. (<50)	mg/kg	
DNP	N.D.(<50)	mg/kg	
DNHP	N.D. (<50)	mg/kg	
DEHP	N.D. (<50)	mg/kg	

Note The viscosity, density and refractive index were measured at 20 °C. Chromaticity measurement method: Pt–Co method. Elemental analysis: atomic adsorption method. N.D. means not detected (lower than the detection limit). Ester content was determined using the area normalization method. DEHP content was measured using GC-MS analysis with an external reference

Table 5.36 DEHCH inspection report

#	Inspection items			Inspection result	Note
1	Volumetric resistivity/(Ώ·cm)		–	4.9 × 10^{10}	
2	Heated weight loss/%	0.1	–		
3	Thermal decomposition temperature/°C	242	–		
4	Thermal stability	Acid value/(mg KOH/kg)	Before test	0.008	180 °C, 2 h
5			After test	0.008	
6		Chromaticity/Hazen	Before test	30	
7			After test	30	
8	Transparancy/%	98.2	–		
9	Extraction/%	2.3	100 °C, 24 h		
10	Migration/%	3.4	70 °C, 72 h		

(5) Drinking water treatment: sterile water dispenser.

2. Test method and steps

(1) Inspection reference: Cosmetic Safety Technical Specifications (2015), "acute oral toxicity test".
(2) Sample treatment and dose preparation: 10.0 g sample mixed with 40 mL salad oil as gavage fluid.
(3) Administration method: oral.
(4) Test steps: after 16 h fasting, the test substance was administered orally at a single dose to 20 small mice (10 males and 10 females). The gavage was 20 mL/kg and the dose was 5000 mg/kg bw. After 14 days, the mortality and poisoning were recorded.

3. Test results

No obvious poisoning was observed after administration of toxicant. No morality was observed after 14 days. After sacrifice, the gross anatomy did not detect obvious abnormal. The small mice acute oral toxicity test results are shown in Tables 5.37 and 5.38.

4. Conclusions

DINCH was administered to male and female small mice in acute oral tests: LD_{50} >5000 mg/kg bw. The toxicity inspection results: according to the toxicity classification standard in Cosmetic Safety Technical Specifications (2015), the small mice acute oral toxicity of DINCH and DEHCH is actually non-toxic.

Table 5.37 Mice acute oral toxicity test results of DINCH (No. HZ2017-0302)

Animal gender	Dose group/(mg/kg)	# of animals	Body weight/g		# of mortality	Mortality rate/%
			Initial Weight	Final weight		
Female	5000	10	18.7 ± 0.7	29.1 ± 1.1	0	0
Male	5000	10	19.9 ± 1.2	33.5 ± 1.5	0	0

Table 5.38 Mice acute oral toxicity test results of DEHCH (No. HZ2017-0136)

Animal gender	Dose group/(mg/kg)	# of animals	Body weight/g		# of mortality	Mortality rate/%
			Initial weight	Final weight		
Female	5000	10	19.2 ± 0.7	28.4 ± 3.3	0	0
Male	5000	10	20.0 ± 1.2	32.9 ± 1.8	0	0

5.6.3 Ring-Hydrogenated Product Pellets and PVC Articles

1. Plasticizer pellets

On February 21, 2017, our team entrusted Shandong Weigao Group Medical Polymer Company Limited to use 50 kg DEHCH to make PVC pellets for use in manufacturing transfusion (infusion) supplies. The PVC pellets were tested with a computer-controlled electronic universal tester following Standard WG JR-SOR-ZL-080. The plasticization performance was compared with that of commercial DINCH (BASF) and TOTM (a plasticizer not yet prohibited or limited in REACH and RoSH). Inspection result: the inspection items are passed. The plasticizer inspection report and its comparison with DINCH and TOTM are shown in Tables 5.39, 5.40, 5.41, 5.42, 5.43, 5.44, 5.45, 5.46, 5.47 and 5.48.

5. PVC articles

The inspection report for DEHCH-plasticized PVC articles is in Table 5.49. Inspection result: the sample was inspected according to WGJR-SOP-ZL-030 (Plasticized polyvinyl chloride (PVC) compounds for transfusion (infusion) equipment/TPE compound test standard); the tested items are passed.

Table 5.39 PVC/DEHCH MT tubing materials test conditions (2017.02.24)

#	Sample specifications			Test speed /(mm/min)	Temperature/°C	Humidity/%
	Thickness *h*/mm	Thickness *h*/mm	Thickness *h*/mm			
1	1.8	6	25	200	24	22
2	1.8	6	25			
3	1.8	6	25			

Table 5.40 PVC/DEHCH MT tubing materials test results (2017.02.24)

#	Yield strength				Maximum strength			Hardness
	Yield force /N	Yield strength/MPa	Deformation/mm	Elastic modulus/MPa	Maximum force /N	Tensile strength /MPa	Elongation at break /%	
1	100.48	9.30	0	7.73	197.78	18.31	576.72	81.4
2	95.23	8.82	0	10.01	189.71	17.57	528.73	
3	91.00	8.43	0	13.07	181.19	16.78	551.38	
Average	95.57	8.85	0	10.27	189.56	17.55	552.28	

Table 5.41 PVC/DEHCH tubing materials test conditions (2017.04.28)

#	Sample Specifications			Test Speed /(mm/min)	Temperature/°C	Humidity/%
	Thickness *h*/mm	Thickness *h*/mm	Thickness *h*/mm			
1	1.8	6	25	200	29	26
2	1.8	6	25			
3	1.8	6	25			

Table 5.42 PVC/DEHCH tubing materials test results (2017.04.28)

#	Yield strength				Maximum strength			Hardness
	Yield force /N	Yield strength /MPa	Deformation /mm	Elastic modulus /MPa	Maximum force/N	Tensile strength /MPa	Elongation at break/%	
1	93.67	8.67	0	6.10	137.08	12.69	670.03	80
2	77.94	7.22	0	7.34	154.57	14.31	791.14	
3	73.00	6.76	0	4.11	146.15	13.53	776.66	
average	81.54	7.55	0	5.85	145.93	13.51	745.94	

Table 5.43 PVC/DEHCH injection dropper materials test conditions (2017.04.28)

#	Sample specifications			Test speed /(mm/min)	Temperature/°C	Humidity/%
	Thickness *h*/mm	Thickness *h*/mm	Thickness *h*/mm			
1	1.8	6	25	200	29	26
2	1.8	6	25			
3	1.8	6	25			

The inspection report for DOP-plasticized PVC articles is in Table 5.50. Inspection result: the sample was inspected according to WGJR-SOP-ZL-030 (Plasticized polyvinyl chloride (PVC) compounds for transfusion (infusion) equipment/TPE compound test standard); the tested items are passed.

The inspection report for TOTM-plasticized PVC articles is in Tables 5.51 and 5.52. Inspection result: the sample was inspected according to WGJR-SOP-ZL-030 (Plasticized polyvinyl chloride (PVC) compounds for transfusion (infusion) equipment/TPE compound test standard); the tested items are passed.

Comparing Tables 5.49 and 5.50, it is concluded that, when used as a plasticizer in PVC compounds for transfusion (infusion) equipment, DEHCH has no obvious difference from DOP in all the performance metrics. Both are qualified products.

Comparing Tables 5.49 with 5.51 and 5.52, it is concluded that, when used as a plasticizer in PVC compounds for transfusion (infusion) equipment, DEHCH has no

Table 5.44 PVC/DEHCH injection dropper materials test results (2017.04.28)

#	Yield strength				Maximum strength			Hardness	Shore hardness (HD)
	Yield force /N	Yield strength /MPa	Deformation /mm	Elastic modulus /MPa	Maximum force/N	Tensile strength /MPa	Elongation at Break/%		
1	87.02	8.06	0	24.29	174.13	16.12	382.92	80	
2	86.51	8.01	0	21.26	172.32	15.96	408.29		42.9
3	95.23	8.82	0	20.68	189.46	17.54	479.49		
Average	89.59	8.30	0	22.08	178.64	16.54	423.57		

Table 5.45 PVC/DINCH plasticized bag test conditions (2017.04.26)

#	Sample Specifications			Test Speed /(mm/min)	Temperature/°C	Humidity/%
	Thickness *h*/mm	Thickness *h*/mm	Thickness *h*/mm			
1	1.8	6	25	200	29	26
2	1.8	6	25			
3	1.8	6	25			

Table 5.46 PVC/DINCH plasticized bag test results (2017.04.26)

#	Yield strength				Maximum strength			Hardness
	Yield force /N	Yield strength /MPa	Deformation/mm	Elastic modulus /MPa	Maximum force /N	Tensile strength /MPa	Elongation at break/%	
1	81.52	7.55	0	7.79	159.26	14.75	581.11	81.3
2	84.39	7.81	0	6.34	167.73	15.53	658.63	
3	74.71	6.92	0	4.99	148.12	13.71	591.75	
Average	80.21	7.43	0	6.37	158.37	14.66	610.50	

Table 5.47 PVC/TOTM Murphy's dropper 1 materials test conditions (2017.04.18)

#	Sample specifications			Test speed /(mm/min)	Temperature/°C	Humidity/%
	Thickness *h*/mm	Thickness *h*/mm	Thickness *h*/mm			
1	1.8	6	25	200	23	38
2	1.8	6	25			
3	1.8	6	25			

Table 5.48 PVC/TOTM injection dropper materials test results (2017.04.18)

#	Yield strength				Maximum strength			Hardness
	Yield force /N	Yield strength /MPa	Deformation /mm	Elastic modulus /MPa	Maximum force /N	Tensile strength /MPa	Elongation at break/%	
1	77.79	7.20	0	4.18	154.62	14.32	653.17	78.1
2	60.04	5.56	0	5.84	120.09	11.12	483.87	
3	67.40	6.24	0	8.46	133.35	12.35	510.93	
Average	68.41	6.33	0	6.16	136.02	12.60	549.32	

obvious difference from TOTM in all the performance metrics. Both are qualified products.

Table 5.49 Shandong Weigao Medical Polymer Co., Ltd. Inspection Center Report

Inspection items	Technical requirements	Inspection results
Appearance	Colorless or near-colorless, transparent pellets with uniform plasticization and no external impurities	Pass
Reductive Substance	Amount of 0.02 mol/L $KMnO_4 \leq 0.3$ mL/20 mL	0.2 mL/20 mL
Acidity	Compared to control, $\Delta pH \leq 1.0$	0.2
Color	Clear, colorless	Pass
Heavy metal	0.3 μg/mL	Pass
UV absorbance	$\leq 0.3(230 - 360$ nm)	0.04
Zn	≤ 0.4 μg/mL	0.2 μg/mL
Non-volatile substances	≤ 2.0 μg/100 mL	0.7 μg/100 mL
Fluorescent substances	No fluorescence when tested under 355 nm UV lamp	Pass

Note Inspectee entity: Zhengzhou University; Inspection environment: the temperature and humidity meet the requirements; production entity: Zhengzhou University, sample # Y2017022365; product name: soft PVC plastics for manufacturing transfusion (infusion) equipment, batch # 170221-22-1 DEHCH; product code MT; sample status: excellent; sample #: 4; inspection type: verification; inspection standard: WGJRs-SOP-ZL-030; inspection time: 2017.02.27–2017.02.28

Table 5.50 Shandong Weigao Medical Polymer Co., Ltd. Inspection Center Report

Inspection items	Technical requirements	Inspection results	Individual judgment
Appearance	Colorless or near-colorless, transparent pellets with uniform plasticization and no external impurities	Pass	Qualified
Reductive substance	Amount of 0.02 mol/L KMnO4 $\leq$ 0.3 mL/20 mL	0.1 mL/20 mL	Qualified
Acidity	Compared to control, $\Delta pH \leq 1.0$	0.2	Qualified
Color	Clear, colorless	Pass	Qualified
Heavy metal	0.3 μg/mL	Pass	Qualified
UV absorbance	$\leq 0.3(230$–360 nm)	0.1	Qualified
Zn	≤ 0.4 μg/mL	0.1 μg/mL	Qualified
Non-volatile substances	≤ 2.0 μg/100 mL	0.7 μg/100 mL	Qualified
Fluorescent substances	No fluorescence when tested under 355 nm UV lamp	Pass	Qualified

Table 5.51 Shandong Weigao Medical Polymer Co., Ltd. Inspection Center Report (injection dropper material)

Inspection items	Technical requirements	Inspection results	Individual judgment
Reductive substance	Amount of 0.02 mol/L $KMnO_4 \leq 0.3$ mL/20 mL	0.4 mL/20 mL	Qualified
Acidity	Compared to control, $\Delta pH \leq 1.0$	0.2	Qualified
Color	clear, colorless	Pass	Qualified
Heavy metal	0.3 μg/mL	Pass	Qualified
UV absorbance	≤0.3(230–360 nm)	0.1	Qualified
Zn	≤0.4 μg/mL	0.4 μg/mL	Qualified
Non-volatile substances	≤2.0 μg/100 mL	0.7 μg/100 mL	Qualified

Table 5.52 Shandong Weigao Medical Polymer Co., Ltd. Inspection Center Report (Murphy's dropper 1 material)

Inspection items	Technical requirements	Inspection results	Individual judgment
Reductive substance	Amount of 0.02 mol/L KMnO4 ≤ 0.3 mL/20 mL	0.2 mL/20 mL	Qualified
Acidity	compared to control, $\Delta pH \leq 1.0$	0.3	Qualified
Color	clear, colorless	Pass	Qualified
Heavy metal	0.3 μg/mL	Pass	Qualified
UV absorbance	≤0.3(230 –360 nm)	0.1	Qualified
Zn	≤0.4 μg/mL	0.4 μg/mL	Qualified
Non-volatile substances	≤2.0 μg/100 mL	0.7 μg/100 mL	Qualified

References

1. Widegern JA, Finke RG. A review of soluble transition-metal nanoclusters as arene hydrogenation catalysts. J Mol Catal A: Chem. 2003;191(2):187–207.
2. Gu JF, Cui HY, Qian SS. Catalytic hydrogenation of DOP liquid phase to DEHCH and its kinetics. Plast Ind. 2012;2:87–91.
3. Zhao KJ, Zhu YH, Feng HQ. Study of hydrogenation of dioctyl phthalate. Shandong Chem Ind. 2012;41(7):31–3.
4. Li YL. Phthalic anhydride production technology and market trends. Petrochem Technol Econ. 2005;21(2):44–8.
5. Yang XP. Patent tracking and analysis of benzyl anhydride production technology. Petrochem Technol Econ. 2011;1:12.
6. Zhao J, Xue M, Huang Y, et al. Hydrogenation of dioctyl phthalate over supported Ni catalysts. Catal Commun. 2011;16(1):30–4.
7. Katsuhiko S, Shigeyasu M, Takeshi K, et al. Studies on the thermal decomposition of phthalic acid esters. IV. Thermal decomposition of mono (2-ethylhexyl) phthalate. Chem Pharma Bull. 1979;27(12):3140–3144.
8. Liu YX, Wei ZJ, Fu J. Selective hydrogenation of phthalein to phthalein over nickel-based catalyst. J Catal. 2008;29(1):52–6.

9. Zhao J, Yu HB, Fei YN. Analysis of hydrogenation products of diisooctyl phthalate by gas chromatography-mass spectrometry. Phys Chem Test Chem. 2015;51(6):745–8.
10. Saido K, Tomono K, Taguchi H, et al. Kinetic study on phthalates. J Oleo Sci. 2001;50(11):875–81.
11. Can H, Metin Ö. A facile synthesis of nearly monodisperse ruthenium nanoparticles and their catalysis in the hydrolytic dehydrogenation of ammonia borane for chemical hydrogen storage. Appl Catal B: Environm. 2012;125:304–10.
12. Ma XF, Lin R, Beuerle C, et al. Effects of surface activation on the structural and catalytic properties of ruthenium nanoparticles supported on mesoporous silica. Nanotechnology. 2014;25(4):45701.
13. Yang J, Sargent E, Kelley S, et al. A general phase-transfer protocol for metal ions and its application in nanocrystal synthesis. Nat Mater. 2009;8(8):683–9.
14. Su F, Lv L, Lee FY, et al. Thermally reduced ruthenium nanoparticles as a highly active heterogeneous catalyst for hydrogenation of monoaromatics. J Am Chem Soc. 2007;129(46):14213–23.
15. Miao S, Liu Z, Han B, et al. Ru nanoparticles immobilized on montmorillonite by ionic liquids: A highly efficient heterogeneous catalyst for the hydrogenation of benzene. Angew Chem Int Ed. 2006;45(2):266–9.

Chapter 6
Pilot Demo and Industrialization

Our research team collaborated with industry to perform pilot demonstrations and industrialization design, developed key facilities, constructed complete sets of facilities and production protocols for catalyst preparation and ring hydrogenation of phthalates, published product industrial standards, and entrusted third parties and SGS-CSTC Standard Technical Services Co. Ltd. to inspect the products. The production protocol is safe, energy-saving, and without waste disposal or environmental pollution. The product parameters meet the requirements of REACH and RoSH.

6.1 Pilot Demo Studies

Our research team collaborated with industry on pilot-scale demonstrations and investigated the catalyst activity, selectivity, repeatability, and stability. The pilot demo facility is shown in Fig. 6.1.

The pilot demo facility includes static mixer M, external circulation reactor R, gravity separator Q, etc.

As shown in Fig. 6.1, phthalates from raw material tank V1 are pressurized via the high-pressure feed pump P1 and are well mixed in the static mixer M with high-pressure H_2 from the H_2 preparation unit and catalyst from the solid–liquid pump P2. Through the inlet b, the reactant mixture reaches the Venturi nozzle f at the bottom of the internal tube d of the external circulation reactor R together with the high-pressure H_2 from gas inlet a. Due to the Venturi effect, a low-pressure adsorption field and Vortex reaction zone are formed to achieve the fast mass transport between the gas, liquid, and solid phases and consequently high conversion of phthalates. The external circulation reactor is heated through the jacket to control the temperature below 130 °C and the pressure at 2–4 MPa. The mixture flow (or residence) time in the internal tube d is carefully controlled to reduce side reactions and achieve high selectivity. Then the reaction mixture overflows the top of the internal tube to the external tube e. The majority of it passes the overflow weir and enters the gravity

Z. Liu, *Green Catalytic Hydrogenation of Phthalate Plasticizers*,
https://doi.org/10.1007/978-981-16-9789-0_6

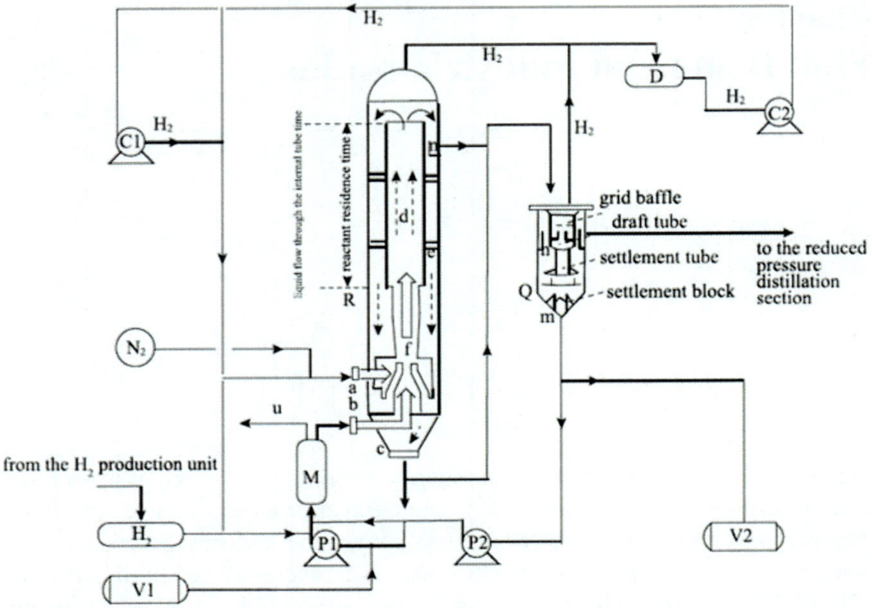

Fig. 6.1 Phthalate hydrogenation pilot demo facility. M is the static mixer, with u on top of it, the purge gas venting port; R is the external circulation reactor, with a the H_2 or N_2 inlet, b the reaction feeding port, c the lower discharge port, d the internal tube, e the external tube, and f the Venturi nozzle; Q is the gravity separator, with m the lower discharge port, n the overflow weir; C1 and C2 are the gas compressors; P1 is the high-pressure feed pump; P2 is the solid–liquid pump; V1 is the raw material storage tank; V2 is the catalyst storage tank; D is the buffer tank.

separator Q, and a small portion containing a liquid–solid mixture of catalyst and cyclohexane-1,2-dicarboxylate enters the gravity separator Q and the static mixer M through the bottom inlet c. The gravity separator has a grid baffle, a draft tube and an overflow weir at its top, a settlement tube at its middle, and a settlement block and settlement core at its bottom to achieve gas–liquid separation, liquid–solid separation, and minimal catalyst clogging. After separation, H_2 exits the top of the gravity separator Q and flows through the heat exchanger, buffer tank D, and gas compressor C2 to be recycled. The solid catalyst is extracted from the bottom of the gravity separator Q via a solid–liquid pump and returns to the static mixer M. The ring-hydrogenated products pass the overflow weir n and enter the reduced pressure distillation section. When the procedure is stopped, the catalyst is stored in the storage tank V2.

Figure 6.2 shows the GC spectra of 4 batches of DOP hydrogenation products generated using the above facility during August 23–28, 2016.

The peak at the retention time 1.2 min is the ethanol diluent. The peaks at 1.3 min and 1.8 min are the low-boiling-point side products. The peak at 5.4 min is the product DEHCH. The peak at 5.7 min is the unreacted DOP.

Table 6.1 shows the DOP conversion ratio C_{DOP}, DEHCH yield Y_{DEHCH}, and the

Fig. 6.2 GC spectra of DOP hydrogenation products

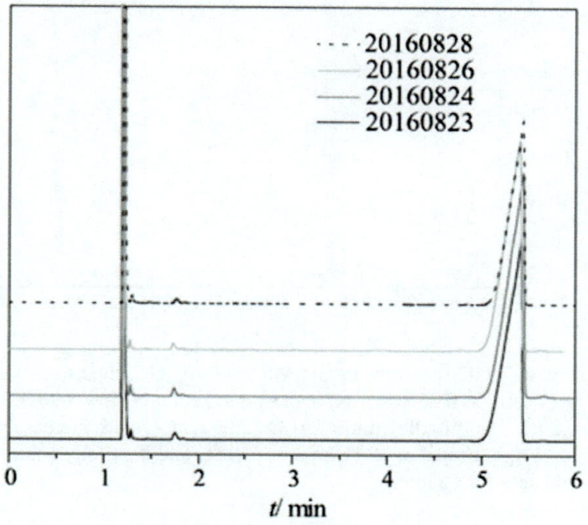

Table 6.1 C_{DOP}, Y_{DEHCH}, and low-boiling-point side product content from pilot demo facility

Batch	Temperature/°C	H_2 Pressure /MPa	C_{DOP}/%	Y_{DEHCH}/%	Side product content/%
2016.08.23	135	2.5	98.78	98.59	1.22
2016.08.24	132	3.0	98.84	98.49	1.14
2016.08.26	130	3.5	99.01	98.72	0.99
2016.08.28	130	4.0	100	99.02	0.98

contents of low-boiling-point side products, obtained using the areal normalization method.

GC-MS analysis of the reaction products is shown in Fig. 6.3.

Table 6.2 shows the MS analysis results of the products.

From Table 6.2, the raw material DOP (peak 9) contains a trace amount of DBP (peak 5). In the hydrogenation process; most DOP is converted to DEHCH (peaks 7 and 8 are cis–trans isomers). The DBP is converted to DBCH (peaks 3 and 4 are cis–trans isomers). There are unconverted DOP (peak 9) and DBP (peak 5). The side products include n-octanol (peak 1), butyl cyclohexanecarboxylate (peak 2), and octyl cyclohexanecarboxylate (peak 6). The products were purified using reduced pressure distillation and nitrogen extraction.

The pilot test indicates that a single DOP conversion ratio reaches as high as 99%. Complete conversion was achieved via second hydrogenation to the hydrogenation product that contained 1% DOP. After the four batches, the catalyst was kept in the reactor for an extended period of time. When the facility was turned back on, the catalyst still exhibited excellent stability and lifetime.

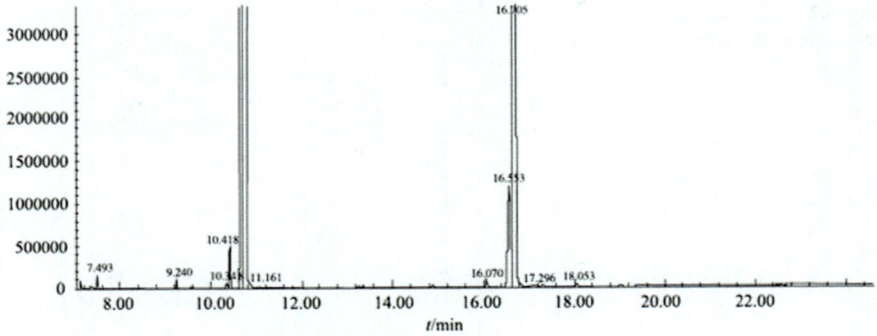

Fig. 6.3 GC spectrum in GC-MS analysis. GC conditions: HP-5MS capillary column (60 m × 0.25 mm × 0.25 μm), N_2 carrier gas, column flow rate 1.5 mL/min, injected sample volume 0.2 μL, splitting flow ratio 5:1. Column temperature increase protocol: initial temperature 100 °C for 1 min, heated at 12 °C/min to 250 °C and kept for 1 min, then heated at 20 °C/min to 270 °C and kept for 15 min.

Table 6.2 MS analysis results of DOP hydrogenation reaction products

Peak #	Compound	Molecular weight	Retention time/min
1	n-Octanol	130	7.493
2	Butyl cyclohexanoate	184	9.240
3	Dibutyl cyclohexane-1,2-dicarboxylate	284	10.679
4	Dibutyl cyclohexane-1,2-dicarboxylate	284	10.750
5	Dibutyl phthalate	278	11.161
6	Octyl cyclohexanoate	268	16.070
7	Dioctyl cyclohexane-1,2-dicarboxylate	396	16.553
8	Dioctyl cyclohexane-1,2-dicarboxylate	396	16.705
9	Dioctyl phthalate	390	17.296

6.2 Industrialization Design of 50 k Ton/y Scale Phthalate Hydrogenation

6.2.1 Design Flowchart

Our research team collaborated with industry to complete the industrialization design of 50 k ton/y scale phthalate hydrogenation. The flowchart is shown in Fig. 6.4.

Figure 6.4 includes H_2 purification, hydrogenation, separation and purification, recycling, and exhaust recycling systems. The protocol is safe, energy-saving, and free of waste disposal.

The H_2 purification system functions to remove the CO, S- and N-containing compounds to avoid catalyst poisoning. The hydrogenation system includes three

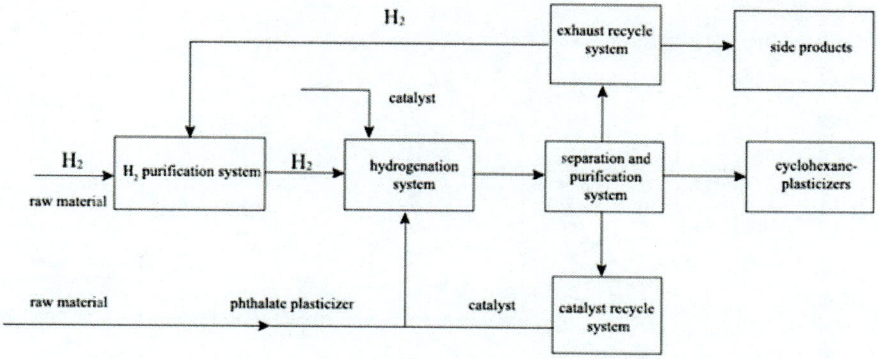

Fig. 6.4 Design flowchart

series-connected autoclave-type reactors to achieve 100% phthalate conversion. The separation and purification system realizes gas–liquid and liquid–solid separations; H_2 returns to the hydrogenation system after the gas–liquid separation; side products, low-boiling-point alcohols, and alkanes are collected; the cyclohexane-1,2-dicarboxylates flow into the product tank. After the liquid–solid separation, the catalyst is returned to the hydrogenation system via the recycling system.

6.2.2 Hydrogenation System

The hydrogenation system is shown in Fig. 6.5, including the feedstock buffer tank, hydrogenation reactor, and catalyst storage tank.

The hydrogenation reactors are the key facilities that use Hastelloy alloy with HC276 liners. Each reactor is equipped with multi-layer stirrers to enhance the gas–liquid–solid mass transport and increase the reaction rate. These reactors have different functions and thus their operational temperature, stir rate, and reactant retention time are not identical. Each reactor is equipped with multiple temperature sensors. If changes in temperature or pressure occur, the H_2 flow will be adjusted or cut to ensure system safety. The design pressure and temperature of the reactors are 6 MPa and 280 °C to leave sufficient space for adjustment. There is a sampling port at the middle of reactors for drawing samples to analyze the conversion and selectivity with GC and monitor the reaction process. The phthalate feedstock flux is defined according to the effective reactor volume so that the overall reactant retention time falls within 60–180 min.

The hydrogenation reaction is completed in the first reactor. Under conditions of 60–110 °C, 3.0–5.0 MPa H_2 pressure, 100–200 r/min stir rate, and 20–60 min reaction time, the phthalate conversion ratio reaches >99%. Because hydrogenation is an exothermal reaction, the released heat can be used to maintain the reaction temperature and meanwhile to avoid overheating to minimize side reactions.

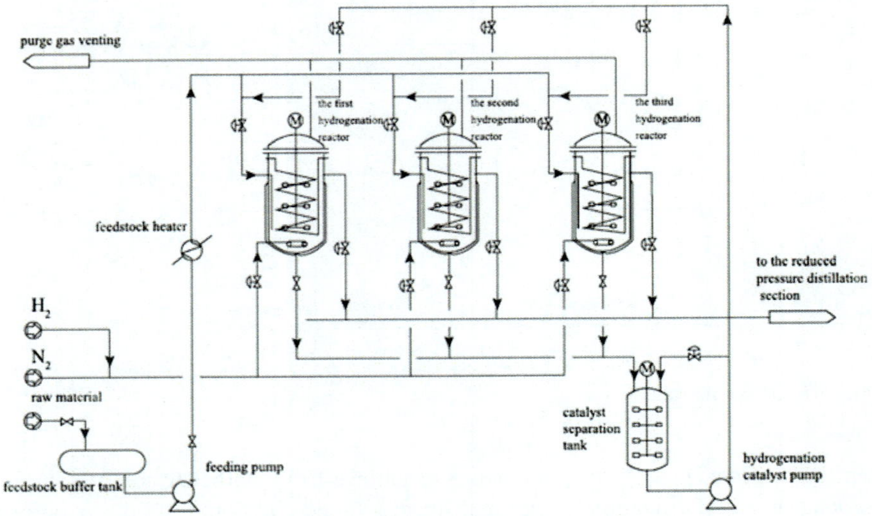

Fig. 6.5 The hydrogenation system of the 50kt/year environmentally friendly plasticizer project

The function of the second reactor is to backmix the small amount of unreacted phthalates residing at the dead ends of the first reactor and adhered to the stirrer shaft, inlet tube, and temperature sensor to achieve 100% conversion under mild conditions. The temperature is maintained at 100 °C and the H_2 pressure at 3–4 MPa with a 20 min reaction time. The third reactor is a backup.

The feedstock buffer tank provides phthalate raw materials that are pressurized using a high-pressure pump, are converged with the catalyst from the catalyst recycling system via a solid–liquid pump, and enter the first reactor. The pressurized H_2 is introduced from the bottom.

6.2.3 *Separation and Purification System*

The separation system, shown in Fig. 6.6, mainly includes feedstock preheater, vacuum distillation tower, extraction tower, solvent recycling tower, light component tank, heavy component tank, vacuum distillation tower condenser, solvent recycling tower condenser, etc. Among them, the vacuum distillation tower is the key facility.

The phthalate hydrogenation products is preheated and then enters the middle of the vacuum distillation tower. From the tower bottom are introduced a certain amount of inert gases such as N_2, CO_2, and superheated steam. The hydrogenation products move downward to exchange mass and heat with the upward-moving inert gas flow to achieve separation of the target product with the side products. After liquid–solid separation, the catalyst returns to the hydrogenation system via a solid–liquid pump. The target product is released from the tower bottom and, after heat exchange with

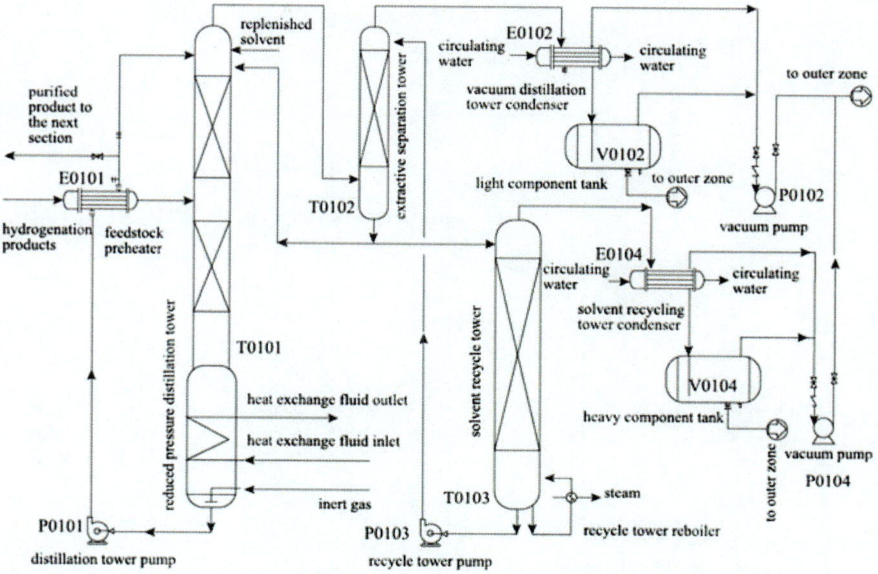

Fig. 6.6 The separation and purification system

the feedstock material, enters the next section. The solvent and side products flow with the inert gas and enter the bottom of the extraction tower. The light components including low-boiling-point alcohols and alkanes enter the vacuum distillation tower condenser with the inert gases, are liquefied, and enter the light component tank. The solvents and cyclohexanecarboxylates enter the solvent recycling tower from the bottom; the former return to the extraction tower for repeated use and the latter enter the recycling tower condenser and then the heavy component tank as liquids. The whole system maintains its temperature at 160–180 °C and its pressure at −0.06 to −0.098 MPa to avoid localized liquid overheating and thermal decomposition of the product.

6.2.4 Catalyst Recycling System

The catalyst recycling system, shown in Fig. 6.7, includes a catalyst separation tank, hydrogenation product relay tank, catalyst recycling tank, membrane filter, etc. Among them, the catalyst separation tank is the key facility.

The catalyst-containing materials from the hydrogenation reactors are separated into the hydrogenation products and catalyst in the catalyst separation tank via gravity settlement. The hydrogenation products enter the relay tank for further separation, and H_2 enters the exhaust cleanup system. The small amount of suspending catalyst is filtered using the membrane filter. The accumulated catalyst is then recoiled by the

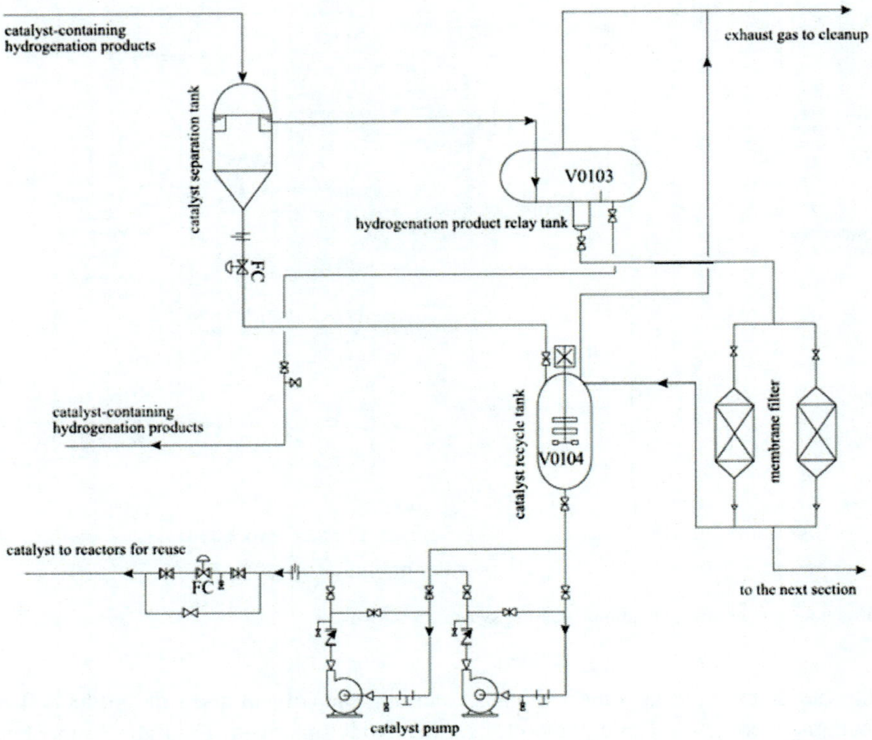

Fig. 6.7 The catalyst recycle system

product into the catalyst recycling tank, which is converged with the catalyst from the catalyst separation tank and sent to the hydrogenation reactors via a catalyst pump for repeated use.

6.2.5 Exhaust Recycling System

The exhaust recycling system, shown in Fig. 6.8, includes the electrical heater, preheater, pretreatment reactor, molecular sieve reactor, and precision filter.

The exhaust from the hydrogenation system first enters the preheater and is heated to 120–160 °C and then enters the pretreatment reactor loaded with supported catalyst. The active composition of the supported catalyst is one or more of Pt, Ni, Co, Mo, Cu, and Zn and can remove the S, N, O-containing impurities from the exhaust at 160–240 °C and 1.0–4.5 MPa H_2 pressure. Then, the treated exhaust successively passes the molecular sieve reactor to separate out H2 in >99% purity under 1.0–4.5 MPa H_2 pressure, and the precision filter to remove any possible catalyst powder and other particles. The recycled H_2 is returned to the hydrogenation system for

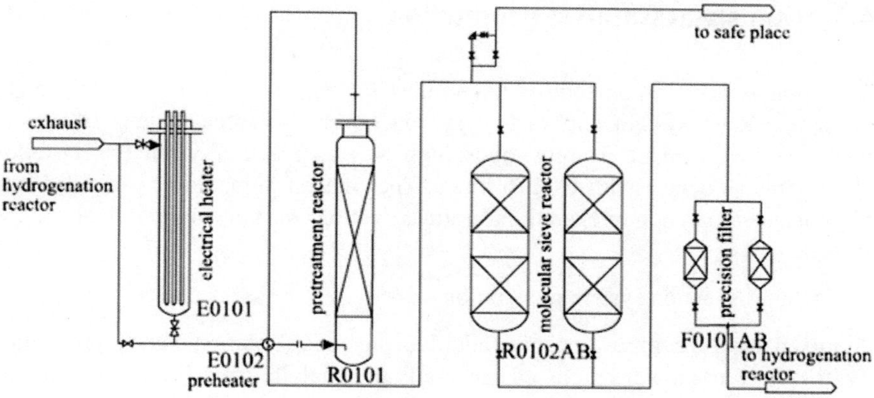

Fig. 6.8 The exhaust recycling system

repeated use. The separated impurity gas includes CH_4, CO, N_2, etc., and can be used as fuel gas.

The design and construction plans of the 50 kt/year phthalate ring hydrogenation industrial facilities, or called 50 kt/year environmentally friendly non-toxic plasticizer project, have passed the government examinations, including.

(1) Feasibility report of 50kt/y environmentally friendly non-toxic plasticizer project (construction agency: Qitaihe Longao Environmental Technology Co., Ltd.).
(2) Environmental impact report of 50kt/y environmentally friendly non-toxic plasticizer project (Certificate #: National Environmental Assessment Certificate A 1702; assessment agency: Harbin Institute of Technology, December 2016).
(3) Safety assessment report of 50kt/y environmentally friendly non-toxic plasticizer project (assessment agency: Daqing Taige Safety Technology Co., Ltd.).
(4) Pre-assessment of occupational hazards of 50kt/y environmentally friendly non-toxic plasticizer project (Certificate #: RXZPY 2016-037; assessment agency: Harbin Runxin Occupational Health Test and Assessment Co., Ltd.).
(5) Safety facility design report of 50kt/y environmentally friendly non-toxic plasticizer project.
(6) Fire-fighting facility design report of 50kt/y environmentally friendly non-toxic plasticizer project.
(7) Occupational disease prevention design report of 50kt/y environmentally friendly non-toxic plasticizer project.

The 50 kt/year phthalate ring hydrogenation industrial facilities strictly follow the design concepts of safe production, energy saving, loss reduction, and environmental protection.

6.3 Industrial Catalyst Production

The catalyst preparation industrial facilities include the batching system, high-pressure reaction system, and washing system, cover a land area of about 100 m^2, as well as an analytical lab. The core of the high-pressure reaction system is a Hastelloy alloy 680L autoclave with HC276 liners. The catalyst preparation is divided into high-dispersion system preparation, in situ reduction, washing and purification, and storage.

1. High-dispersion system preparation

A mixed active composition and additive precursor solution and a precipitant hydroxide solution were prepared separately in equal volumes. Their total volume was 70–80% of the effective autoclave volume. The mass fraction was 10–20% for the mixed active composition and additive precursor solution and was 5–10% for the precipitant hydroxide solution. The hydroxide solution was added to the autoclave first and heated to 40–80 °C at a stir rate of 100–200 r/min. Then, the mixed precursor solution was added to start the reaction at an elevated stir rate of 200–300 r/min. The reaction lasted for 30–100 min to form the highly dispersed active composition and additive system.

2. In situ reduction

High purity N_2 replaced the air in the autoclave to a pressure of 3.0–5.0 MPa kept for 5 min and then released. The operation was repeated 3 times. Then, high purity H_2 replaced the N_2 using the same method. The reduction reaction lasted for 4–8 h under conditions of 100–150 °C, 3.0–5.0 MPa H_2 pressure, and 300–400 r/min stir rate.

3. Washing and purification

After the reduction reaction, the system was cooled naturally or with cooling water. When the autoclave temperature dropped below 95 °C, the H_2 was emptied from the autoclave and the autoclave was connected to air. The materials in the autoclave were guided into the washing and purification system. The measured pH value should fall within the required range. The washing and purification system used polypropylene plastics to avoid Cr, Fe, Ni contaminations via corrosion. High purity water was used to wash out the residues in the autoclave. The catalyst was washed under stirring, until the pH of the supernatant reached 7.

The catalyst of each batch was directly used in the solvent-free DOP hydrogenation reaction at the catalyst/DOP mass ratio of 0.25%, 120 °C, 4.0 MPa H_2 pressure, 60 min reaction time, and nanocrystal size 3.5–6 nm. The conversion ratio and selectivity were comparable to the lab test.

The catalyst preparation is the core technology of the catalytic ring hydrogenation of phthalate plasticizers.

6.4 Industrial Standards of Cyclohexane-Based Plasticizers

In 2017, Qitaihe Longao Environmental Technology Co., Ltd., Zhengzhou Green Catalytic Hydrogenation Technology Key Laboratory, and Henan Kemingjunde Chemical Technology Co., Ltd. co-drafted and co-published 6 industrial standards for cyclohexane-based plasticizers.

Our standards were formatted and written following "Directives for the work of standardization Part 1: Format and writing of standards" (GB/T 1.1-2009). These standards implement the national standard "Toy safety" (GB 6675-2014) reinforcing the 0.1% limit of 6 plasticizers including DBP, BBP, DEHP, DOP, DINP, and DIDP and took into effect on January 1, 2016. The inspection methods involved in these standards use corresponding national standards.

6.4.1 Scope

Our standards stipulate the requirements, test methods, inspection rules, signs, packaging, transportation, and storage of industrial cyclohexane-1,2-dicarboxylates.

Our standards are applicable to cyclohexane-1,2-dicarboxylates prepared via hydrogenation of phthalate plasticizers. The products are mainly used in industries such as plasticizers for PVC and other polar polymers, plastics, rubber, paint, emulsifier, etc., as well as toxicologically sensitive applications.

6.4.2 Normative References

The terms in the following standard documents become the terms of our standards via being cited. For dated references, their revisions (not including errata) and revised versions do not apply to our standards, but all parties using our standards are encouraged to explore the use of the newest versions of these documents. The newest versions of undated references apply to our standards.

GB/T 601 Chemical reagent—Preparations of reference titration solutions;

GB/T 603 Chemical reagent—Preparations of reagent solutions for use in test methods (GB/T 603—2002, ISO 6353-1: 1982, NEQ);

GB/T 8170—2008 Rules of rounding off for numerical values & expression and judgment of limiting values;

GB/T 1664 Plasticizers—Determination of color;

GB/T 4472—2011 Determination of density and relative density for chemical products;

GB/T 6488—2008 Liquid chemicals. Determination of refractive index at 20 °C;

GB/T 1668—2008 Plasticizers—Determination of acid value and acidity;

GB/T 11133 Standard test method for determination of water in petroleum products, lubricating oils, and additives—Coulometric Karl Fischer titration method;

GB/T 9722—2006 Chemical reagent—General rules for the gas chromatography;

GB/T 1671—2008 Plasticizers—Determination of flash point-Cleveland open cup method;

DZ/T 0223—2001 Analysis method for inductively coupled plasma mass spectrometry (ICP-MS);

GB/T 6678—2003 General principles for sampling chemical products;

GB/T 6680—2003 General rules for sampling liquid chemical products;

GB/T 6682 Water for analytical laboratory use—Specification and test methods (GB/T 6682—2008, NEQ, ISO 3696: 1987).

6.4.3 Test Methods

The standard solutions, reagents, and products used in our standards followed GB/T 601 and GB/T 603, otherwise specified. Analytical tests used only analytical purity reagents and Level 3 water defined in GB/T 6682.

The determination of experimental data and test results in our standards followed the rounding value comparison method defined in GB/T 8170–2008.

1. Determination of appearance

Visual inspection in natural light.

2. Determination of color

Measurement following GB/T 1664.

3. Determination of density

Measurement following Webster's balance method in GB/T 4472-2011.

Allowable error: the mathematical average value of two parallel measurements was used as the result with four decimals. The difference between the two parallel measurements was no higher than 0.0005.

4. Determination of refractive index

Measurements following GB/T 6488-2008.

Allowable error: the mathematical average value of two parallel measurements was used as the result with four decimals. The difference between the two parallel measurements was no higher than 0.0005.

5. Determination of ester content

(1) Mechanism: the sample was injected in a GC spectrometer and flowed on the column with the carrier gas to be separated. The eluents were detected using an H_2 flame ionization detector, which were recorded in the spectrum. The

contents of cyclohexane-based esters were obtained using the area normalization method. The total content of all cyclohexane-based esters is the ester content.

(2) reagents and materials: carrier and auxiliary gases. (i) N_2, > 99.99%; (ii) H_2, > 99.99%; (iii) compressed air, purified.

(3) Instruments: GC spectrometer, H_2 flame ionization detector.

(4) Chromatographic columns: (i) quartz capillary column: 0.32 mm × 30 m × 0.5 μm, commercial; (ii) aging of quartz capillary column: the capillary column was installed in FC spectrometer and the gas tightness was checked. With N_2 connected, the column temperature was increased stepwise from 100 °C to 270 °C for > 7 h until the baseline was stable.

(5) Sample injector: 1μL glass micro syringe.

(6) Chromatography workstation.

(7) Operation conditions: the instrument was operated as follows. Appropriate adjustments could be made for different instruments to obtain decent separation. (i) Column temperature: programed heating, initial temperature 150 °C for 2 min, 10 °C/min, final temperature 260 °C for 10 min; (ii) gasification chamber temperature: 260 °C; (iii) detector chamber temperature: 260 °C; (iv) N_2 flow rate: 30 mL/min; (v) H_2 flow rate: 30 mL/min; (vi) airflow rate: 300 mL/min.

(8) Analytical steps: the instrument was adjusted as above until the baseline was stable. 0.2μL sample was injected using the micro syringe. The chromatography workstation automatically showed the areal percentages of compositions. The linear range of the instrument should meet the requirements for the normalization quantitative analysis. The retention times of the compositions on the column are listed in Table 6.3.

(9) Allowable error: the mathematical average value of two parallel measurements was used as the result with three decimals. The difference between the two parallel measurements was no higher than 0.150%.

(10) Detection limit: the lowest detection concentration is 0.002%.

6. Determination of acid value

Measurements following GB/T 1668-2008.

Allowable error: the mathematical average value of two parallel measurements was used as the result with three decimals. The difference between the two parallel measurements was no higher than 0.003.

7. Determination of water content

Measurements following GB/T 11133.

Allowable error: the mathematical average value of two parallel measurements was used as the result with three decimals. The difference between the two parallel measurements was no higher than 0.003%.

Table 6.3 Retention time of each component on the GC column

Peak #	Component	Retention time/min
1	Octane	2.157
2	Methyl cyclohexane	2.334
3	3-Methyl heptane	2.594
4	2-Ethylhexanol	4.682
5	1,4-Dimethyl cyclohexane	8.167
6	Cyclohexyl isoamyl ester	9.228
7	Diisobutyl cyclohexane-1,2-dicarboxylate	9.386
8	Dibutyl cyclohexane-1,2-dicarboxylate	9.54
9	Hexyl cyclohexanoate	11.03
10	2-Ethylbutyl cyclohexanoate	11.137
11	Butyl octyl cyclohexane-1,2-dicarboxylate	14.828
12	Octyl cyclohexanoate	15.569
13	Diisooctyl cyclohexane-1,2-dicarboxylate	16.025
14	Dioctyl cyclohexane-1,2-dicarboxylate	16.821
15	Dioctyl phthalate	17.331
16	Di(2-ethylhexyl) cyclohexane-1,2-dicarboxylate	18.027

8. Determination of phthalate content

Measurements following GB/T 9722-2006.

Allowable error: the mathematical average value of two parallel measurements was used as the result with three decimals. The difference between the two parallel measurements was no higher than 0.001%.

9. Determination of flashing point

Measurements following GB/T 1671-2008.

Allowable error: the mathematical average value of two parallel measurements was used as the result with zero decimals. The difference between the two parallel measurements was no higher than 5 °C.

10. Determination of metal contents

Measurements following DZ/T 0223-2001.

Allowable error: the mathematical average value of two parallel measurements was used as the result with two decimals. The difference between the two parallel measurements was no higher than 0.03%.

6.4.4 Inspection Rules

1. Categories of inspections

Table 6.3 lists the factory inspection items.

2. Batching rule

One batch is defined as the uniform product from one storage tank or one package.

3. Manufacturer inspection

The product can be released only after inspection by the manufacturer's quality control department, with the formatted quality certificate attached. The certificate contains the product name, manufacturer name, standard number, grade, lot number, production date, and inspector code.

4. Sampling

The number of samples is determined according to the requirements in GB/T 6678-2003. The sampling is performed following GB/T 6680-2003. A clean, dry sampling tube is inserted in the upper, middle, and bottom part of the container to draw samples of >1000 mL. The samples are well blended and fed into two clean, dry ground bottles. The bottles are then sealed and specify the producer, product name, lot number, sampling date, and sampler name. One bottle is subjected to inspection. The other bottle will be kept for 3 months in case of re-inspection.

5. Re-inspection

If one of the factory inspection items fails to meet our standards, samples will be drawn from the double amounts of the product and re-inspected. Once an item fails the re-inspection, this product batch is regarded as unqualified.

6. Sign, packaging, transportation, and storage

(1) Sign

Each product pack should have a clear, firm sign containing the product name, standard number, grade, manufacturer name, address, contact number, trademark, net weight, production date, lot number, etc.

(2) Packaging

This product should be packaged in clean, dry, robust, impurity-free 200 mL zinc-plated iron or steel buckets. The screw socket on the bucket cap should be sealed with clean polyethylene to avoid leaking loss. If demanded by users, another packaging can also be used. Table 6.3 lists the factory inspection items. Bulk containers also must satisfy the above requirement.

(3) Transportation

Sun, rain, and violent impact should be avoided during product transportation and moving.

(4) Storage

This product should be stored in a well-vented, dry warehouse or shed. With appropriate transportation and storage conditions, the storage time is 6 months after the production date, after which the product can still be used if the inspection according to our standards is passed.

6.4.5 Industrial Standard Numbers and Technical Index

The industrial standards and technical index of DEHCH, DBCH, DIBCH, DINCH, DPHCH, and DIDCH are listed below.

(1) The industrial standard of DEHCH is Q/QL A001-2017. Its technical index is shown in Table 6.4.
(2) The industrial standard of DBCH is Q/QL A002-2017. Its technical index is shown in Table 6.5.
(3) The industrial standard of DIBCH is Q/QL A003-2017. Its technical index is shown in Table 6.6.
(4) The industrial standard of DINCH is Q/QL A004-2017. Its technical index is shown in Table 6.7.
(5) The industrial standard of DPHCH is Q/QL A005-2017. Its technical index is shown in Table 6.8.
(6) The industrial standard of DIDCH is Q/QL A006-2017. Its technical index is shown in Table 6.9.

Table 6.4 Technical specifications of DEHCH

Items	Metrics		
	Premium grade	First class	Qualified grade
Appearance	Transparent, oily liquid with no visible impurity		
Chromaticity (Pt–Co)/Hazen	≤25	≤40	≤60
Density (20 °C)/(g/cm^3)	0.94–0.958		
Refractive index n_D^{20}	1.455–1.466	1.455–1.466	–
Ester content/%	≥99.5	≥99.0	≥98.0
Acid value/(mg KOH/g)	≤0.08	≤0.15	≤ 0.20
Water content /%	≤0.10	≤0.15	≤0.20
Phthalate content/%	≤0.01	≤ 0.02	≤0.05
Flash point (open cup method)/°C	≥196	≥192	≥192
(Sb, As, Ba, Sn, Cr, Pb, Hg, Se) content /(mg/kg)	≤1	≤1	≤1.5
Cd content/(mg/kg)	≤0.6	≤0.6	≤1

Note the above inspection items were determined according to user request and through coordination between provider and consumer

Table 6.5 Technical specifications of DBCH

Items	Metrics		
	Premium grade	First class	Qualified grade
Appearance	transparent, oily liquid with no visible impurity		
Chromaticity (Pt–Co)/Hazen	≤20	≤30	≤40
Density (20 °C)/(g/cm^3)	1.044–1.066		
Refractive index n_D^{20}	1.455–1.466	1.455–1.466	–
Ester content/%	≥99.5	≥99.0	≥98.0
Acid value/(mg KOH/g)	≤0.08	≤0.15	≤0.20
Water content /%	≤0.10	≤0.15	≤0.20
Phthalate content/%	≤0.01	≤0.02	≤0.05
Flash point (open cup method)/°C	≥160	≥155	≥155
(Sb, As, Ba, Sn, Cr, Pb, Hg, Se) content /(mg/kg)	≤1	≤1	≤1.5
Cd content/(mg/kg)	≤0.6	≤0.6	≤1

Note the above inspection items were determined according to user request and through coordination between provider and consumer

Table 6.6 Technical specifications of DIBCH

Items	Metrics		
	Premium grade	First class	Qualified grade
Appearance	transparent, oily liquid with no visible impurity		
Chromaticity (Pt–Co) /Hazen	≤25	≤35	≤60
Density (20 °C)/(g/cm^3)	1.034–1.056		
Refractive index n_D^{20}	1.455–1.466	1.455–1.466	–
Ester content/%	≥99.5	≥99.0	≥98.0
Acid value/(mg KOH/g)	≤0.08	≤0.15	≤0.20
Water content /%	≤0.10	≤0.15	≤0.20
Phthalate content/%	≤0.01	≤0.02	≤0.05
Flash point (open cup method)/°C	≥160	≥155	≥155
(Sb, As, Ba, Sn, Cr, Pb, Hg, Se) content /(mg/kg)	≤1	≤1	≤1.5
Cd content/(mg/kg)	≤0.6	≤0.6	≤1

Note The above inspection items were determined according to user request and through coordination between provider and consumer

Table 6.7 Technical specifications of DINCH

Items	Metrics		
	Premium grade	First class	Qualified grade
Appearance	transparent, oily liquid with no visible impurity		
Chromaticity (Pt–Co) /Hazen	≤30	≤35	≤60
Density (20 °C)/(g/cm^3)	0.944 –0.966		
Refractive index n_D^{20}	1.455–1.466	1.455–1.466	–
Ester content/%	≥99.5	≥99.0	≥98.0
Acid value/(mg KOH/g)	≤0.08	≤0.15	≤0.20
Water content /%	≤0.10	≤0.15	≤0.20
Phthalate content/%	≤0.01	≤0.02	≤0.05
Flash point (open cup method)/°C	≥210	≥205	≥205
(Sb, As, Ba, Sn, Cr, Pb, Hg, Se) content /(mg/kg)	≤1	≤1	≤1.5
Cd content/(mg/kg)	≤0.6	≤0.6	≤1

Note The above inspection items were determined according to user request and through coordination between provider and consumer

Table 6.8 Technical specifications of DPHCH

Items	Metrics		
	Premium grade	First class	Qualified grade
Appearance	Transparent, oily liquid with no visible impurity		
Chromaticity (Pt–Co)/Hazen	≤25	≤40	≤60
Density (20 °C)/(g/cm^3)	0.944–0.966		
Refractive index n_D^{20}	1.445–1.476	1.445–1.476	–
Ester content/%	≥99.5	≥99.0	≥98.0
Acid value/(mg KOH/g)	≤0.08	≤0.15	≤0.20
Water content /%	≤0.10	≤0.15	≤0.20
Phthalate content/%	≤0.01	≤0.02	≤0.05
Flash point (open cup method)/°C	≥210	≥205	≥205
(Sb, As, Ba, Sn, Cr, Pb, Hg, Se) content /(mg/kg)	≤1	≤1	≤1.5
Cd content/(mg/kg)	≤0.6	≤0.6	≤1

Note the above inspection items were determined according to user request and through coordination between provider and consumer

Table 6.9 Technical specifications of DIDCH

Items	Metrics		
	Premium grade	First class	Qualified grade
Appearance	Transparent, oily liquid with no visible impurity		
Chromaticity (Pt–Co) /Hazen	$\leq$30	$\leq$35	$\leq$ 60
Density (20 °C)/(g/cm^3)	0.946–0.969		
Refractive index n_D^{20}	1.445–1.476	1.445 –1.476	–
Ester content/%	$\geq$99.5	$\geq$99.0	$\geq$ 98.0
Acid value/(mg KOH/g)	$\leq$0.08	$\leq$0.15	$\leq$ 0.20
Water content /%	$\leq$0.10	$\leq$0.15	$\leq$ 0.20
Phthalate content/%	$\leq$0.01	$\leq$0.02	$\leq$ 0.05
Flash point (open cup method)/°C	$\geq$219	$\geq$215	$\geq$ 215
(Sb, As, Ba, Sn, Cr, Pb, Hg, Se) content /(mg/kg)	$\leq$1	$\leq$1	$\leq$ 1.5
Cd content/(mg/kg)	$\leq$0.6	$\leq$0.6	$\leq$ 1

Note The above inspection items were determined according to user request and through coordination between provider and consumer.

6.5 Third-Party Inspections

1. Hangzhou C&K Testing Technic Co., Ltd.

On March 28, 2017, our research team entrusted Hangzhou C&K Testing Technic Co., Ltd. to inspect the SVHC contents in Zhengzhou University product DEHCH. The SVHC includes 173 substances of very high concern (SVHC) defined in EU No. 1907/2006 (i.e., REACH). The mass fraction of SVHC does not exceed 0.1%.

Inspection result: the mass fraction of the 173 SVHC is lower than 0.1%. The report is shown in Table 6.10.

Among the 173 SVHC defined by REACH #57, phthalate plasticizers and their inspection results are shown in Table 6.11.

2. SGS-CSTC Standards Technical Services (Tianjin) Co. Ltd.

Introduction of SGS: SGS certification is a certificate of conformity service SGS offers based on standards, laws, and user requests. To provide a certificate, examination, inspection, identification, and certification must be performed (SGS certificate itself is not a "certificate" but refers to SGS's certification service). SGS is an abbreviation of Societe Generale de Surveillance S. A., which was established in 1878 and now is the globally largest and oldest non-governmental third-party multinational corporation offering quality control and technical appraisal.

On April 1, 2017, our research team entrusted SGS-CSTC Standards Technical Services (Tianjin) Co. Ltd. to test DEHCH following the content limit requirements defined by EU RoSH 2011/65/EU Appendix I Revision (EU) 2015/863. Here are the results: based on the tests performed on the products, the contents

Table 6.10 DEHCH inspection report (No. TR20171630)

#	Inspection item	CAS#	Detection limit	Inspection result
1	Anthracene	120-12-7	100	N.D
2	4,4'-Methylenedianiline (MDA)	101-77-9	100	N.D
3	Musk xylene	81-15-2	100	N.D
4	Hexabromocyclododecane	25637-99-4, 3194-55-6 (134237-50-6) (134237-51-7) (134237-52-8)	100	N.D
5	C_{10-13} chlorcosane	85535-84-8	100	N.D
6	Dibutyl phthalate (DBP)	84-74-2	10	N.D
7	Di(2-ethylhexyl) phthalate (DEHP)	117-81-7	10	20
8	Butyl benzyl phthalate (BBP)	85-68-7	10	N.D
9	Cobalt chloride ($CoCl_2$)	7646-79-9	100	N.D
10	Bis(tributyltin) oxide (TBTO)	56-35-9	100	N.D
11	Sodium dichromate	7789-12-0, 10588-01-9	100	N.D
12	Lead hydrogen arsenate	7784-40-9	100	N.D
13	As_2O_3	1327-53-3	100	N.D
14	As_2O_3	1303-28-2	100	N.D
15	Triethyl arsenate	15606-95-8	100	N.D
16	Anthracene oil	90640-80-5	100	N.D
17	Anthracene oil, anthracene paste, distn lights	91995-17-4	100	N.D
18	Anthracene oil, anthracene paste, anthracene fraction	91995-15-2	100	N.D

(continued)

Table 6.10 (continued)

#	Inspection item	CAS#	Detection limit	Inspection result
19	Anthracene oil, anthracene-low	90640-82-7	100	N.D
20	Anthracene oil, anthracene paste	90640-81-6	100	N.D
21	Coal tar pitch, high temperature	65996-93-2	100	N.D
22	Acrylamide	79-06-1	100	N.D
23	2,4-Dinitrotoluene	121-14-2	100	N.D
24	Diisobutyl phthalate (DIBP)	84-69-5	10	N.D
25	Tri(2-chloroethyl) phosphate	121-14-2	100	N.D
26	Lead chromate	7758-97-6	100	N.D
27	Lead chromate molybdate sulfate red (C.I. Pigment Red 104)	12656-85-8	100	N.D
28	Lead sulfochromate yellow (C.I. Pigment Yellow 34)	1344-37-2	100	N.D
29	Trichloroethene	79-01-6	100	N.D
30	Boric acid	10043-35-3, 11113-50-1	100	N.D
31	Disodium tetraborate, anhydrous	1303-96-4, 1303-43-4, 12179-04-3	100	N.D
32	Disodium tetraborate, heptahydrate	12267-73-1	100	N.D
33	Sodium chromate	7775-11-3	100	N.D
34	Potassium chromate	7789-00-6	100	N.D
35	Ammonium dichromate	7789-09-5	100	N.D
36	Potassium dichromate	7778-50-9	100	N.D

(continued)

Table 6.10 (continued)

#	Inspection item	CAS#	Detection limit	Inspection result
37	Chromium trioxide	1333-82-0	100	N.D
38	2-Ethoxyethanol	110-80-5	100	N.D
39	2-Methoxyethanol	109-86-4	100	N.D
40	Cobalt acetate	71-48-7	100	N.D
41	Cobalt carbonate	513-79-1	100	N.D
42	Cobalt nitrate	10141-05-6	100	N.D
43	Cobalt sulfate	10124-43-3	100	N.D
44	Derivatives from chromium trioxide such as chromic acid, dichromic acid, oligomers of chromic acid	7738-94-5, 13530-68-2	100	N.D
45	2-Ethoxyethylacetate	111-15-9	100	N.D
46	Strontium chromate	7789-06-2	100	N.D
47	Di(C_{7-11} branched or straight alkyl) phthalate	68515-42-4	100	N.D
48	Hydrazine	7803-58-7, 302-01-2	100	N.D
49	1-Methyl-2-pyrrolidone	872-50-4	100	N.D
50	1,2,3-Trichloropropane	96-18-4	100	N.D
51	Di(C_{6-8} branched alkyl, C_7 rich) phthalate	71888-89-6	100	N.D
52	Calcium arsenate	7778-44-1	100	N.D
53	Diethylene glycol dimethyl ether	111-96-6	100	N.D
54	Potassium hydroxyoctaoxodizincatedichromate	11103-86-9	100	N.D
55	Lead dipicrate	6477-64-1	100	N.D
56	*N,N*-Dimethylacetamide	127-19-5	100	N.D
57	Arsenic acid, orthoarsenic acid	7778-39-4	100	N.D

(continued)

Table 6.10 (continued)

#	Inspection item	CAS#	Detection limit	Inspection result
58	2-Methoxyaniline	90-04-0	100	N.D
59	Lead arsenate	3687-31-8	100	N.D
60	1,2-Dichloroethane	107-06-2	100	N.D
61	Pentazinc chromate octahydroxide	49663-84-5	100	N.D
62	4-(1,1,3,3-Tetramethylbutyl)phenol	140-66-9	100	N.D
63	Polymers of formaldehyde with aniline	25214-70-4	100	N.D
64	Di(2-methoxyethyl) phthalate	117-82-8	100	N.D
65	Lead azide	13424-46-9	100	N.D
66	Lead 2,4,6-trinitroresorcinate	15245-44-0	100	N.D
67	3,3′-Dichloro-4,4′-methylenedianiline	101-14-4	100	N.D
68	Phenolphthalein	77-09-8	100	N.D
69	Chromium chromate	24613-89-6	100	N.D
70[a]	Aluminosilicate refractory ceramic fibers	–	100	N.D
71	Zirconia aluminosilicate refractory ceramic fibers	–	100	N.D
72	Triethylene glycol dimethyl ether	112-49-2	100	N.D
73	1,2-Dimethoxyethane	110-71-4	100	N.D
74	Diboron trioxide	1303-86-2	100	N.D
75	Formamide	75-12-7	100	N.D
76	Lead (II) methanesulfonate	17570-76-2	100	N.D
77	1,3,5-Tris(oxiranylmethyl)-1,3.5-triazine-2,4,6(1H,3H,5H)-trione	2451-62-9	100	N.D
78	Teroxirone	59653-74-6	100	N.D

(continued)

Table 6.10 (continued)

#	Inspection item	CAS#	Detection limit	Inspection result
79	4.4′-Bis(dimethylamino)benzophenone	90-94-8	100	N.D
80	*N*,*N*,*N*′,*N*′-Tetramethyl-4,4′-methylenedianiline	101-61-1	100	N.D
81[b]	Crystal violet	548-62-9	100	N.D
82[b]	Basic Blue 26	2580-56-5	100	N.D
83[b]	Solvent Blue 4	6786-83-0	100	N.D
84[b]	4,4′-Bis(dimethylamino)-4”-(methylamino)tritylalcohol	561-41-4	100	N.D
85	Decabromodiphenyl ether	1163-19-5	100	N.D
86	Pentacosafluorotridecanoic acid	72629-94-8	100	N.D
87	Tricosafluorododecanoic acid	307–55-1	100	N.D
88	Henicosafluoroundecanoic acid	2058–94-8	100	N.D
89	Heptacosafluorotetradecanoic acid	376-06-7	100	N.D
90	Diazene-1,2-dicarboxamide	123-77-3	100	N.D
91	Hexahydrophthalic anhydride, 1,3-isobenzofurandione, trans-cyclohexane-1,2-dicarboxylates	85-42-7, 13149-00-3, 14166-21-3	100	N.D
92	Hexahydromethylphathalic anhydride, hexahydro-4-methylphathalic anhydride, hexahydro-1-methylphathalic anhydride, hexahydro-3-methylphathalic anhydride	2550-51-0, 19438-60-9, 48122-14-1, 57110-29-9	100	N.D
93	4-Nonylphenol, branched and linear	–	100	N.D
94	4-(1,1,3,3-Tetramethylbutyl)phenol	–	100	N.D
95	Methoxyacetic acid	625-45-6	100	N.D
96	*N*,*N*-Dimethylformamide	68-12-2	100	N.D
97	Dibutyltin dichloride	683-18-1	100	N.D

(continued)

Table 6.10 (continued)

#	Inspection item	CAS#	Detection limit	Inspection result
98	Lead oxide	1317-36-8	100	N.D
99	Lead tetroxide	1314-41-6	100	N.D
100	Lead tetrafluoroborate	13814-96-5	100	N.D
101	Lead carbonate basic	1319-46-6	100	N.D
102	Lead titanium trioxide	12060-00-3	100	N.D
103	Lead titanium zirconium oxide	12626-81-2	100	N.D
104	Lead silicate	11120-22-2	100	N.D
105	Lead-doped barium silicate	68784-75-8	100	N.D
106	1-Bromopropane	106-94-5	100	N.D
107	Propylene oxide	75-56-9	100	N.D
108	Dipentyl phthalate, branched and linear	84777-06-0	100	N.D
109	Diisopentyl phthalate (DIPP)	605-50-5	100	N.D
110	n-Pentyl iso-pentyl phthalate	776297-69-9	100	N.D
111	1,2-Diethoxyethane	629-14-1	100	N.D
112	Lead acetate basic	51404-69-4	100	N.D
113	Lead sulfate basic	12036-76-9	100	N.D
114	Dibasic lead phthalate	69011-06-9	100	N.D
115	Dioxobis(stearato)trilead	12578-12-0	100	N.D
116	Fatty acids (C_{16-18}), lead salt	91031-62-8	100	N.D
117	Lead cynamidate	20837-86-9	100	N.D
118	Lead nitrate	10099-74-8	100	N.D
119	Pentalead tetraoxide sulphate	12065-90-6	100	N.D

(continued)

Table 6.10 (continued)

#	Inspection item	CAS#	Detection limit	Inspection result
120	C. I. pigment yellow 41	8012-00-8	100	N.D
121	Pentalead tetraoxide sulphide	62229-08-7	100	N.D
122	Tetraethyllead	48-00-2	100	N.D
123	Tetralead trioxide sulphate	12202-17-4	400	N.D
124	Lead phosphite tribasic	12141-20-7	100	N.D
125	Furan	110-00-9	100	N.D
126	Diethyl sulphate	64-67-5	100	N.D
127	Dimethyl sulphate	77-78-1	100	N.D
128	3-Ethyl-2-methyl-2-(3-methylbutyl)-1,3- oxazolidine	143860-04-2	100	N.D
129	dinoseb	88-85-7	100	N.D
130	4,4′-Methylenedi-o-toluidine	838-88-0	100	N.D
131	4,4′-oxydianiline	101-80-4	100	N.D
132	4-Phenylazoaniline	60-09-3	100	N.D
133	2,4-Diaminotoluene	95-80-7	100	N.D
134	2-Methoxy-5-toluidine	120-71-8	100	N.D
135	4-Aminobiphenyl	92-67-1	100	N.D
136	o-Aminoazotoluene	97-56-3	100	N.D
137	o-Toluidine	95-53-4	100	N.D
138	*N*-Methylacetamide	79-16-3	100	N.D
139	Cadmium	7440-43-9	100	N.D
140	Cadmium oxide	1306-19-0	100	N.D
141	Ammonium pentadecafluorooctanoate (APFO)	3285-26-1	100	N.D

(continued)

Table 6.10 (continued)

#	Inspection item	CAS#	Detection limit	Inspection result
142	Pentadecafluorooctanoic acid (PFOA)	335-67-1	100	N.D
143	Dipentyl phthalate (DPP)	131-18-0	100	N.D
144	4-Nonylphenol, branched and linear, including all individual C_9 alkyl isomers and all UVCB substances with branched or linear C_9 alkyl groups)	–	100	N.D
145	Cadmium sulphide	1306-23-6	100	N.D
146	Dihexyl phthalate (DHXP)	84-75-3	100	N.D
147	C. I. direct red 28	573–58-0	100	N.D
148	C. I. direct black 38	1937-37-7	400	N.D
149	2-Imidazolidinethione	96-45-7	100	N.D
150	Lead (II) acetate	301-04-2	100	N.D
151	Trixylyl phosphate	25155-23-1	100	N.D
152	Chromium chloride	10108-64-2	100	N.D
153	Di(hexyl, branched or linear) phthalate	68515-50-4	100	N.D
154	Sodium peroxometaborate	7632–04-4	100	N.D
155	Sodium perborate	–	100	N.D
156	Cadmium fluoride	7790-79-6	100	N.D
157	Cadmium sulphate	10124-36-4, 31119-53-6	100	N.D
158	UV-320	3846-71-7	100	N.D
159	UV-328	25973-55-1	100	N.D
160	2-Ethylhexyl 10-ethyl-4,4-dioctyl-7-oxo-8-oxa-3,5-dithia-4-stannatetradecanoate (DOTE)	15571-58-1	100	N.D

(continued)

Table 6.10 (continued)

#	Inspection item	CAS#	Detection limit	Inspection result
161	Reaction products of DOTE and MOTE	–	100	N.D
162	Di(C_{6-10} alkyl) phthalates, mixed decyl, hexyl and octyl phthalates with diehexyl phthalate content ≥0.3%	68515-51-5, 68648-93-1	100	N.D
163	2-(2′,4′-Dimethyl-3′-cyclohexen-1′-yl)-5-methyl-5-(1″-methylpropyl)-1,3-dioxane (karanal), 2-(2′,6′-dimethyl-3′-cyclohexen-1′-yl)-1,3-dioxane, and their mixtures at any ratios	–	100	N.D
164	Nitrobenzene	98-95-3	100	N.D
165	UV-327	3864-99-1	100	N.D
166	UV-350	36437-37-3	100	N.D
167	1,3-Propanesultone	1120-71-4	100	N.D
168	perfluorononan-1-oic acid and its sodium and ammonium salts	375-95-1, 21049-39-8, 4149-60-4	100	N.D
169	Benzo[a]pyrene	50-32-8	100	N.D
170	Bisphenol A(BPA)	80-05-7	100	N.D
171	Perfluorodecan-1-oic acid and its sodium and ammonium salts	335-76-2, 3108-42-7, 3830-45-3	100	N.D
172	4-Heptyl phenol, branched and linear	–	100	N.D
173	4-(1′,1′-dimethylpropyl)phenol	80-46-6	100	N.D

Note The units in the table were mg/kg; N.D. represents not detected (lower than the detection limit)
[a]This substance can be found in "Categorization, labeling and packaging of chemicals and their mixtures" Appendix VI, Part 3, Table 3.1, Index # 650-017-00-8;
[b] Only when the content of Michler's ketone or Michler's base is ≥0.1 wt% can these four substances meet the requirements for carcinogenic substances 1A or 1B in the REACH, Article 57.

Table 6.11 Phthalate plasticizers included in REACH and their inspection results

#	Phthalate plasticizer	CAS #	Detection Limit	Inspection result
1	DBP (#6 in Table 6.10)	84-74-2	10	N.D
2	DEHP (#7 in Table 6.10)	117-81-7	10	20
3	BBP (#8 in Table 6.10)	85-68-7	10	N.D
4	DIBP (#24 in Table 6.10)	84-69-5	10	N.D
5	Di(C_{7-11} branched and linear alkyl) phthalate (#47 in Table 6.10)	68515-42-4	100	N.D
6	Di(C_{6-8} branched alkyl, C_7 rich) phthalate (#51 in Table 6.10)	71888-89-6	100	N.D
7	Di(methoxyethyl) phthalate (#64 in Table 6.10)	117-82-8	100	N.D
8	Dipentyl phthalate, branched and linear (#108 in Table 6.10)	84777-06-0	100	N.D
9	Diisopentyl phthalate (DIPP) (#109 in Table 6.10)	605-50-5	100	N.D
10	n-Pentyl iso-pentyl phthalate (#110 in Table 6.10)	776297-69-9	100	N.D
11	Dipentyl phthalate (DPP) (#143 in Table 6.10)	131-18-0	100	N.D
12	Dihexyl phthalate (DHXP) (#146 in Table 6.10)	84-75-3	100	N.D
13	Di(hexyl, branched and linear) phthalate (#153 in Table 6.10)	68515-50-4	100	N.D
14	Di(C_{6-10} alkyl) phthalates, mixed decyl, hexyl and octyl phthalates with diehexyl phthalate content $\geq$0.3%	68515-51-5, 68648-93-1	100	N.D

of Pb, Cd, Hg, Cr(VI), polybrominated biphenyls (PBBs), polybrominated diphenyl ethers (PBDEs), and phthalates (such as DBP, BBP, DEHP, and DIBP) meet the limits defined by EU RoSH 2011/65/EU Appendix II Revision (EU) 2015/863. The inspection report is shown in Table 6.12.

Table 6.12 DEHCH inspection report (No. TSNEC 1700513402)

Inspection item	Maximum allowed limit	Unit	Detection limit	Inspection result
Cadmium (Cd)	100	mg/kg	2	N.D
Lead (Pb)	1000	mg/kg	2	N.D
Mercury (Hg)	1000	mg/kg	2	N.D
Cr(VI)	1000	mg/kg	2	N.D
Total of polybromobiphenyls (PBBs)	1000	mg/kg	–	N.D
Bromobiphenyl	–	mg/kg	5	N.D
Dibromobiphenyl	–	mg/kg	5	N.D
Tribromobiphenyl	–	mg/kg	5	N.D
Tetrabromobiphenyl	–	mg/kg	5	N.D
Pentabromobiphenyl	–	mg/kg	5	N.D
Hexabromobiphenyl	–	mg/kg	5	N.D
Heptabromobiphenyl	–	mg/kg	5	N.D
Octabromobiphenyl	–	mg/kg	5	N.D
Nonabromobiphenyl	–	mg/kg	5	N.D
Decabromobiphenyl	–	mg/kg	5	N.D
Total of polybromo diphenyl ethers (PBDEs)	1000	mg/kg	–	N.D
Bromodiphenyl ether	–	mg/kg	5	N.D
Dibromodiphenyl ether	–	mg/kg	5	N.D
Tribromodiphenyl ether	–	mg/kg	5	N.D
Tetrabromodiphenyl ether	–	mg/kg	5	N.D
Pentabromodiphenyl ether	–	mg/kg	5	N.D
Hexabromodiphenyl ether	–	mg/kg	5	N.D
Heptabromodiphenyl ether	–	mg/kg	5	N.D
Octabromodiphenyl ether	–	mg/kg	5	N.D
Nonabromodiphenyl ether	–	mg/kg	5	N.D
Decabromodiphenyl ether	–	mg/kg	5	N.D

(continued)

Table 6.12 (continued)

Inspection item	Maximum allowed limit	Unit	Detection limit	Inspection result
Dibutyl phthalate (DBP)	1000	mg/kg	50	N.D
Butyl benzyl phthalate (BBP)	1000	mg/kg	50	N.D
Di(2-ethylhexyl) phthalate (DEHP)	1000	mg/kg	50	N.D
Diisobutyl phthalate (DIBP)	1000	mg/kg	50	N.D

Note The maximum allowed limit is cited from RoSH (EU) 2015/863 and the result is the content in wet sample mass; "–" represents "not defined"
(EU) 2015/863, revision of 2011/65/EU Appendix II, the test method:
(1) See IEC 62321-5: 2013, Cd and Pb contents measured with ICP-OES.
(2) See IEC 62321-4: 2013, Hg content measured with ICP-OES.
(3) See IEC 62321: 2008, Cr(VI) content measured with UV-vis spectrophotometry.
(4) See IEC 62321-6: 2015, PBBs and PBDEs contents measured with GC-MS.
(1) See IEC 62321-8 (111/321/CD): 2013, phthalate contents measured with GC-MS.